Multiplication and Division with Rod Patterns and Graph Paper

by
Patricia S. Davidson
Robert E. Willcutt

Cuisenaire Company of America
12 Church Street
New Rochelle, NY 10805

ISBN 0-914040-82-0

Printed in U.S.A.

Table of Contents

Making One-Color Trains

Here are all the one-color trains which match the dark green rod.

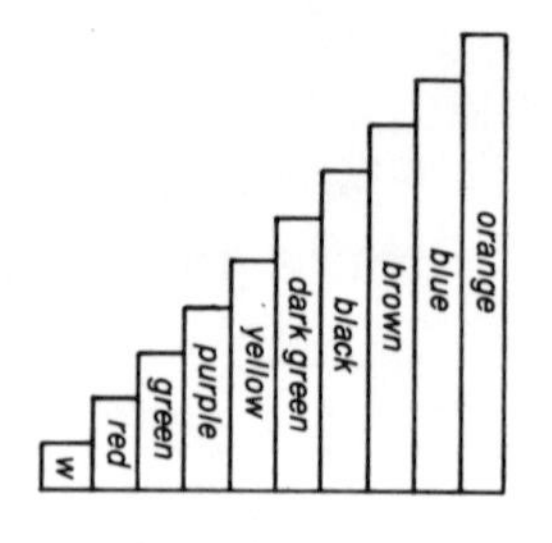

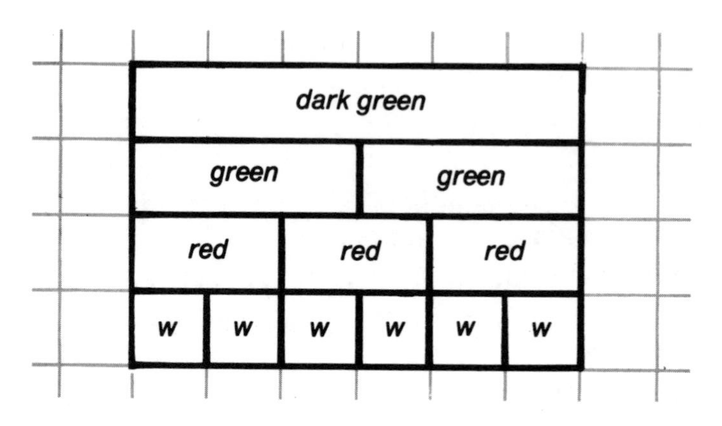

Place your rods on the graph paper and make all the one-color trains which match the brown rod. Now remove the trains and draw them on the graph paper.

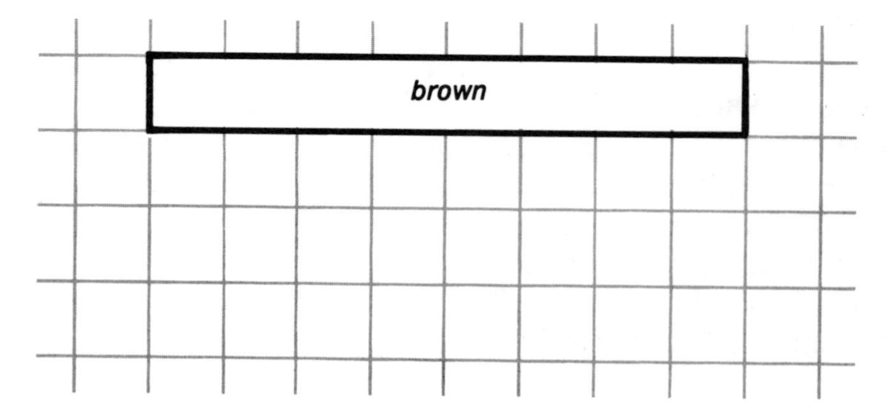

Making One-Color Trains

Place your rods on the graph paper and make <u>one-color trains</u>. Remove the trains and draw them on the graph paper.

w red green purple yellow dark green black brown blue orange

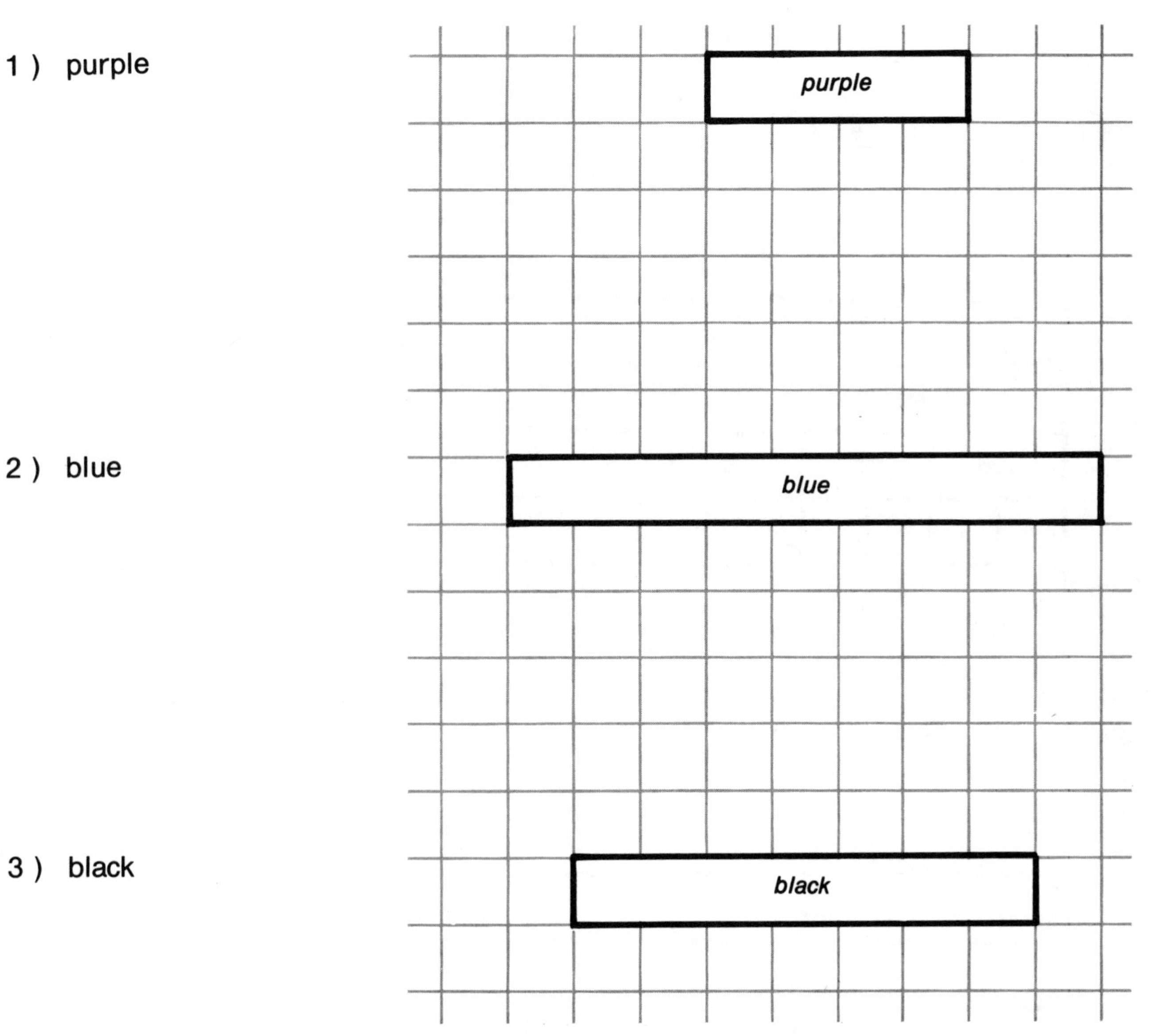

Matching Trains and Facts

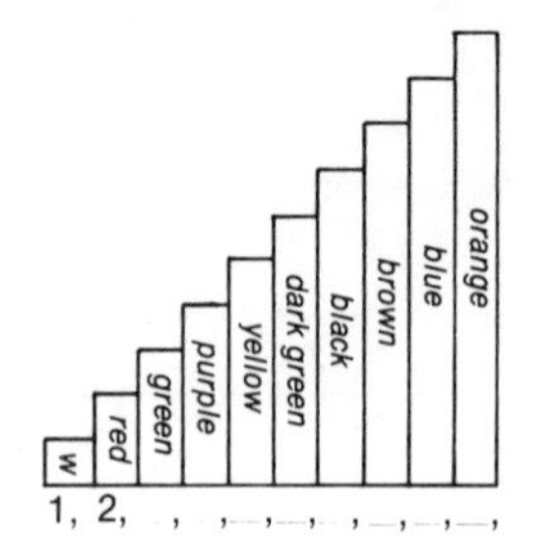

Let white = 1. Then red = 2 because red = 2 whites.

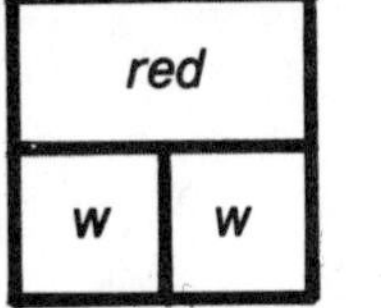

By matching each rod in the staircase with white rods, give each rod a number name.

white = ____ red = ____ green = ____ purple = ____ yellow = ____ dark green = ____
black = ____ brown = ____ blue = ____ orange = ____

Here is a way to build the multiplication facts for 10 with rods. Place your rods on the pattern to build the facts for 10.

orange — 1×10
yellow yellow — 2×5
red red red red red — 5×2
w w w w w w w w w w — 10×1

Place rods on the pattern to build the multiplication facts for 8. Match the multiplication facts with your one-color trains.

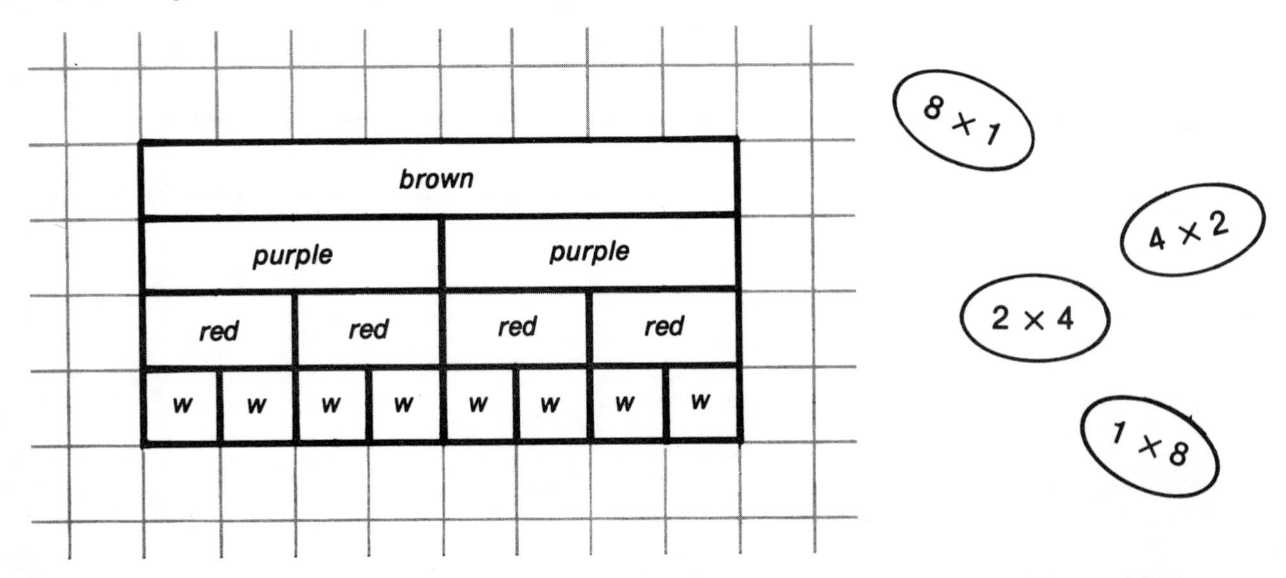

Finding Facts for One-Color Trains

(Let white = 1)

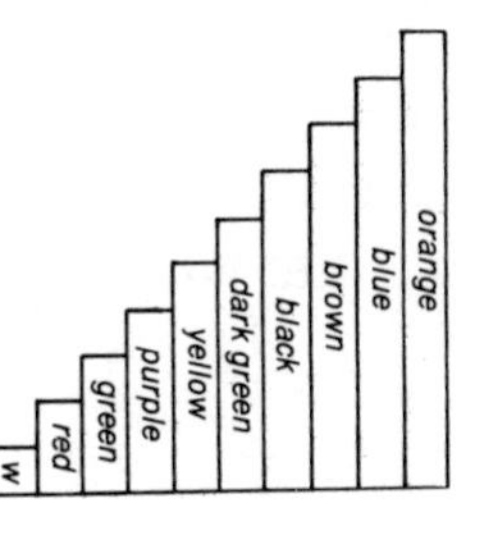

Place your rods on each pattern and build the multiplication facts for each family of one-color trains.

1) blue

blue								
green			green			green		
w	w	w	w	w	w	w	w	w

1 × 9
3 × 3
9 × 1

2) dark green

dark green					
green			green		
red		red		red	
w	w	w	w	w	w

3) black

black						
w	w	w	w	w	w	w

4) purple

purple			
red		red	
w	w	w	w

Finding Trains and Facts

(Let white = 1)

1.) Place rods on the pattern and make the multiplication facts for 15. When making numbers larger than 10 with rods, use orange-plus one other rod. "Orange-plus" trains show numbers by "tens" and "ones."

1×15

3×5

2) Complete the rod pattern for the multiplication facts for 12. When making numbers larger than 10 with rods, use orange-plus one other rod. "Orange-plus" trains show numbers by "tens" and "ones."

1×12

2×6

Finding Trains and Facts

(Let white = 1)

Complete the rod pattern for the multiplication facts for 16 and 18. When making numbers larger than 10 with rods, use orange-plus one other rod. ''Orange-plus'' trains show numbers by ''tens'' and ''ones''.

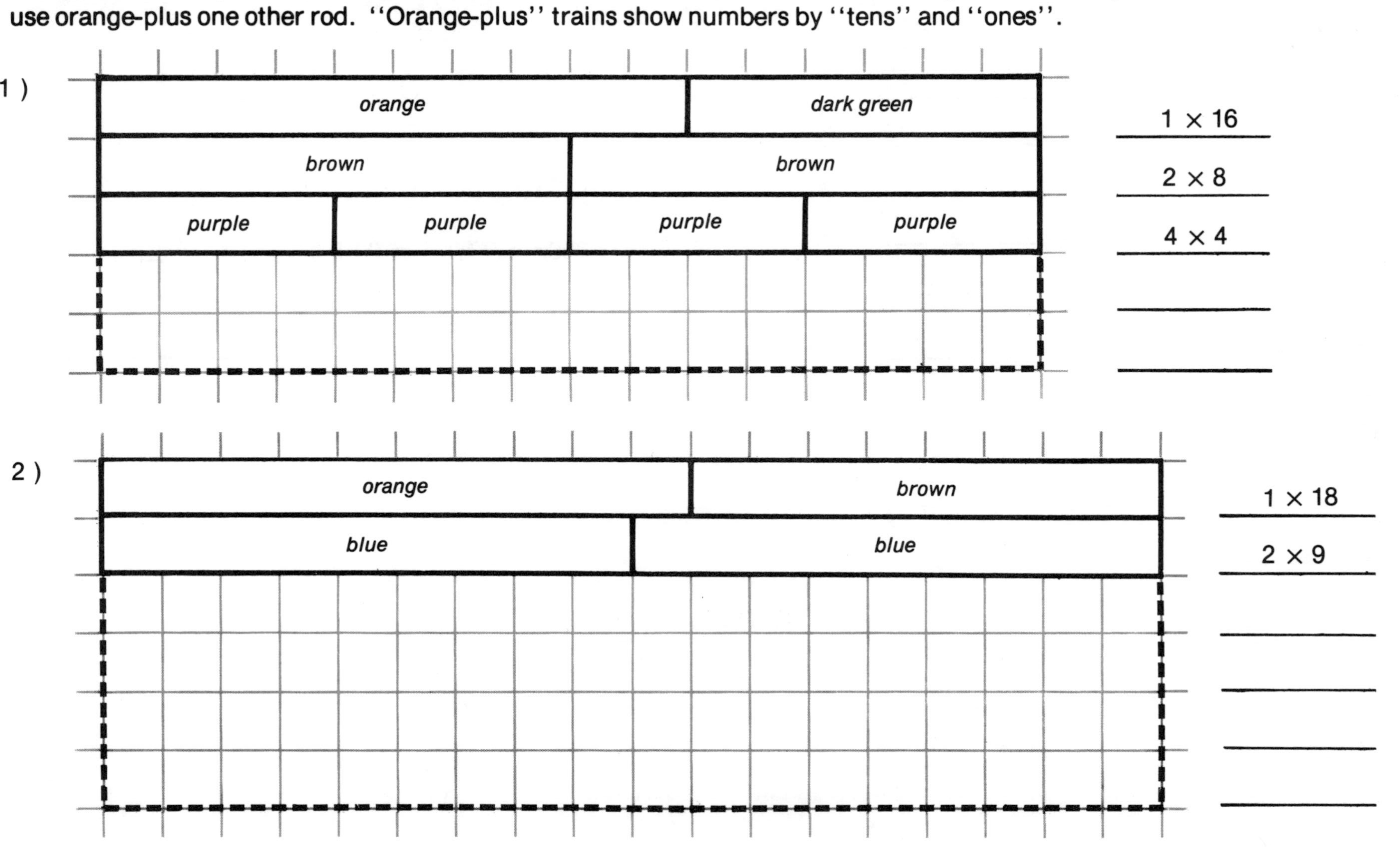

Finding All the Facts

(Let white = 1)

Use your rods on the rod work area below to find all the multiplication facts for each number.

1) 20 (orange + orange) 1×20

______
______
______
______
______

2) 21 (orange + orange + white) 1×21

______
______
______

3) 19 (orange + blue) 1×19

______

4) 24 (orange + orange + purple) 1×24

______
______
______
______
______
______
______

ROD WORK AREA

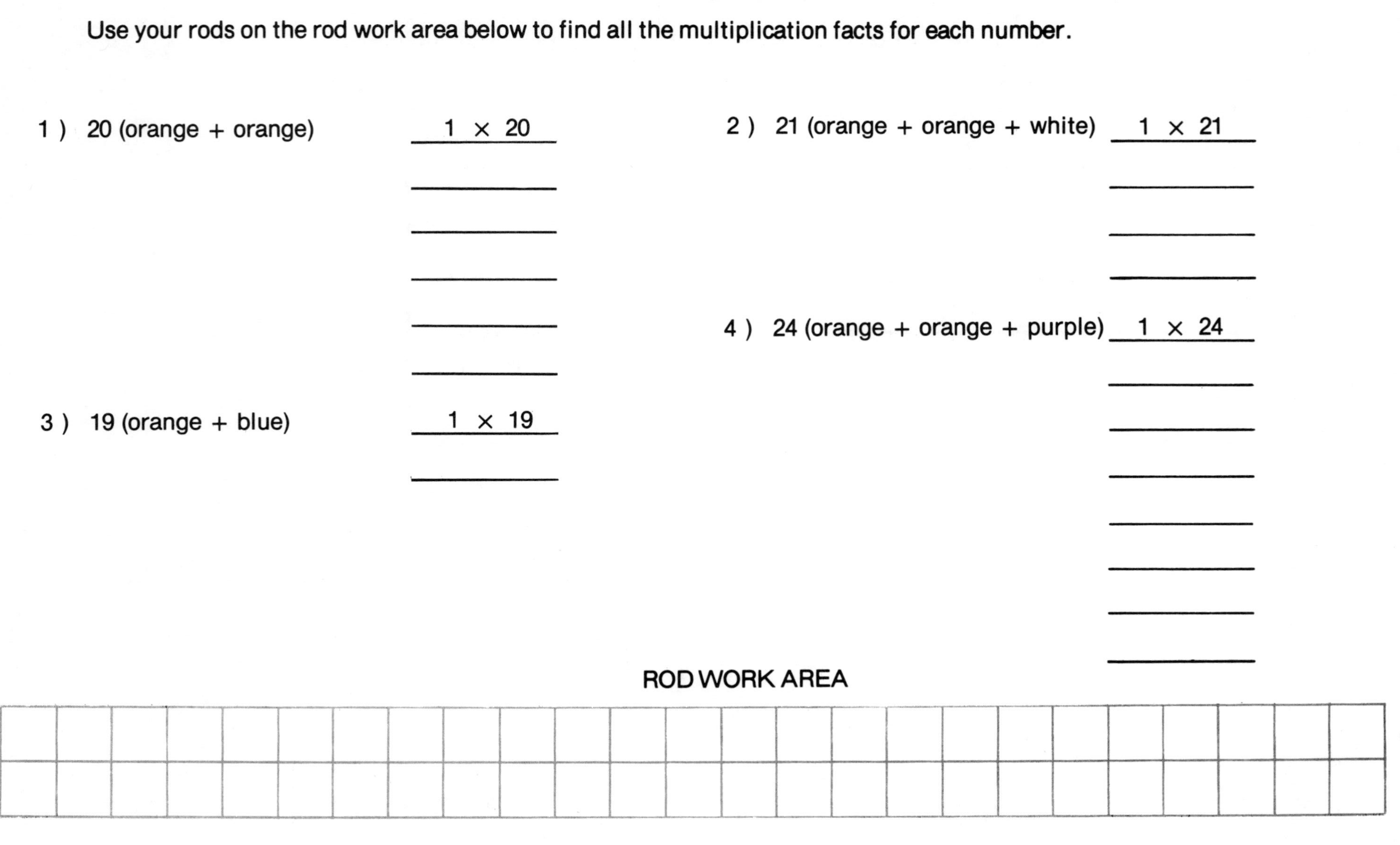

Finding All the Facts

(Let white = 1)

Use your rods on the rod work area below to find all the multiplication facts for each number.

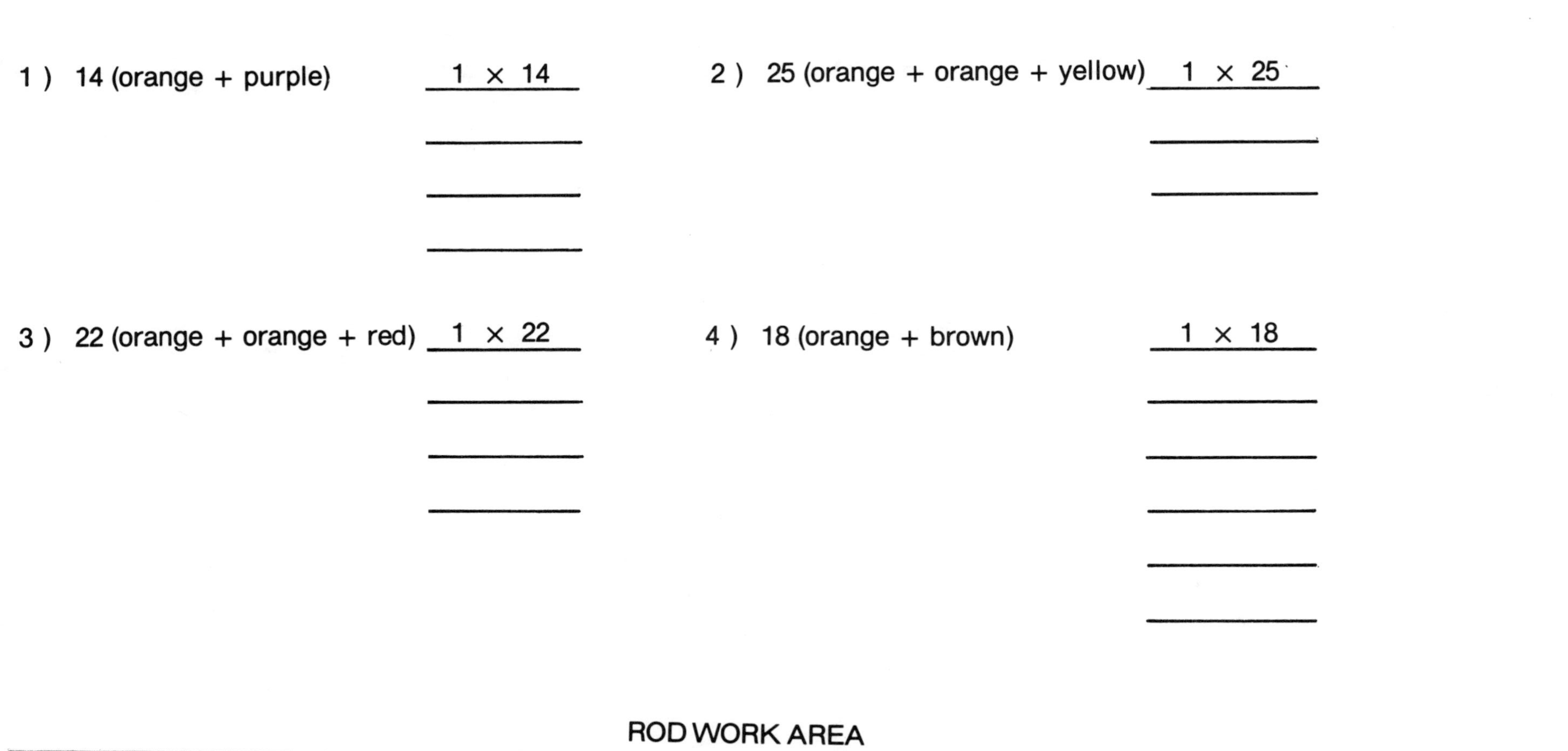

1) 14 (orange + purple) 1×14 ____ ____ ____

2) 25 (orange + orange + yellow) 1×25 ____ ____

3) 22 (orange + orange + red) 1×22 ____ ____ ____

4) 18 (orange + brown) 1×18 ____ ____ ____ ____ ____

ROD WORK AREA

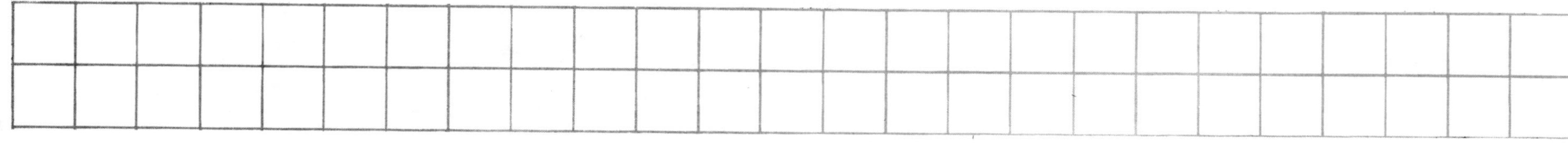

Hunting for Products

(Let white = 1)

Use your rods to match the multiplication facts with the rod pictures. Fill in the answers (products).

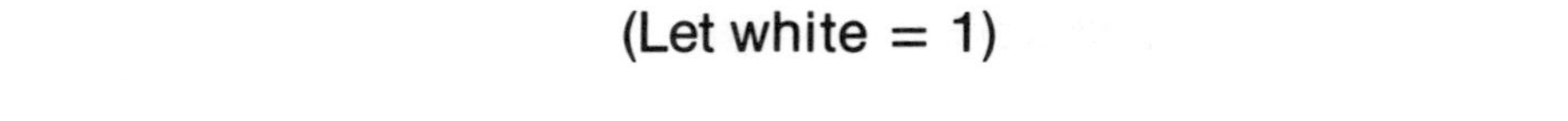

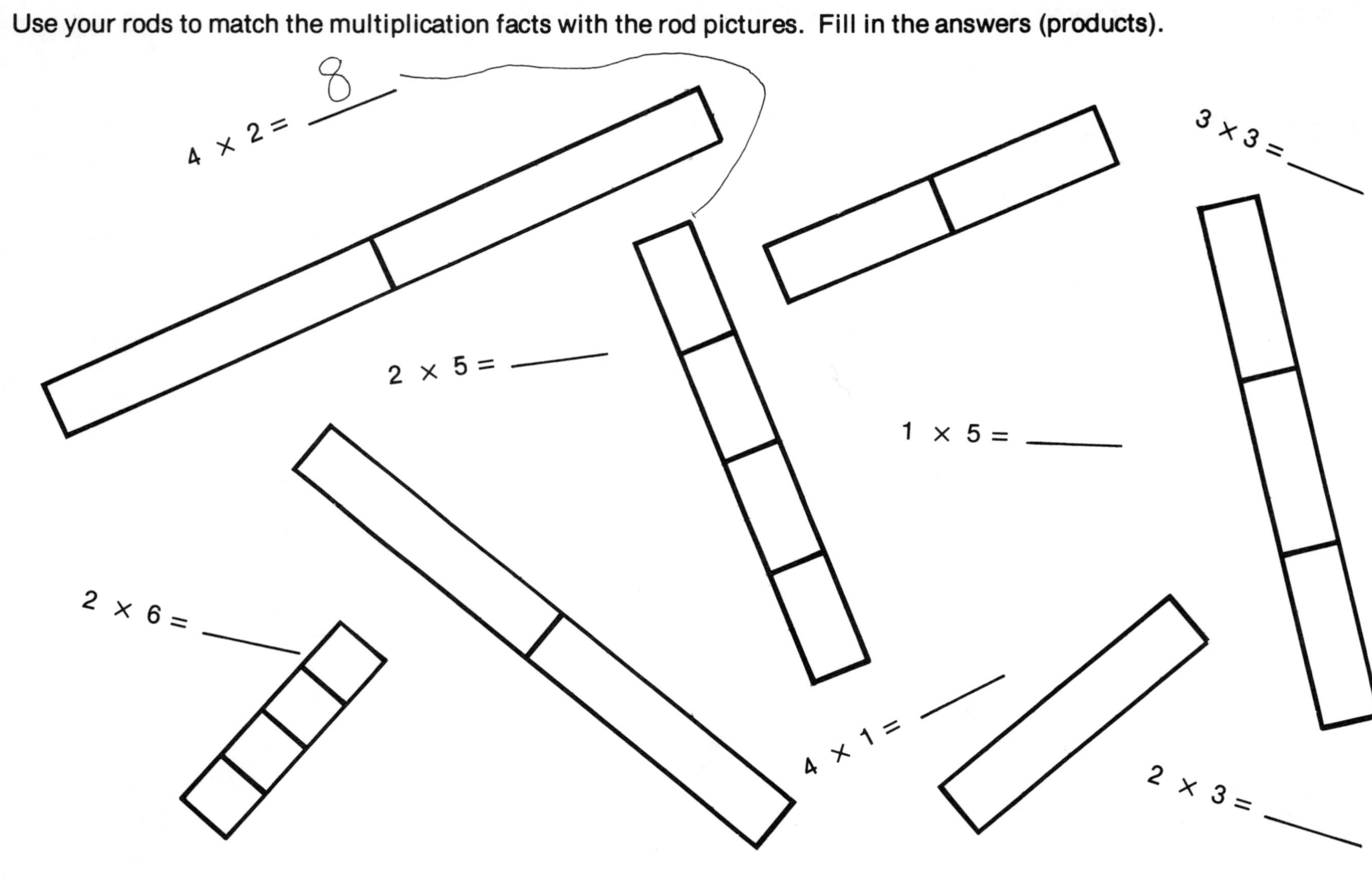

Hunting for Products

(Let white = 1)

Use your rods to match the multiplication facts with the rod pictures. Fill in the answers (products).

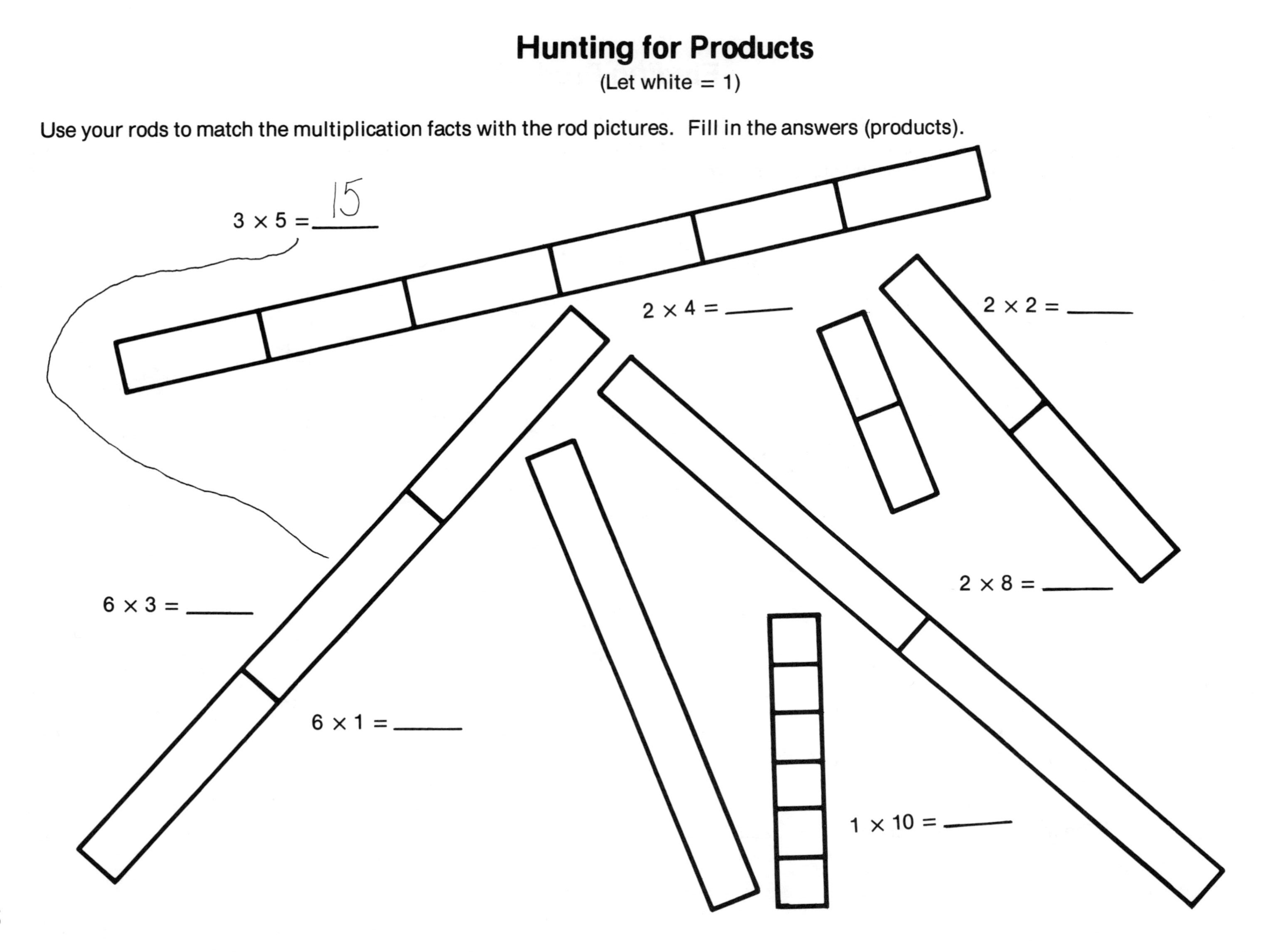

Finding Products

(Let white = 1)

Match the multiplication facts with the rod pictures. Fill in the products. One fact is missing!
Make rod trains beside the worksheet to check your answers.

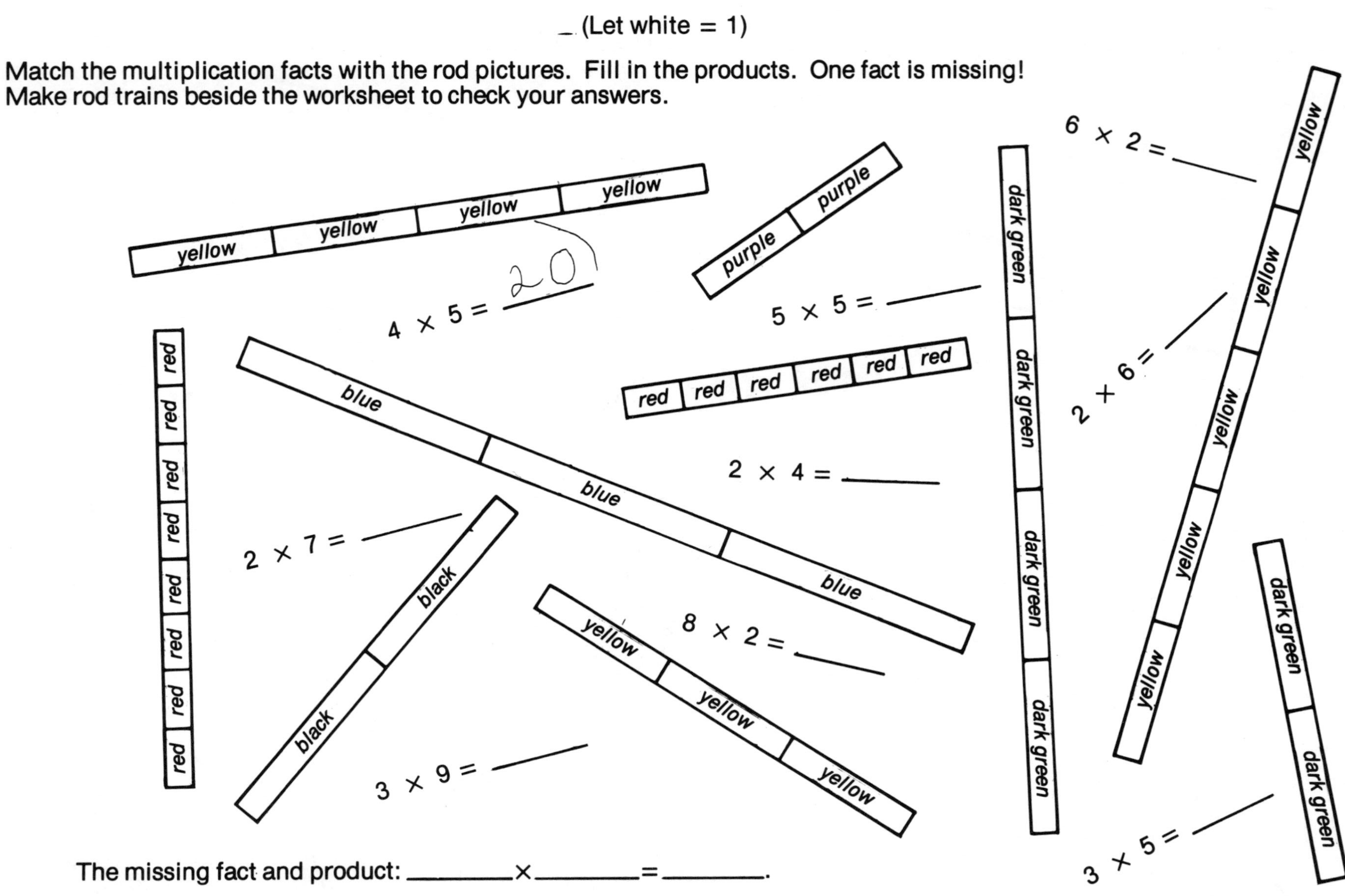

The missing fact and product: ______ × ______ = ______.

Finding Products

(Let white = 1)

Match the multiplication facts with the rod pictures. Fill in the products.
One fact is missing! Make rod trains beside the worksheet to check your answers.

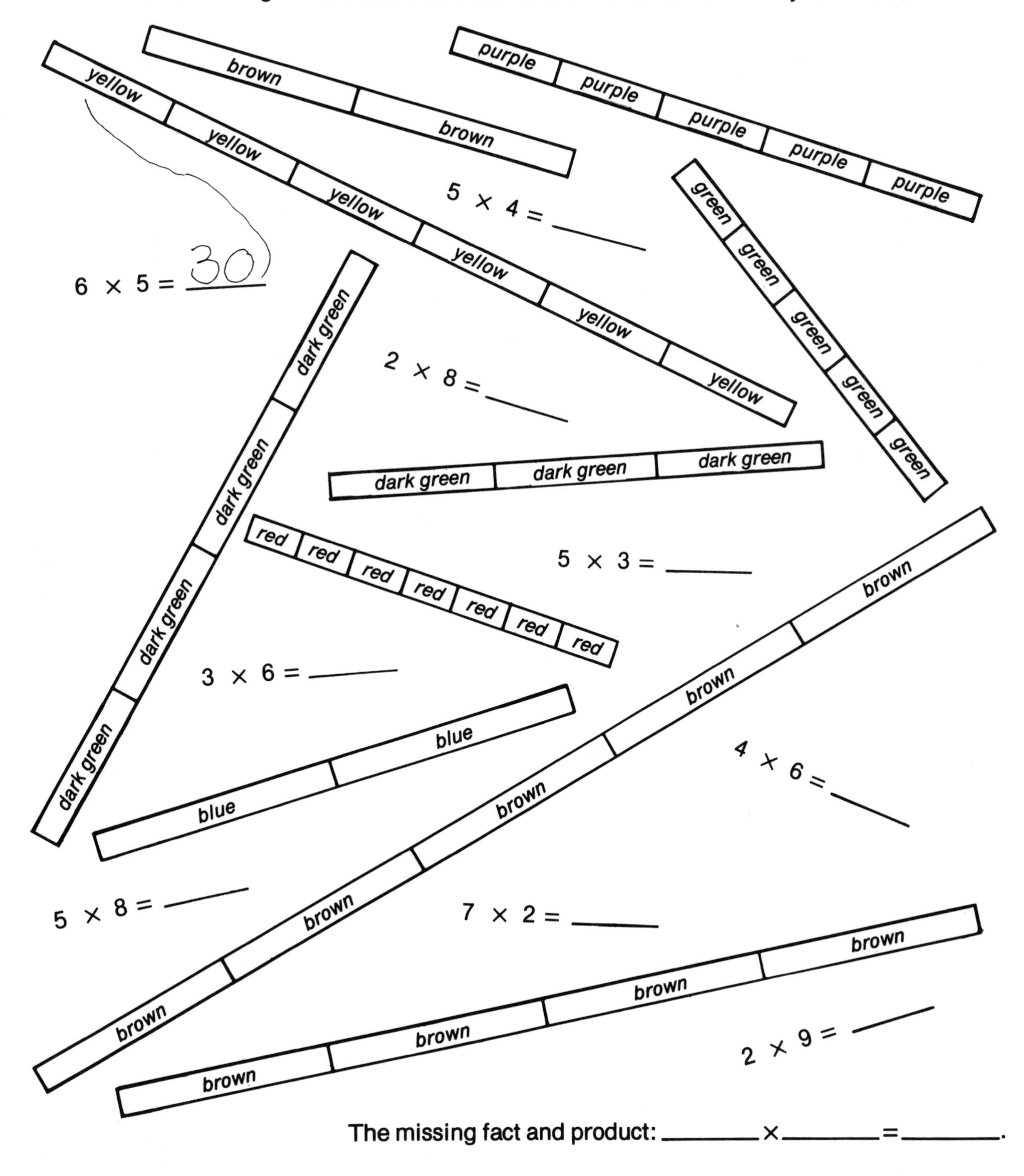

The missing fact and product: ______ × ______ = ______.

Matching Products and Rods

(Let white = 1)

Match the products with the rod pictures. Write in the multiplication facts. One product is missing! Make rod trains beside the worksheet to check your answers.

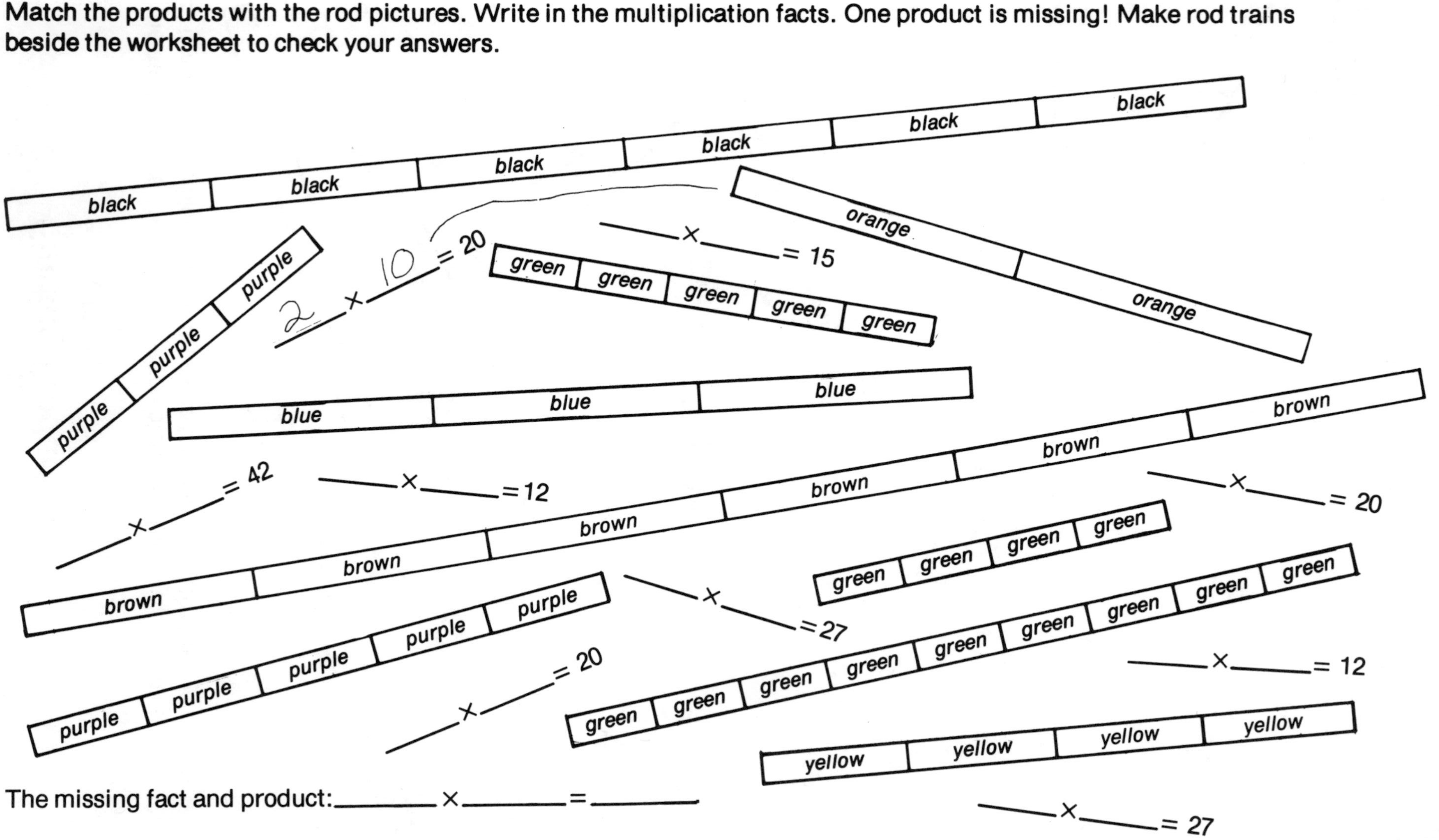

The missing fact and product: ______ × ______ = ______

Matching Products and Rods

(Let white = 1)

Match the products with the rod pictures. Write in the multiplication facts. One product is missing! Make rod trains beside the worksheet to check your answers.

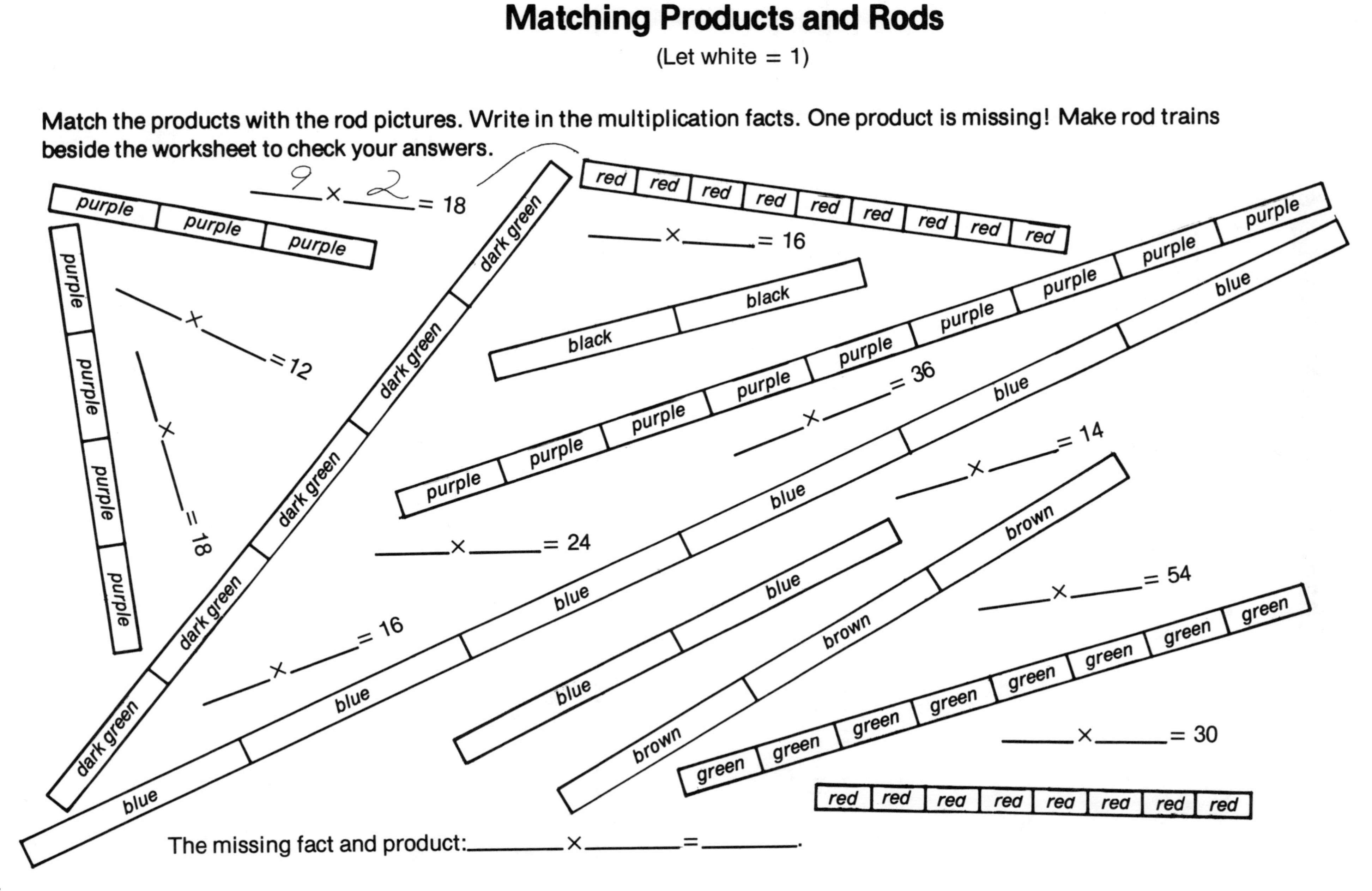

The missing fact and product: ______ × ______ = ______.

Changing Trains to Rectangles

(Let white = 1)

1) The brown rod can be matched with 4 red rods.

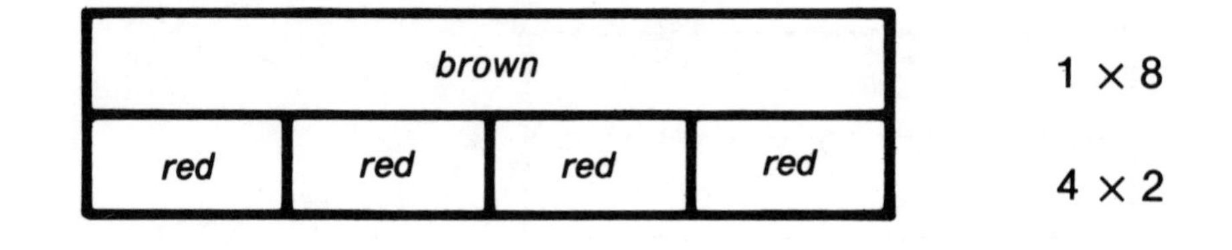

1×8

4×2

The 4 red rods can be made into a rectangle. The dimensions of the rectangle are 4×2.

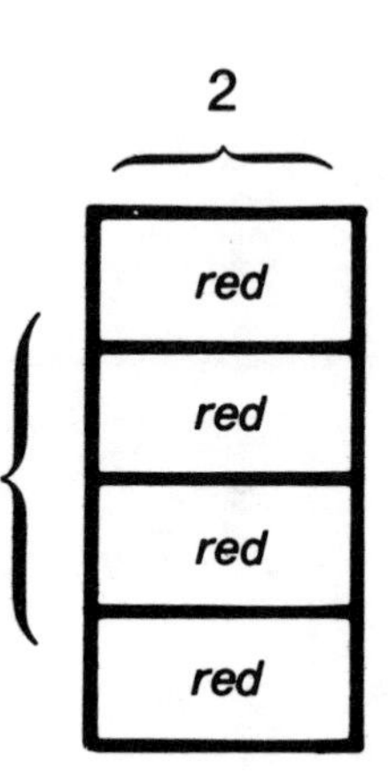

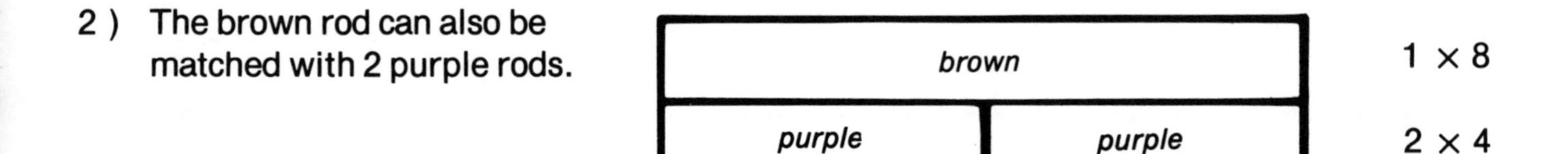

2) The brown rod can also be matched with 2 purple rods.

1×8

2×4

Make and draw a rectangle for the 2 purple rods. The dimensions are 2×4.

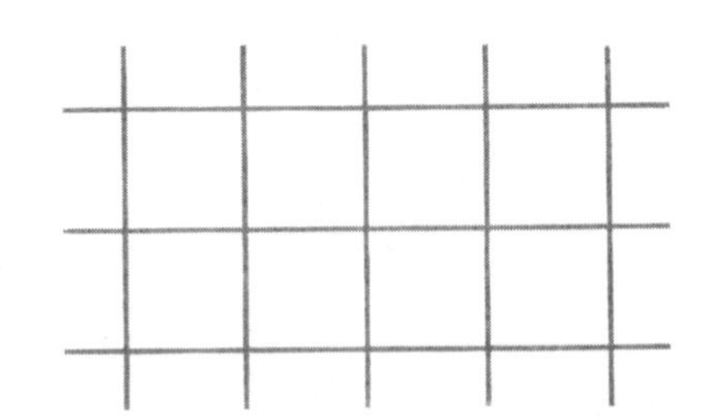

Changing Trains to Rectangles

(Let white = 1)

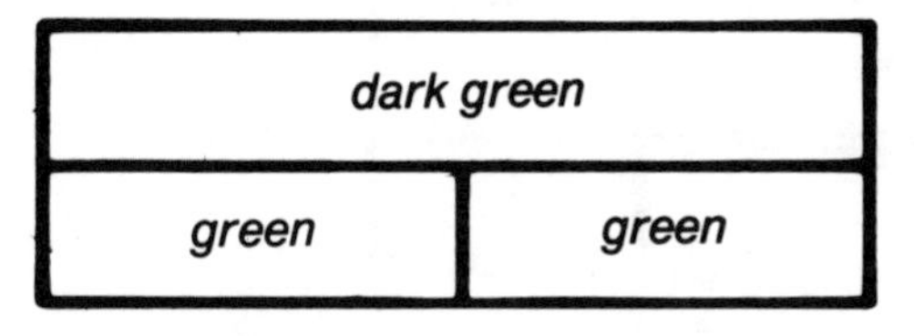

1) The dark green rod can be matched with 2 light green rods.

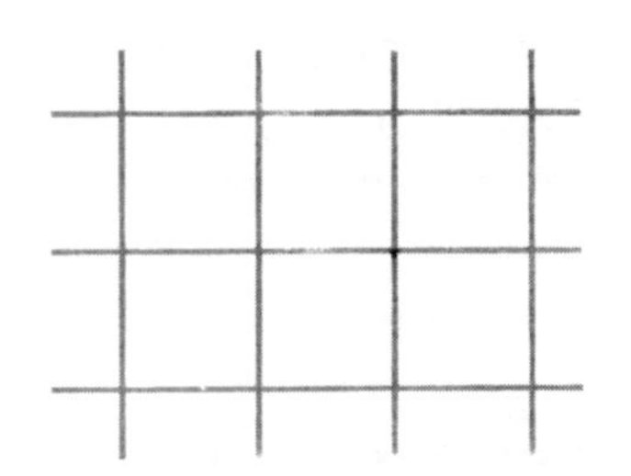

Make and draw a rectangle for the 2 light green rods. The dimensions are

_______ × _______.

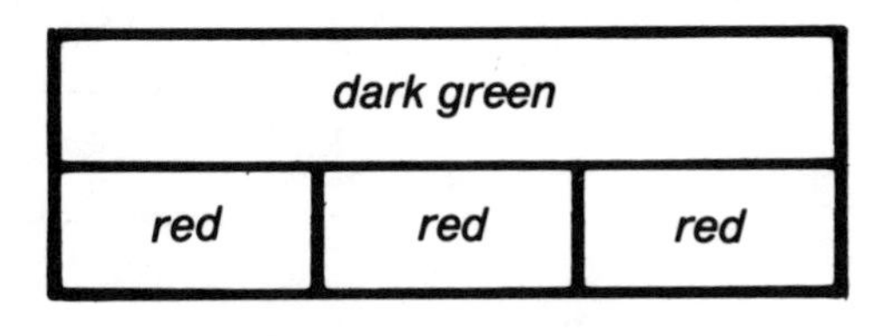

2) The dark green rod can be matched with 3 red rods.

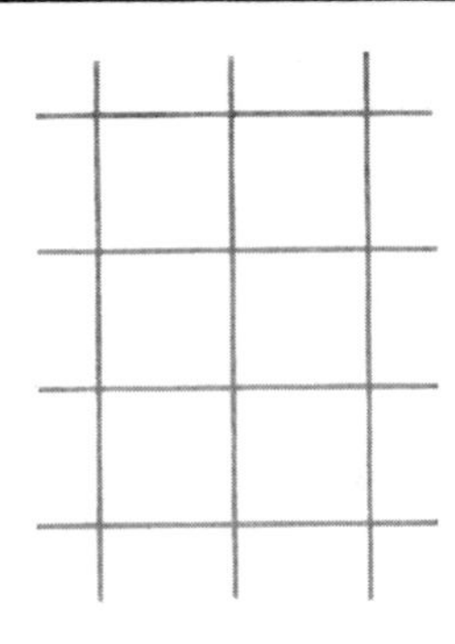

Make and draw a rectangle for the 3 red rods. The dimensions are

_______ × _______.

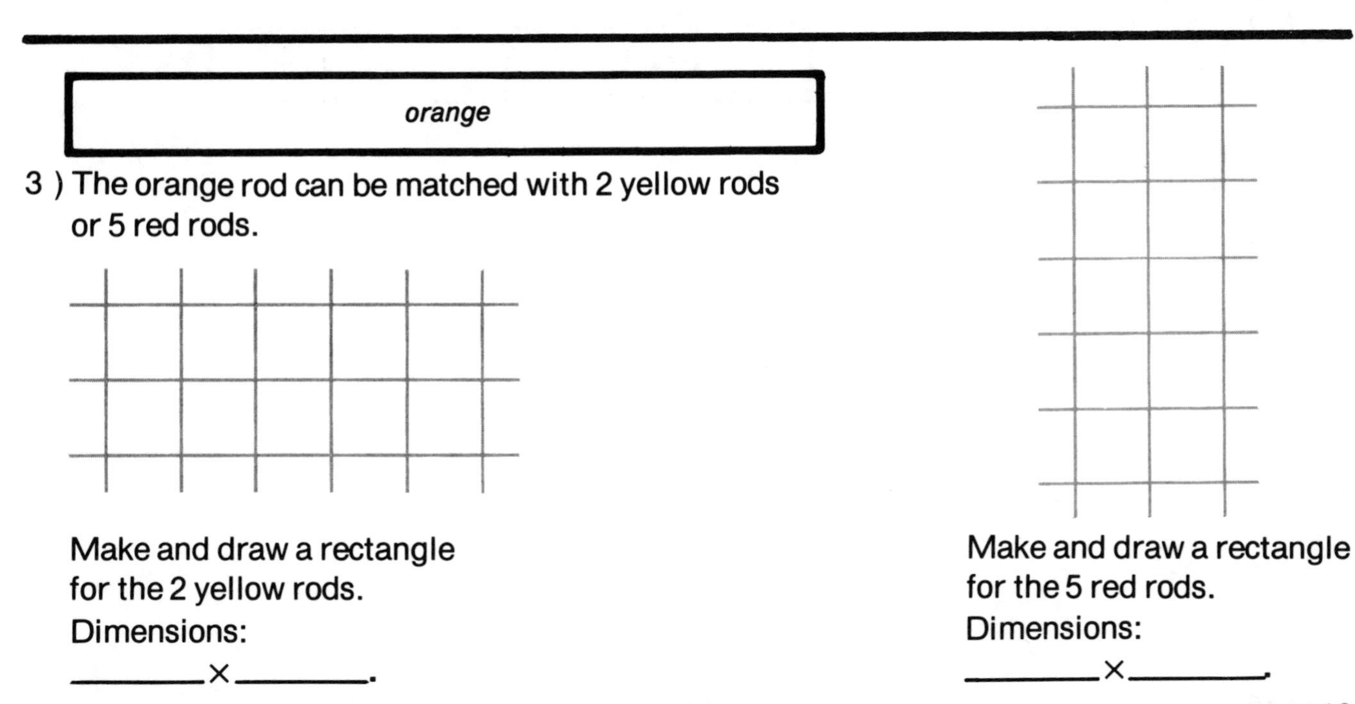

3) The orange rod can be matched with 2 yellow rods or 5 red rods.

Make and draw a rectangle for the 2 yellow rods. Dimensions:

_______ × _______.

Make and draw a rectangle for the 5 red rods. Dimensions:

_______ × _______.

Building Rectangles

(Let white = 1)

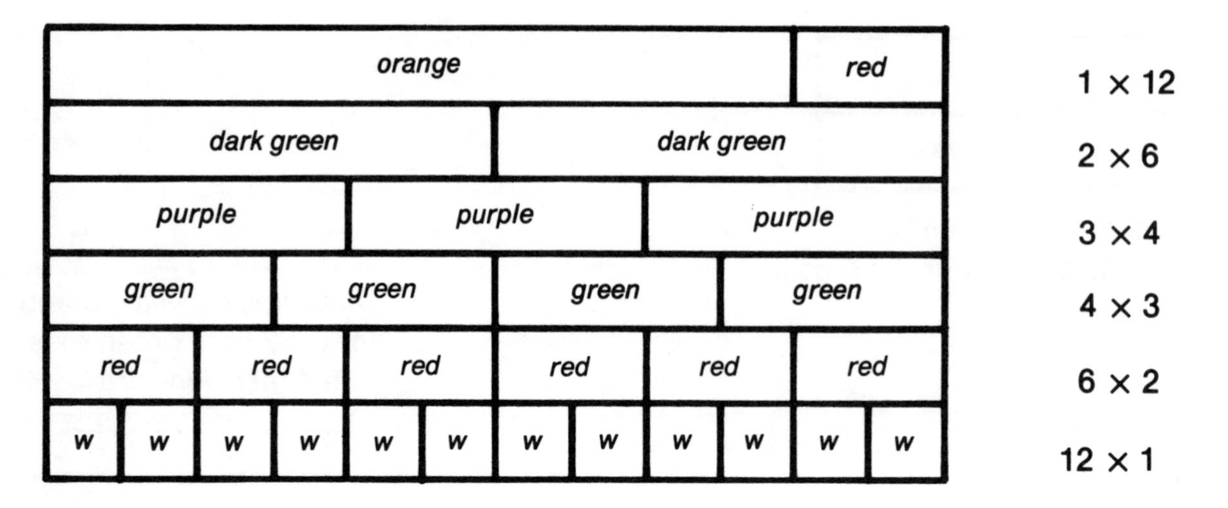

Make and draw a rod rectangle for each of these multiplication facts.

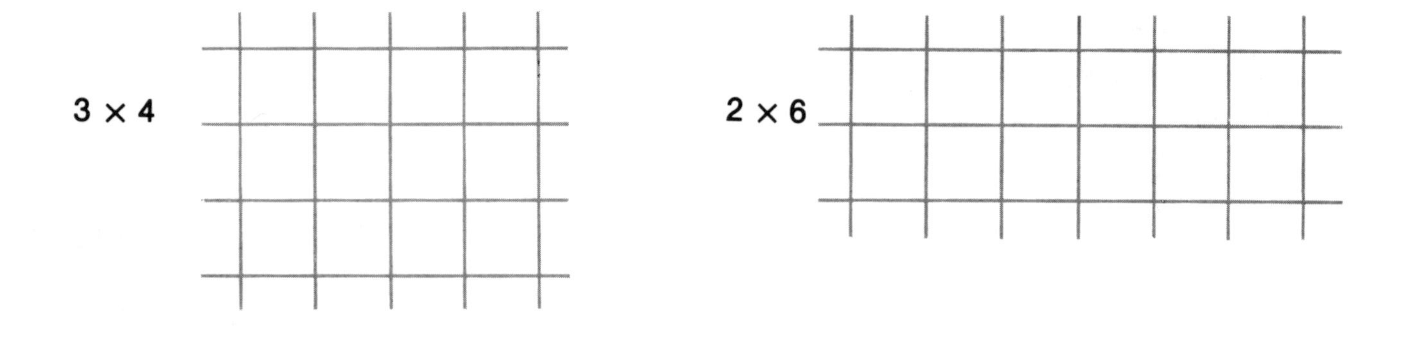

6 × 2

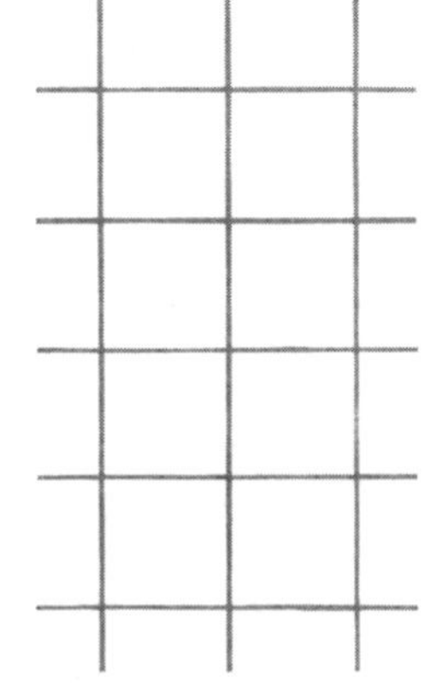

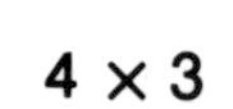

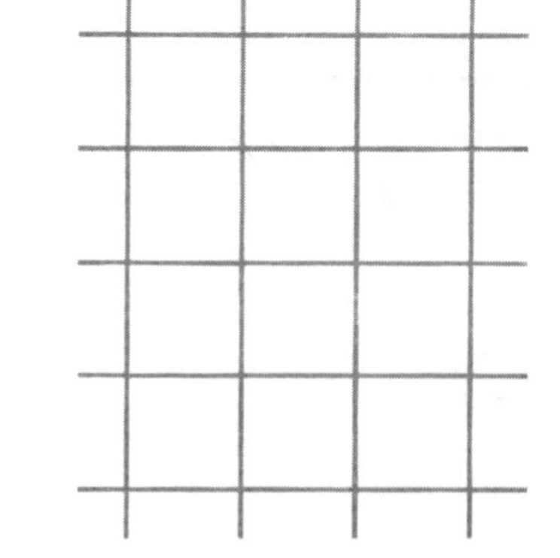

Building Rectangles

(Let white = 1)

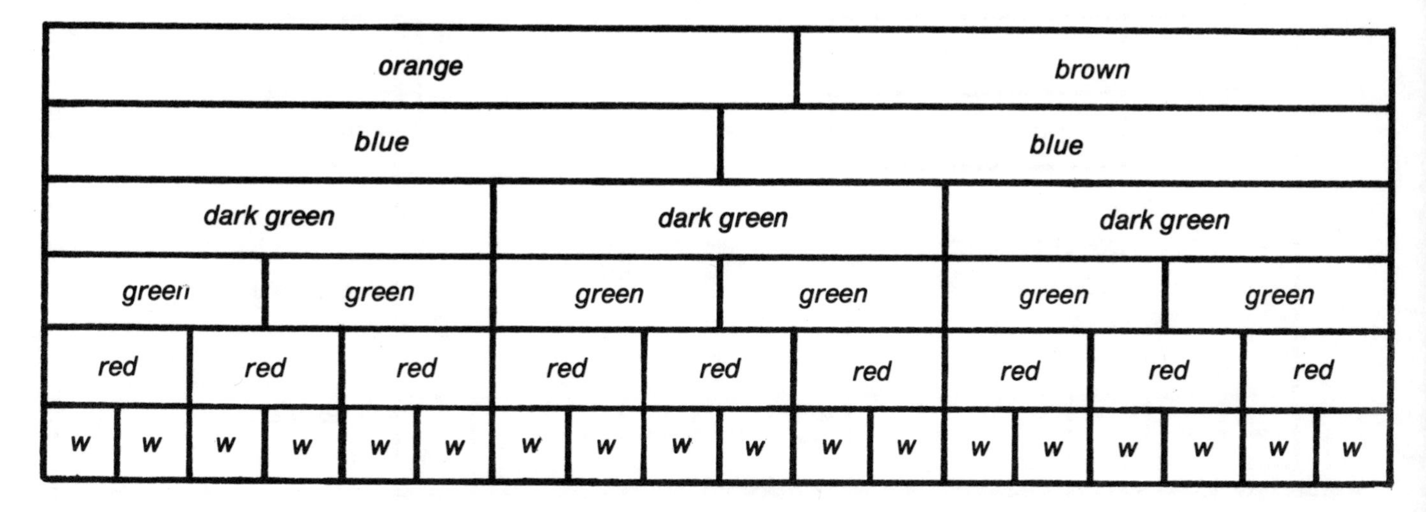

Make and draw a rod rectangle for each of these multiplication facts.

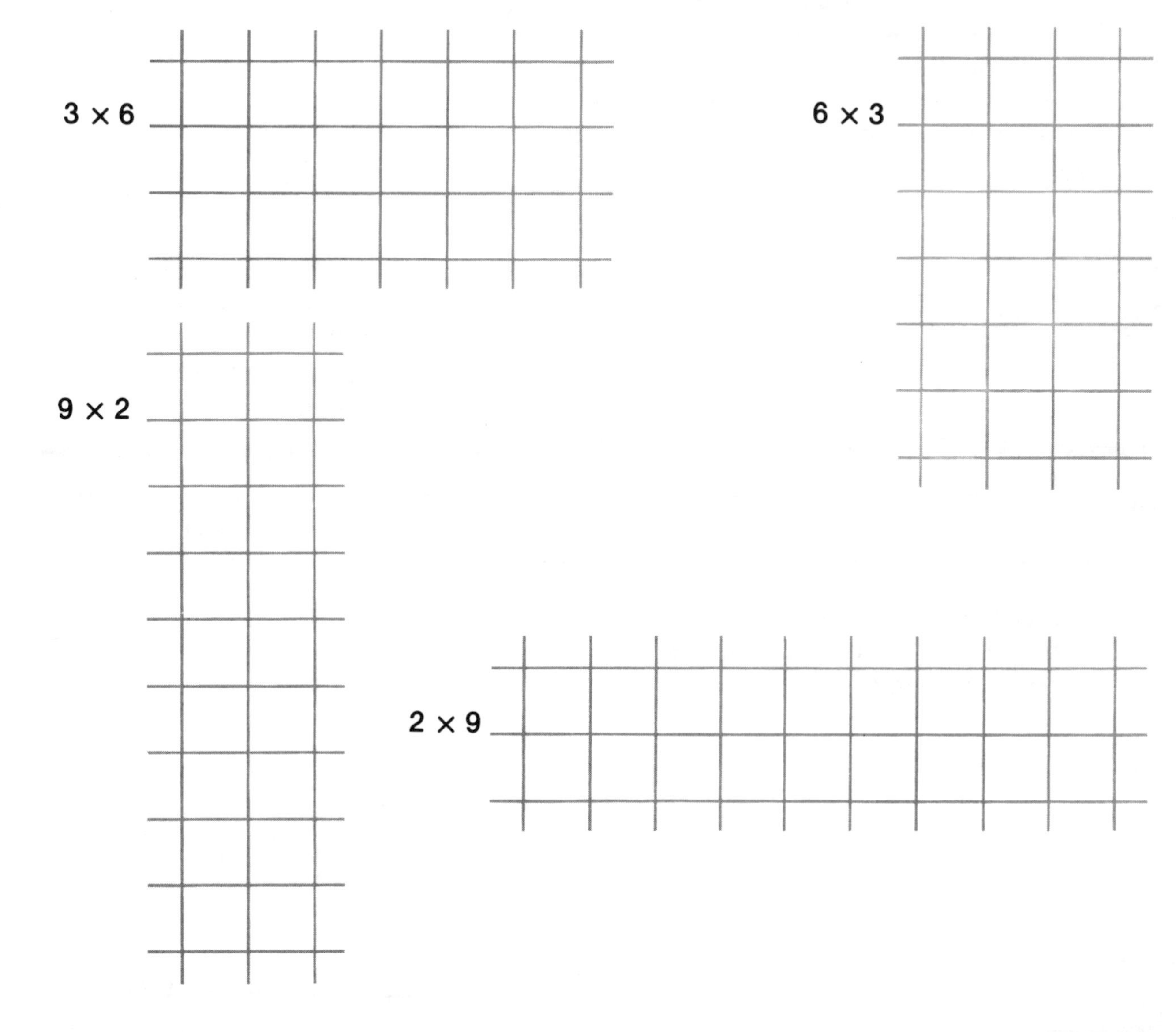

Finding Square Numbers

(Let white = 1)

1)

purple	
red	red

The purple rod can be matched by 2 red rods.

red
red

The 2 red rods can be made into a square, 2 × 2.
2 × 2 = 4.
4 is a <u>square</u> number.

2.)

blue		
green	green	green

The blue rod can be matched by 3 light green rods.

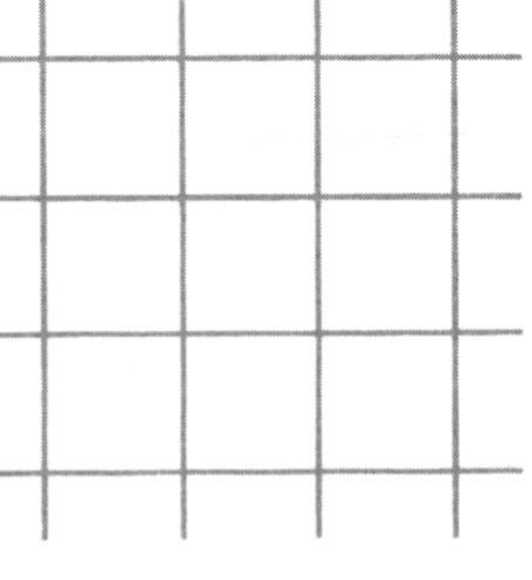

Make and draw the square using the 3 light green rods.
______ is a square number.

---

3)

orange	dark green

Match orange and dark green with purple rods. Make and draw the square using the 4 purple rods.

______ is a square number.

Finding More Square Numbers

(Let white = 1)

Make and draw the next three square numbers.

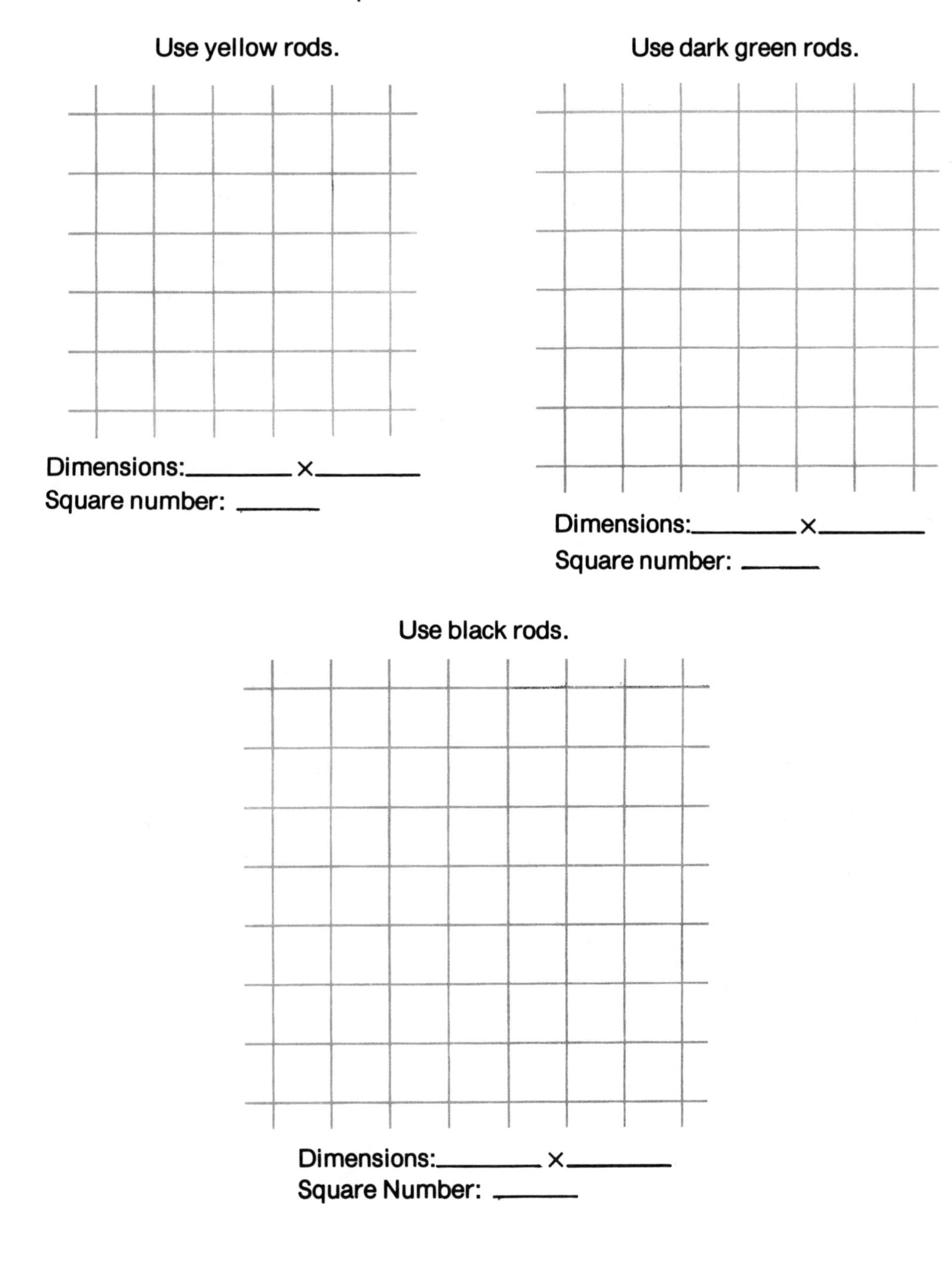

The three square numbers that come after these are ________, ________, and ________.

Seeing Rectangles in Tables

(Let white = 1)

Place the rod rectangle for 3 × 4 on this multiplication table as shown.

×	1	2	3	4	5
1	*purple*				
2	*purple*				
3	*purple*				
4					
5					

3 × 4 can be seen as
4 + 4 + 4 or 12.

Write the product 12 in the lower right corner, as if you were an artist "signing" your rod picture.

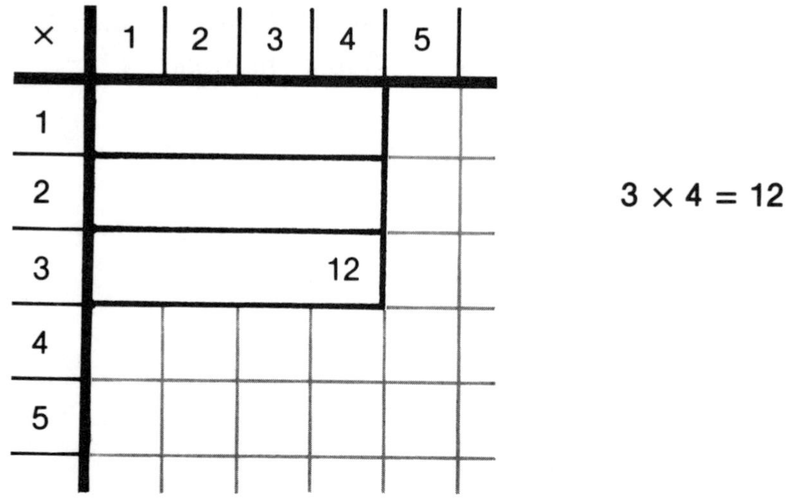

3 × 4 = 12

Now find the product and "sign" this rod rectangle in the lower right corner.

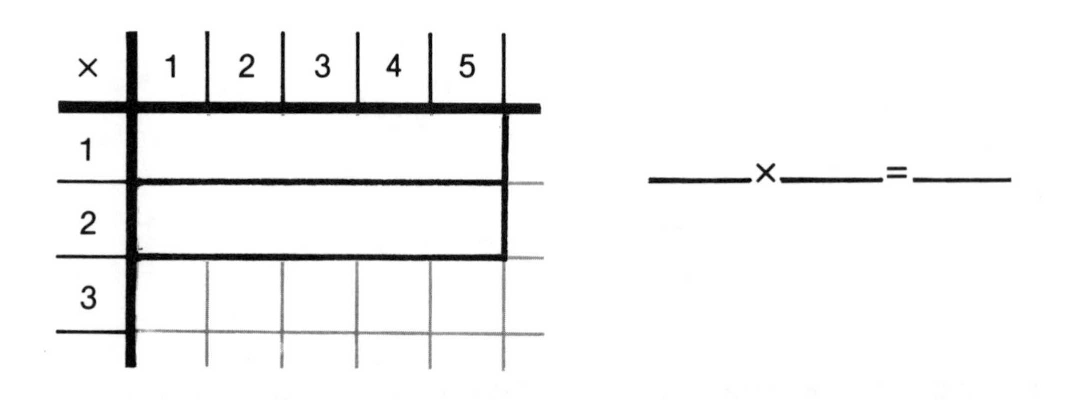

_____ × _____ = _____

Seeing Rectangles in Tables

(Let white = 1)

Use the product to sign each rectangle in the lower right corner. Fill in the multiplication sentence.

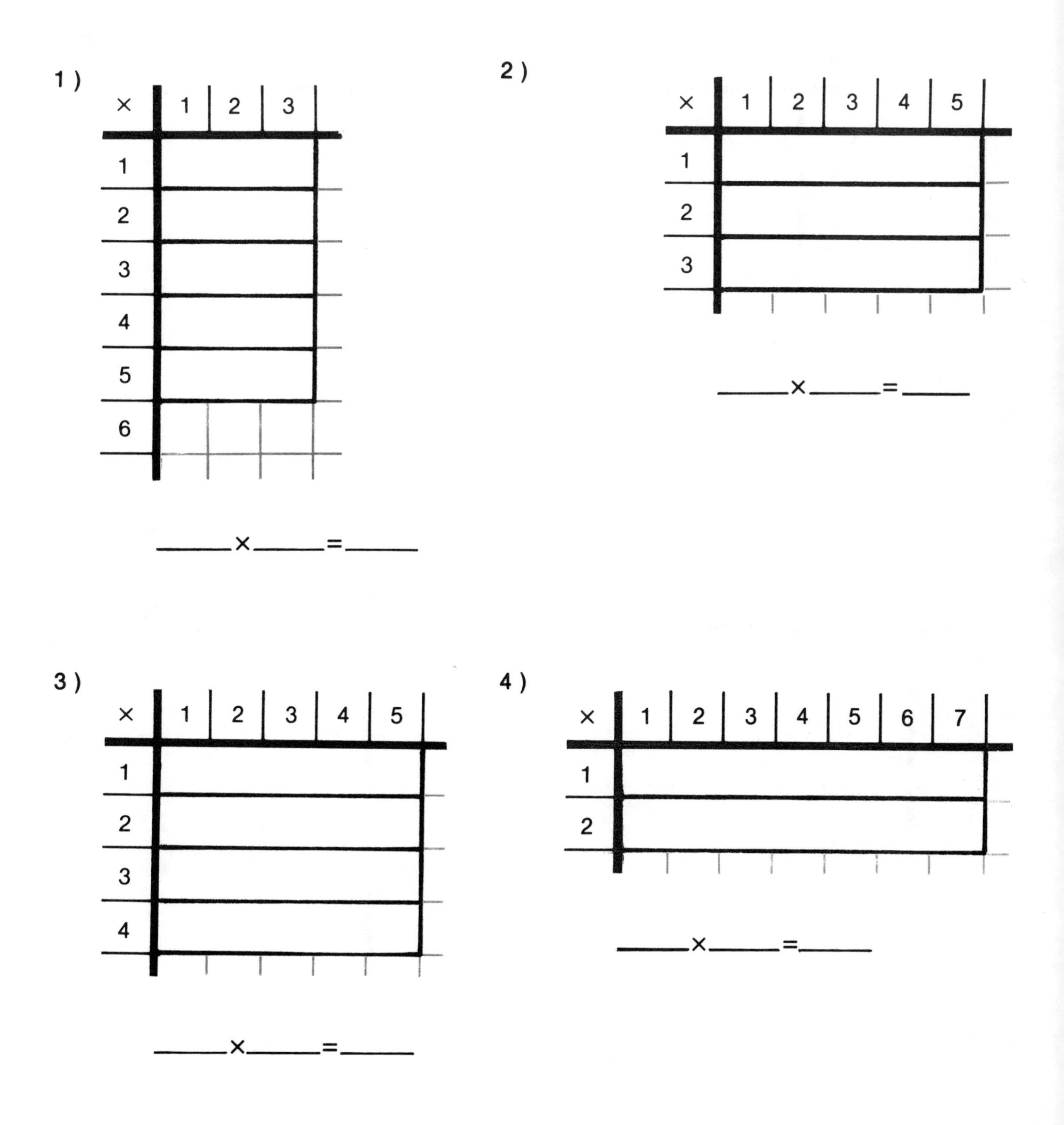

Building Rectangles on Tables

(Let white = 1)

Make and draw the rod rectangle for each multiplication fact. Then sign each rod picture in the lower right corner.

1)

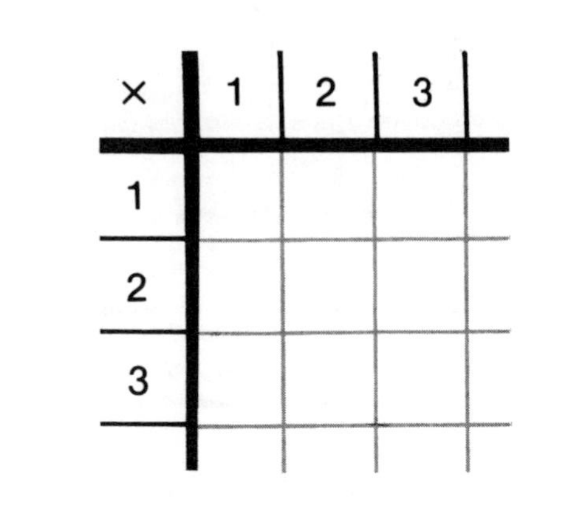

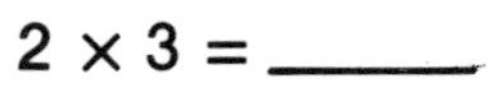

$2 \times 3 =$ ______

2)

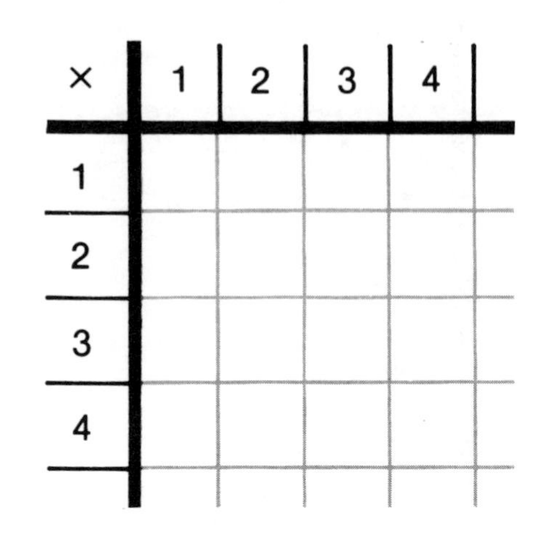

$4 \times 3 =$ ______

3)

4)

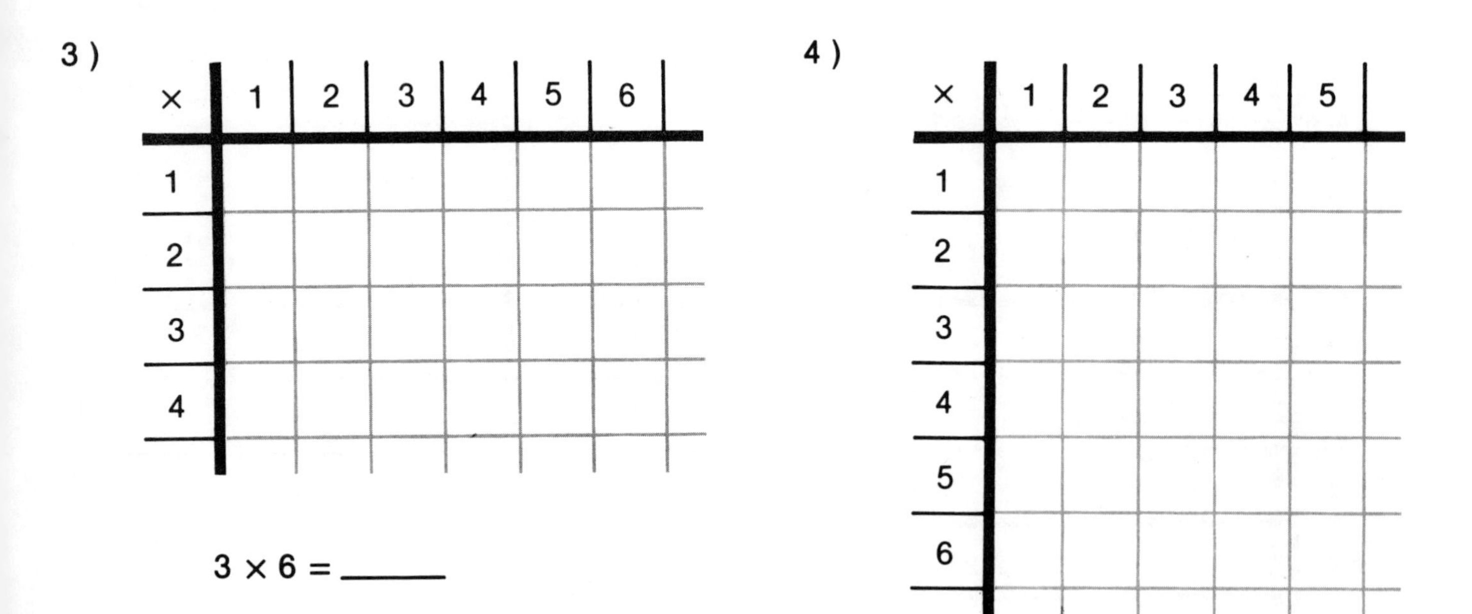

$3 \times 6 =$ ______

$5 \times 4 =$ ______

Building Rectangles on Tables

(Let white = 1)

Make and draw the rod rectangle for each multiplication fact. Use the product to sign each rod picture in the lower right corner.

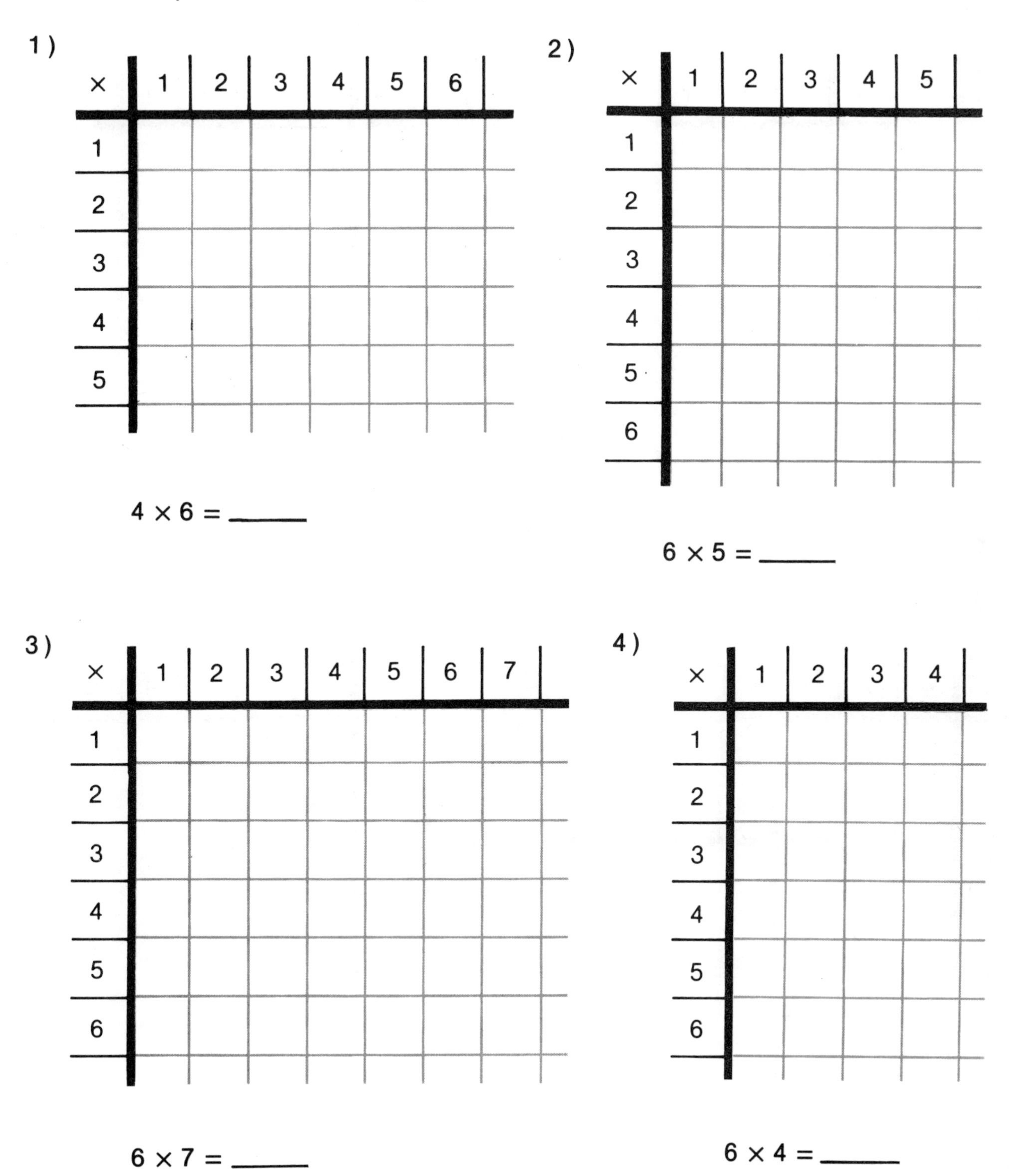

Building Rectangles on Tables

(Let white = 1)

Make and draw the rod rectangle for each multiplication fact. Then sign each rod picture in the lower right corner.

1)

2)

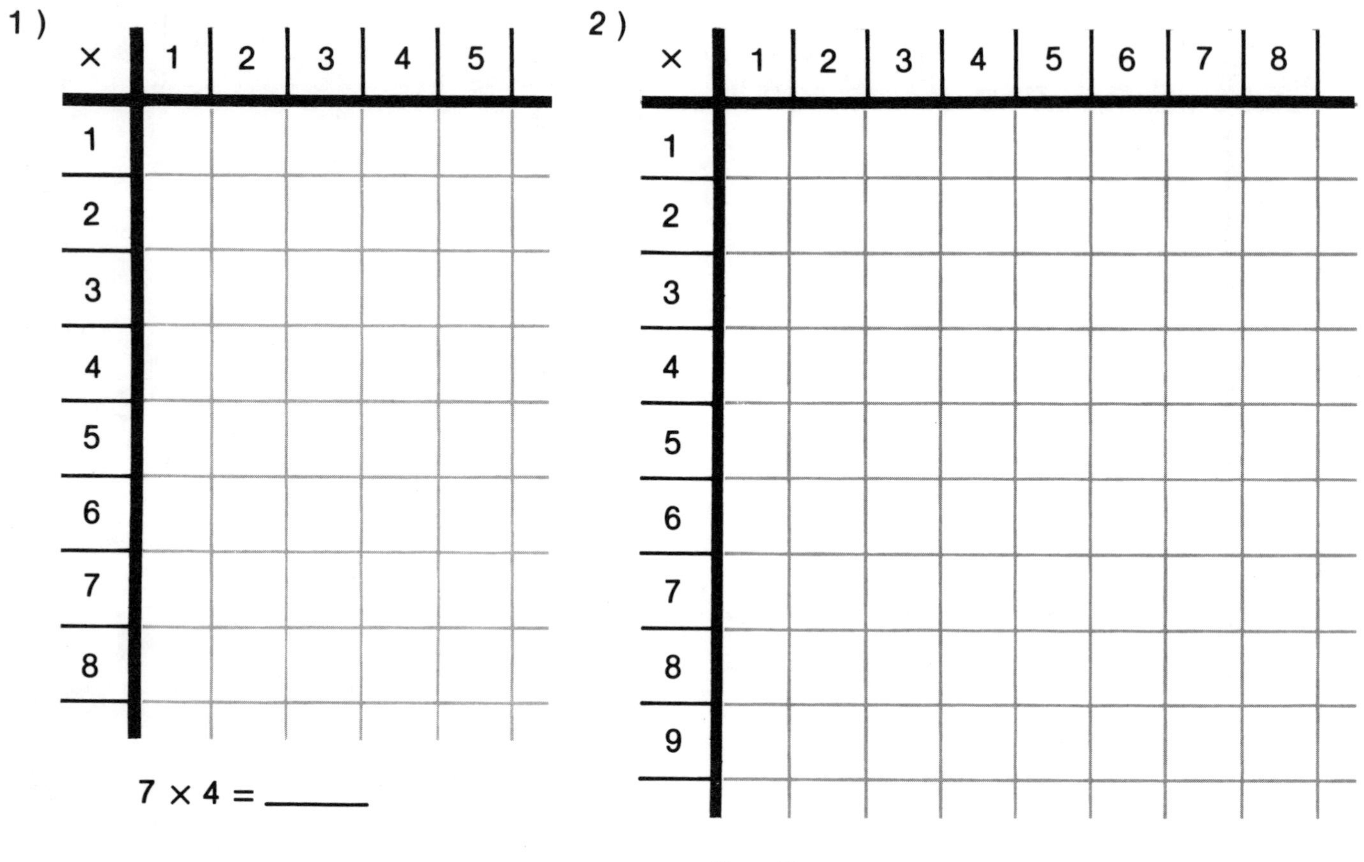

$7 \times 4 =$ ______

$8 \times 6 =$ ______

3)

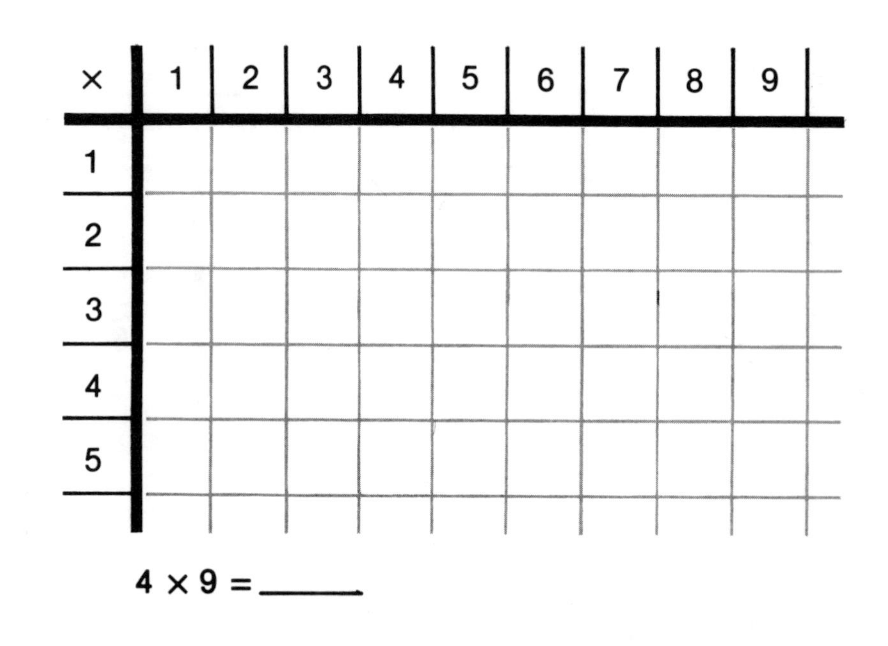

$4 \times 9 =$ ______

Building Rectangles on Tables

(Let white = 1)

Make and draw the rod rectangle for each multiplication fact. Then sign each rod picture in the lower right corner.

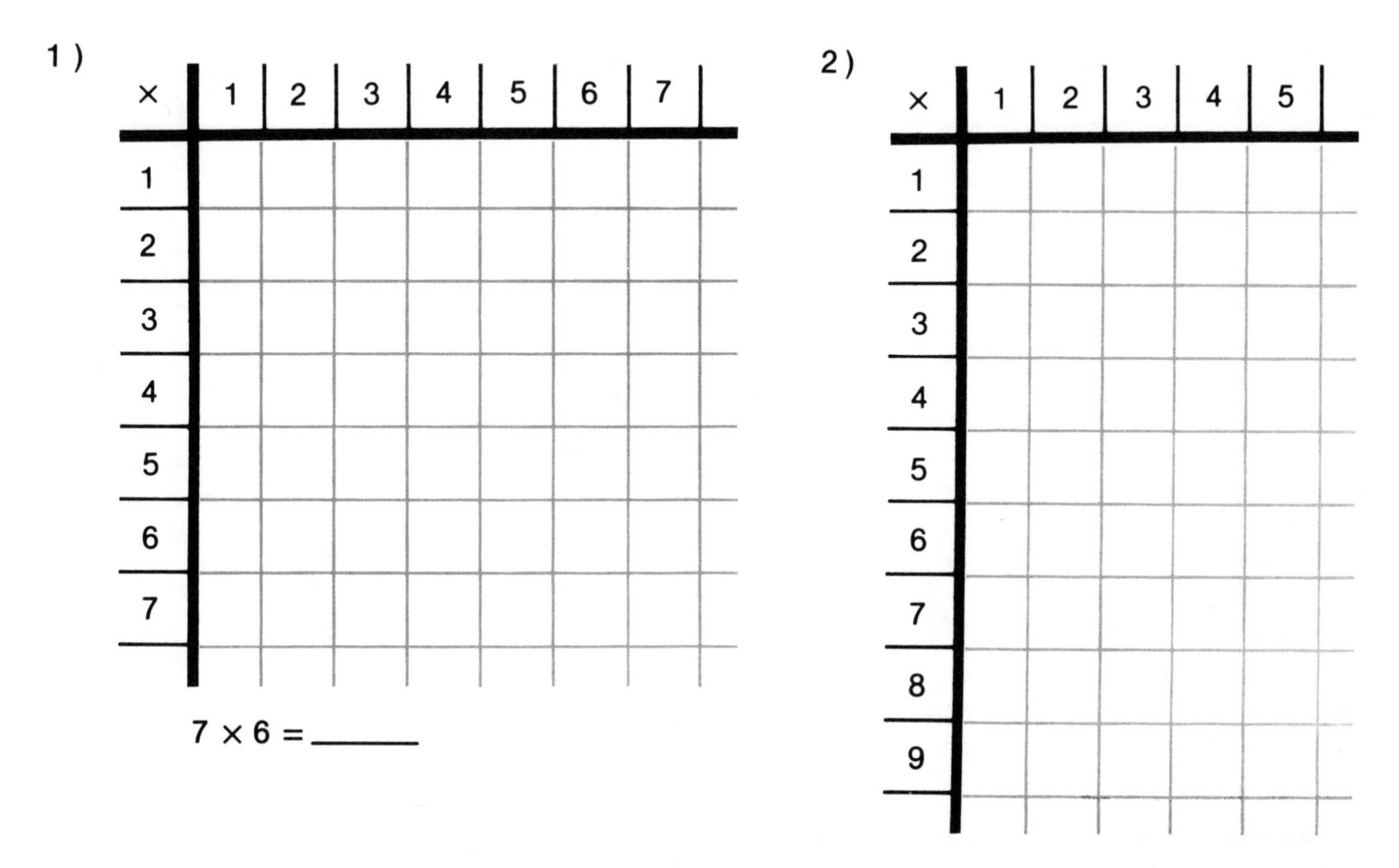

1) $7 \times 6 =$ ______

2) $9 \times 5 =$ ______

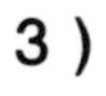

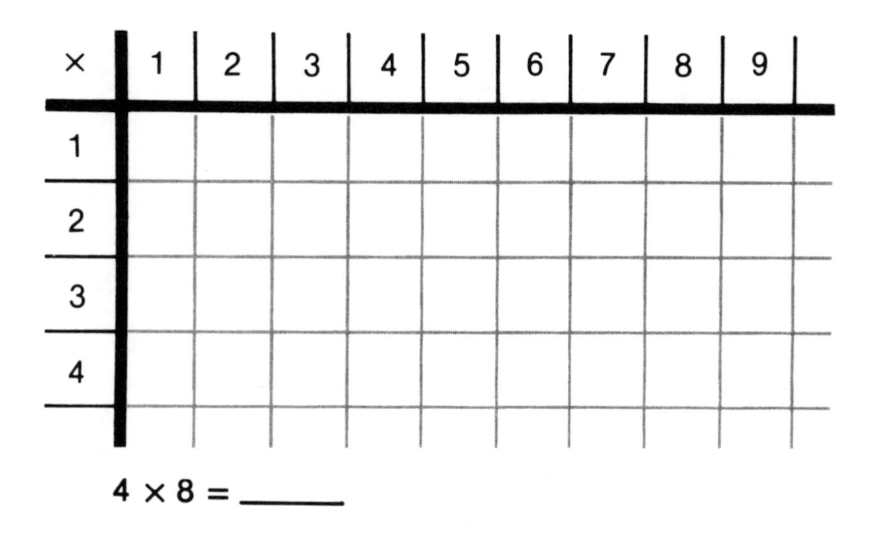

3) $4 \times 8 =$ ______

Finding Facts from Tables

(Let white = 1)

Make and draw the rod rectangle and write the multiplication fact for each product.

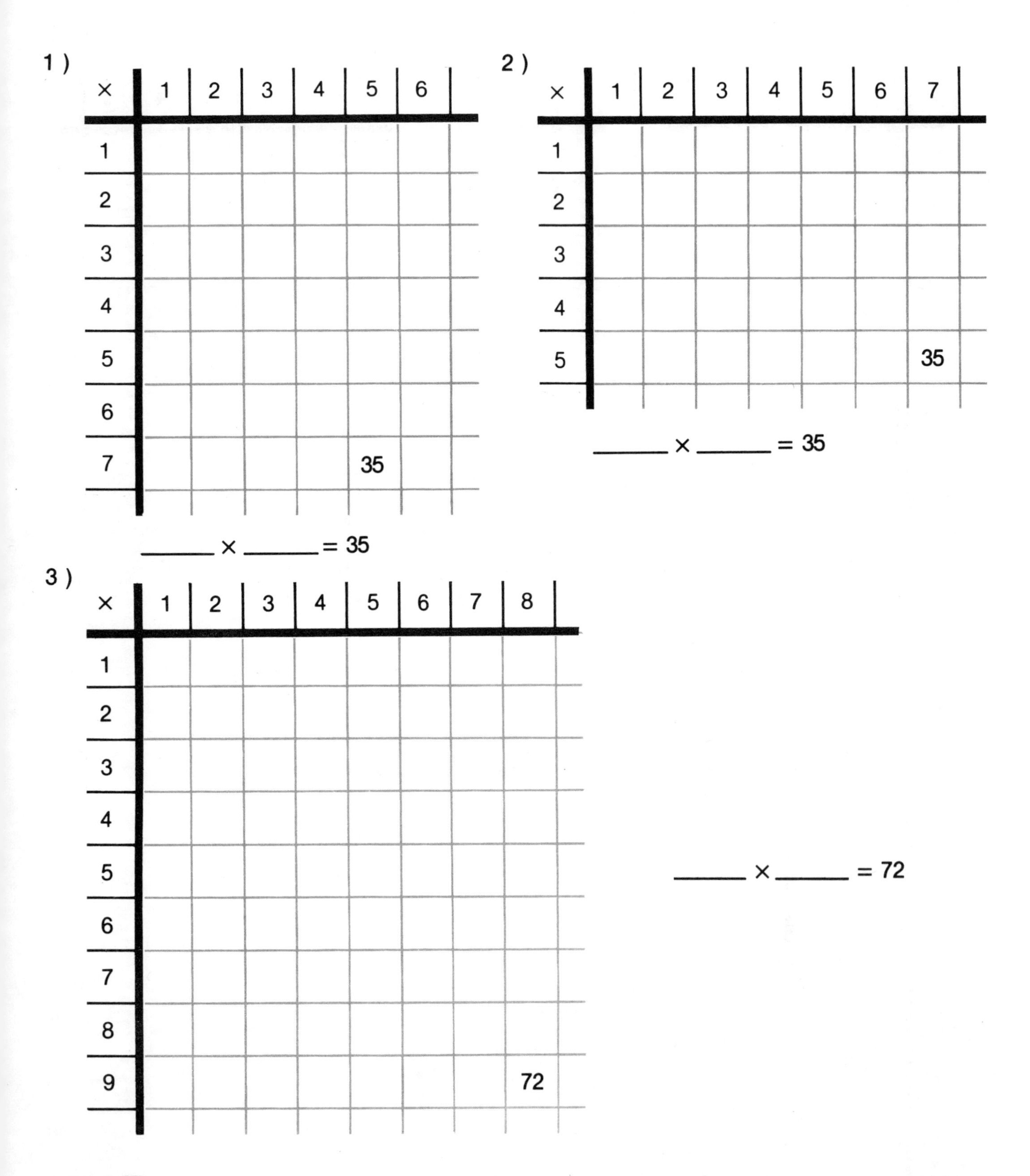

Finding Facts from Tables

(Let white = 1)

Make and draw the rod rectangle and write the multiplication fact for each product.

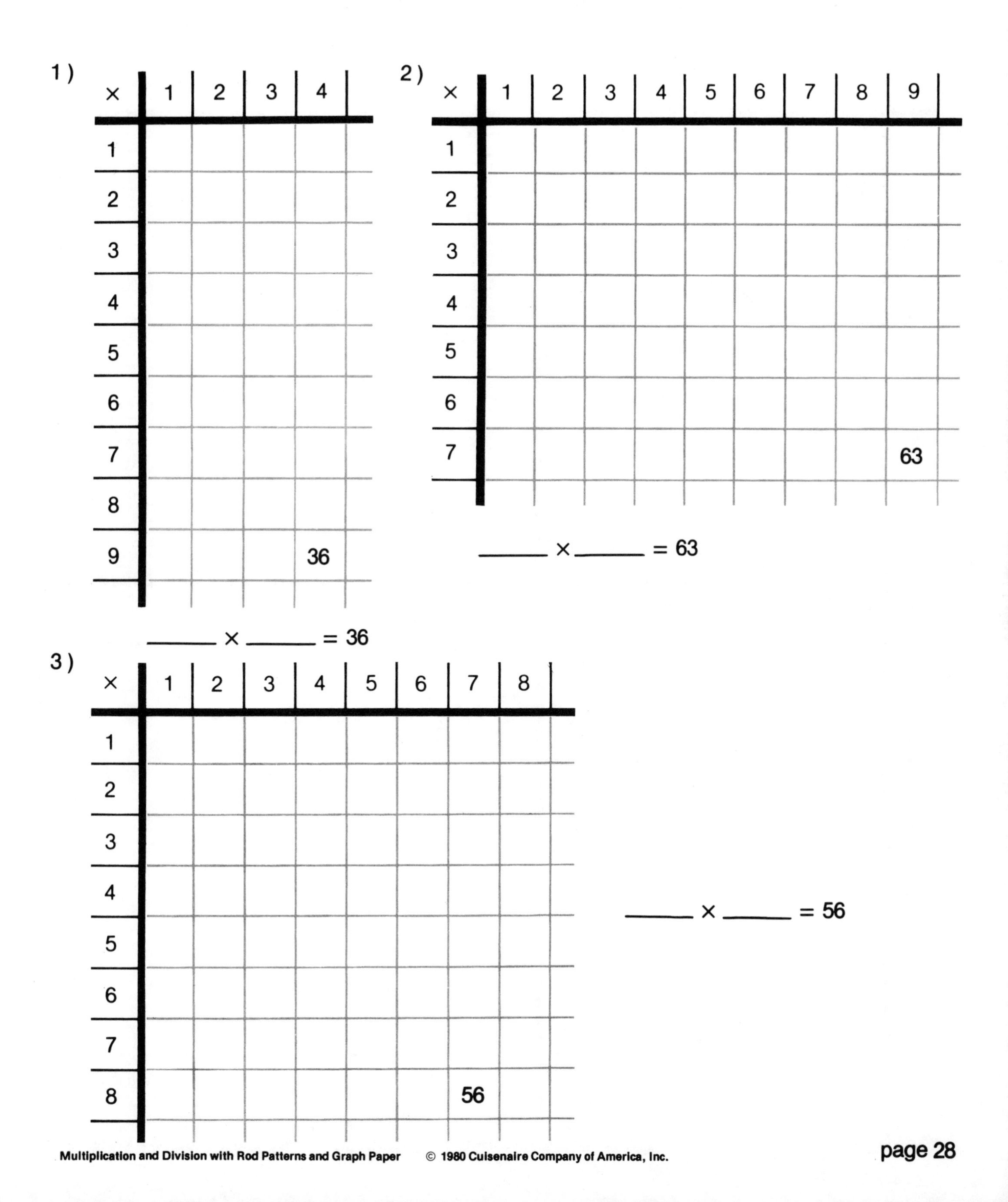

Creating Words

Solve each multiplication fact. Write and circle your answers in the table. If all of the answers are correct, you will see a word.

1)

5 × 5	6 × 5
3 × 8	5 × 4
4 × 2	5 × 8
6 × 8	4 × 5
3 × 5	6 × 2
5 × 3	7 × 5
7 × 2	5 × 2
4 × 8	7 × 8
3 × 2	

×	1	2	3	4	5	6	7	8	9
1									
2									
3									
4									
5					(25)				
6									
7									
8									
9									

2)

4 × 1	7 × 3
5 × 6	4 × 9
7 × 1	3 × 2
3 × 9	5 × 3
4 × 6	6 × 1
3 × 1	7 × 6
6 × 4	6 × 9
3 × 4	3 × 3
7 × 8	7 × 7
5 × 1	3 × 6
3 × 7	7 × 2
5 × 9	3 × 8
6 × 6	5 × 4
7 × 4	7 × 9

×	1	2	3	4	5	6	7	8	9
1									
2									
3									
4									
5									
6									
7									
8									
9									

Creating Words

Solve each multiplication fact. Write and circle your answers in the table. If all of the answers are correct, you will see a word.

1)

4 × 4
1 × 3
4 × 8
5 × 5
6 × 7
7 × 6
3 × 2
3 × 5
7 × 9
6 × 4
5 × 2
8 × 7
4 × 9
2 × 1

6 × 8
5 × 4
2 × 2
8 × 8
7 × 4
3 × 6
5 × 7
4 × 2
8 × 9
7 × 5
1 × 1
3 × 4
2 × 3

×	1	2	3	4	5	6	7	8	9
1									
2									
3									
4									
5									
6									
7									
8									
9									

2)

4 × 8
5 × 3
8 × 2
7 × 5
2 × 7
3 × 5
7 × 3
6 × 4

1 × 7
6 × 3
3 × 7
7 × 1
9 × 3
3 × 6
7 × 2

×	1	2	3	4	5	6	7	8	9
1									
2									
3									
4									
5									
6									
7									
8									
9									

Creating Shapes

Solve each multiplication fact. Write and circle your answers in the table. If all of the answers are correct, you will see a geometric shape. Name the shape.

1)

6 × 8	6 × 4
5 × 5	9 × 7
7 × 9	5 × 7
4 × 6	8 × 6
8 × 8	7 × 5

Shape: ________

×	1	2	3	4	5	6	7	8	9
1									
2									
3									
4									
5									
6									
7									
8									
9									

2)

5 × 8	5 × 5
7 × 4	4 × 7
3 × 6	9 × 3
8 × 5	7 × 7
6 × 9	

Shape: ________

×	1	2	3	4	5	6	7	8	9
1									
2									
3									
4									
5									
6									
7									
8									
9									

Creating Shapes

Solve each multiplication fact. Write and circle your answers in the table. If all of the answers are correct, you will see a geometric shape. Name the shape.

1)

5 × 7	3 × 6
9 × 4	4 × 4
7 × 3	8 × 6
5 × 2	7 × 8

Shape: __________

×	1	2	3	4	5	6	7	8	9
1									
2									
3									
4									
5									
6									
7									
8									
9									

2)

6 × 6	6 × 2
9 × 5	8 × 4
7 × 3	6 × 3
6 × 4	7 × 7
9 × 7	6 × 5
8 × 8	9 × 6
9 × 8	9 × 9

Shape: __________

×	1	2	3	4	5	6	7	8	9
1									
2									
3									
4									
5									
6									
7									
8									
9									

Solving Picture Riddles

Solve each multiplication fact. Write and circle your answers in the table. Your picture will solve the riddle.

Don't lose me, I'm important.
Keep me close by your side.
You can use me to enter,
Or when you go for a ride.

3 × 6
4 × 9
5 × 7
6 × 4
3 × 9
8 × 2
5 × 5
8 × 4
7 × 3
2 × 8

5 × 8
6 × 5
9 × 1
7 × 5
4 × 6
2 × 7
7 × 4
9 × 2
5 × 6

×	1	2	3	4	5	6	7	8	9
1									
2									
3									
4									
5									
6									
7									
8									
9									

Who am I ? ____________________

Solving Picture Riddles

Solve each multiplication fact. Write and circle your answers in the table. Your picture will solve the riddle.

Fluffy and quiet,
Bouncy and quick.
Always so silent,
Even in a hat trick.

4×5	8×8
1×8	7×5
9×7	5×7
6×5	3×6
2×8	6×8
8×5	9×6
5×6	1×5
3×7	4×8
7×8	2×5

×	1	2	3	4	5	6	7	8	9
1									
2									
3									
4									
5									
6									
7									
8									
9									

Who am I ? ______________

Solving Picture Riddles

Solve each multiplication fact. Write and circle your answers in the table. Your picture will solve the riddle.

Get some exercise,
And have fun too.
Learn how to balance
And away with you.

7×3
9×7
5×2
8×4
4×7
6×3
8×2
5×8
9×3
6×5

5×3
6×6
4×9
8×8
6×7
5×4
6×4
8×6
7×7

×	1	2	3	4	5	6	7	8	9
1									
2									
3									
4									
5									
6									
7									
8									
9									

Who am I ? ____________________

Solving Picture Riddles

Solve each multiplication fact. Write and circle your answers in the table. Your picture will solve the riddle.

A computer is my master.
You cannot see my face.
I speak another language
I come from outer space.

9 × 5	7 × 7	3 × 4
7 × 3	2 × 5	9 × 7
5 × 5	4 × 3	4 × 7
3 × 7	9 × 6	8 × 4
10 × 4	5 × 4	5 × 6
1 × 5	3 × 5	7 × 8
8 × 7	9 × 4	6 × 5
4 × 8	4 × 4	7 × 4
7 × 5	5 × 7	10 × 7
3 × 6	7 × 6	2 × 6
6 × 6	1 × 6	

×	1	2	3	4	5	6	7	8	9
1									
2									
3									
4									
5									
6									
7									
8									
9									
10									

Who am I ?______________________

Making Larger Rectangles

(Let white = 1)

Here is the rod rectangle for $\begin{array}{r} 12 \\ \times 4 \\ \hline \end{array}$

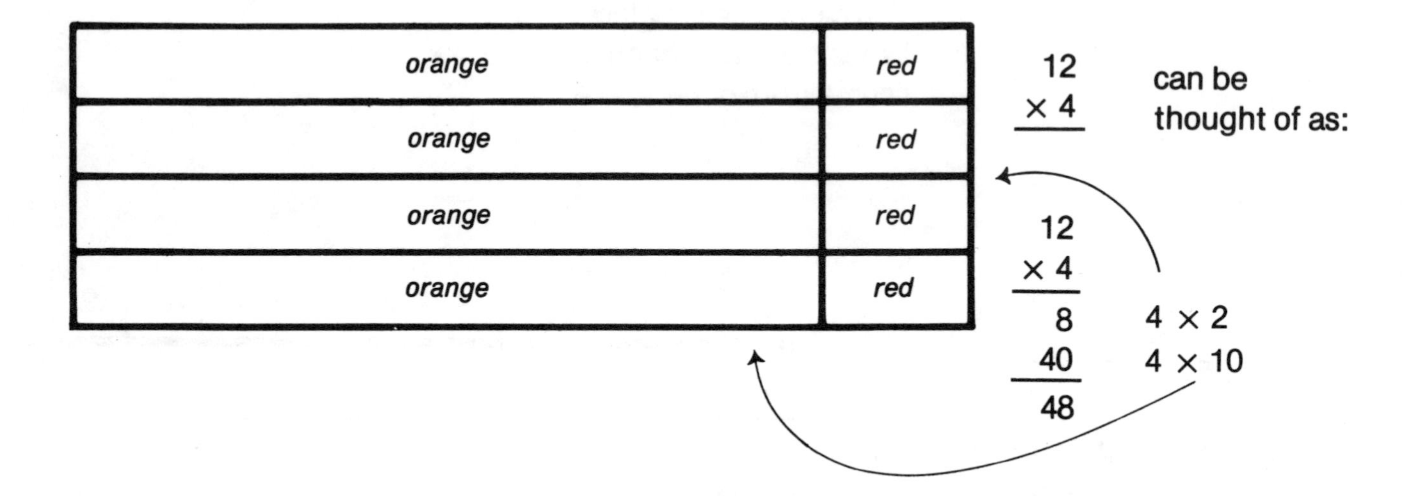

Make and draw the rod rectangle for 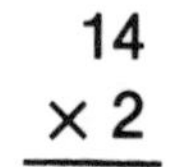

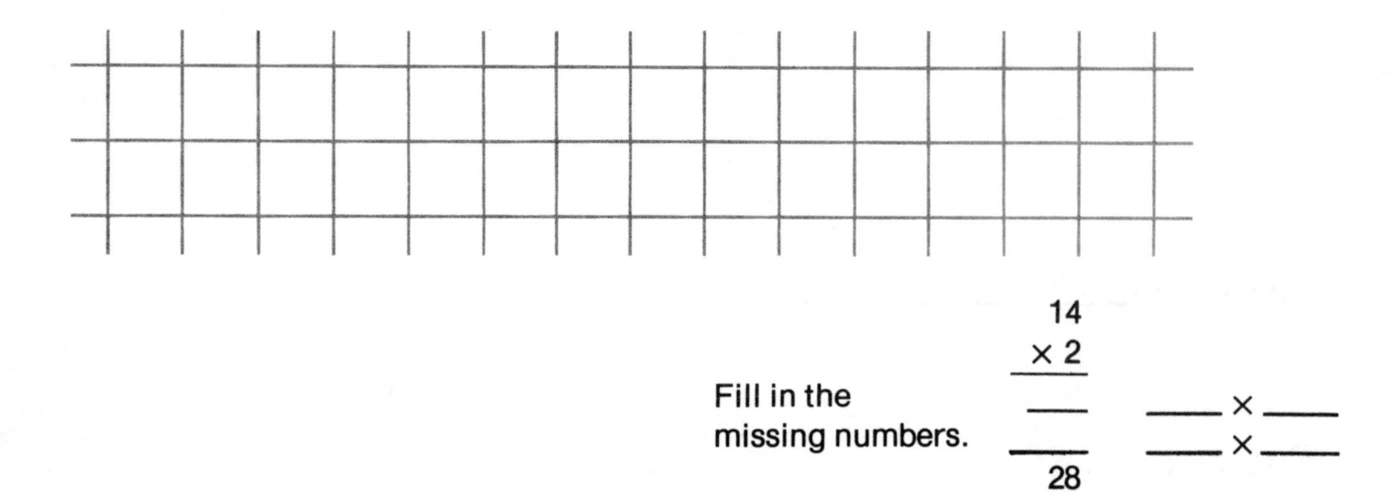

Extending the Table

(Let white = 1)

1) The rod rectangle for 4 × 13 is shown on the table. Find the product and write the answer in the lower right corner of the rod picture.

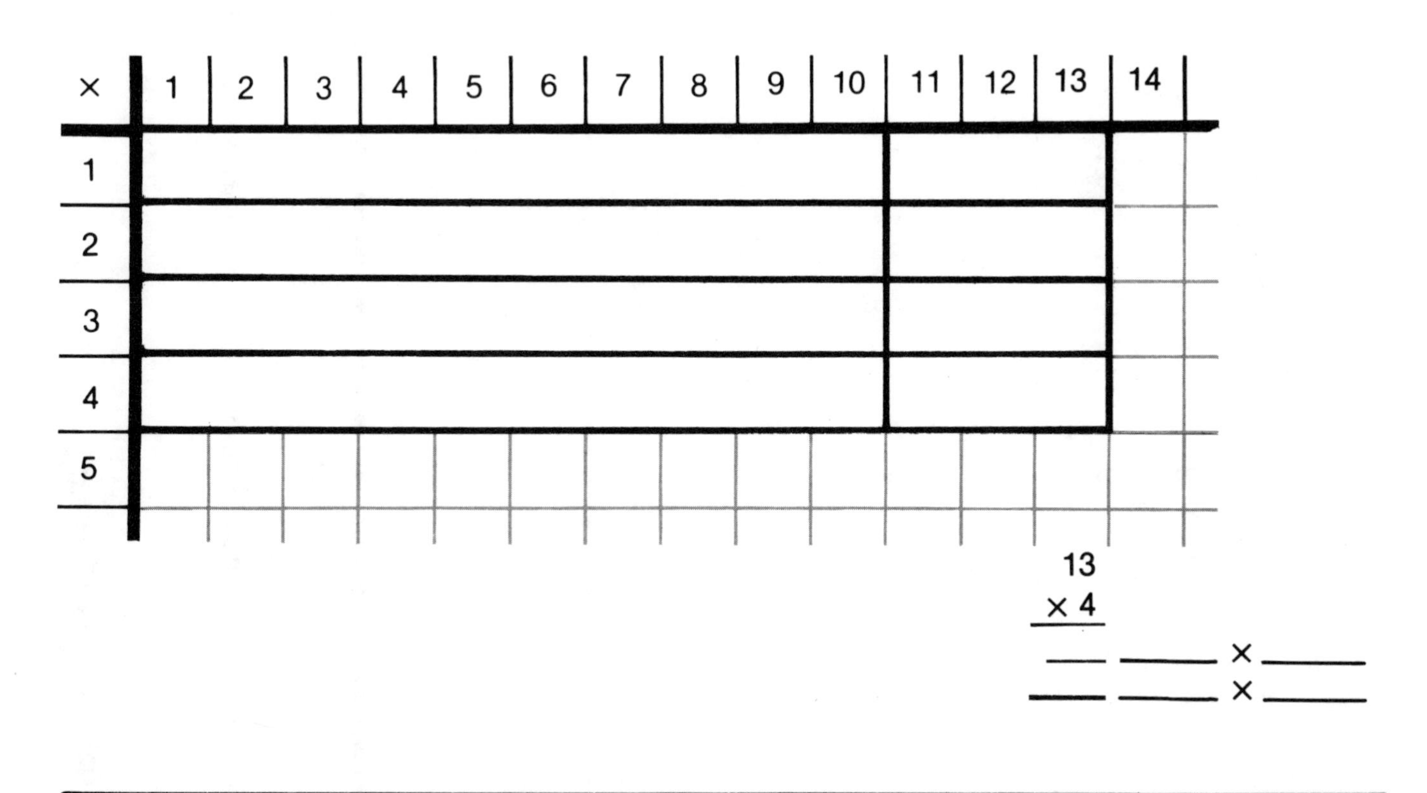

$$\begin{array}{r} 13 \\ \times\ 4 \\ \hline \end{array}$$

___ _____ × _____

___ _____ × _____

---

2) Make and draw the rod rectangle for 3 × 14. Then find the product and write the answer in the lower right corner of the rod picture.

×	1	2	3	4	5	6	7	8	9	10	11	12	13	14
1														
2														
3														
4														

$$\begin{array}{r} 14 \\ \times\ 3 \\ \hline \end{array}$$

___ _____ × _____

___ _____ × _____

Creating Longer Words

Solve each multiplication fact. Write and circle your answers in the table. If all of the answers are correct, you will see a word.

×	1	2	3	4	5	6	7	8	9	10	11	12	13	14	15	16	17	18	19	20
1																				
2																				
3																				
4																				
5																				
6																				
7																				
8																				
9																				

4 × 6	5 × 13	8 × 14	6 × 18	5 × 9	4 × 19	8 × 17	4 × 13	6 × 2
8 × 9	7 × 4	4 × 9	4 × 2	8 × 19	8 × 4	4 × 11	8 × 18	4 × 17
6 × 11	8 × 3	7 × 11	8 × 11	6 × 3	6 × 13	6 × 6	6 × 4	8 × 13
5 × 2	6 × 9	6 × 17	5 × 7	4 × 4	5 × 11	5 × 8	5 × 17	7 × 17
7 × 9	4 × 18	5 × 6	7 × 6	7 × 13	4 × 3	8 × 15	8 × 2	8 × 6

Creating Longer Words

Solve each multiplication fact. Write and circle your answers in the table. If all of the answers are correct, you will see a word.

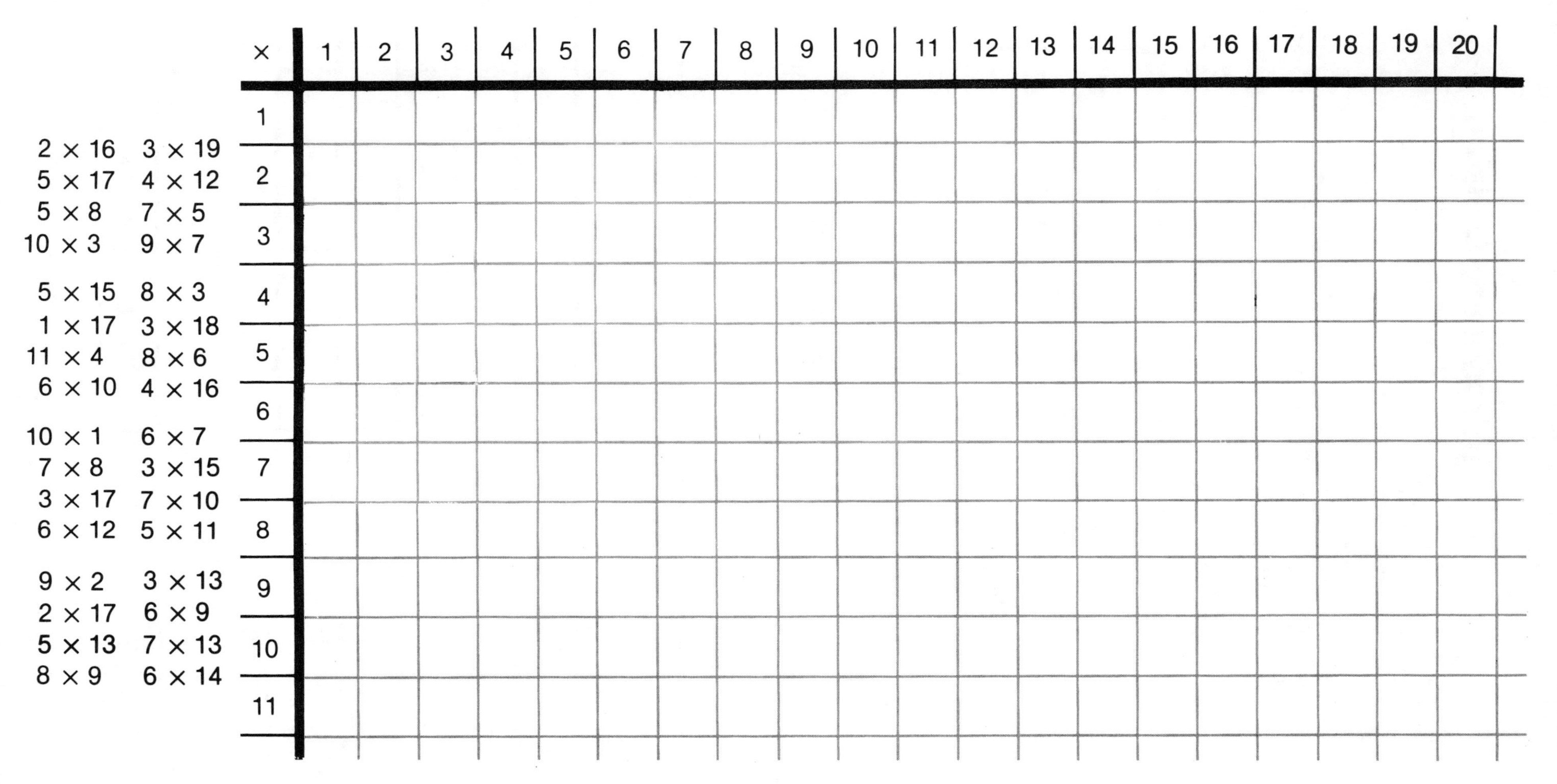

2×16 3×19
5×17 4×12
5×8 7×5
10×3 9×7

5×15 8×3
1×17 3×18
11×4 8×6
6×10 4×16

10×1 6×7
7×8 3×15
3×17 7×10
6×12 5×11

9×2 3×13
2×17 6×9
5×13 7×13
8×9 6×14

Introducing Division

(Let white = 1)

$8 \div 2 = ?$

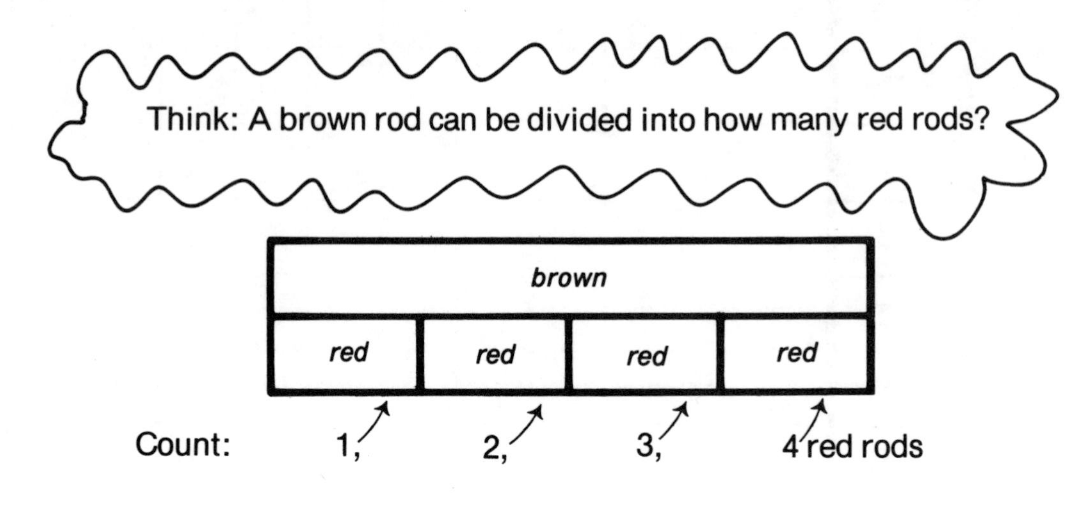

Write this result as:

$8 \div 2 = 4$ (Read as ''8 divided by 2 equals 4.'')

---

1) $10 \div 5 =$ ______

2) $9 \div 3 =$ ______

3) $6 \div 1 =$ ______

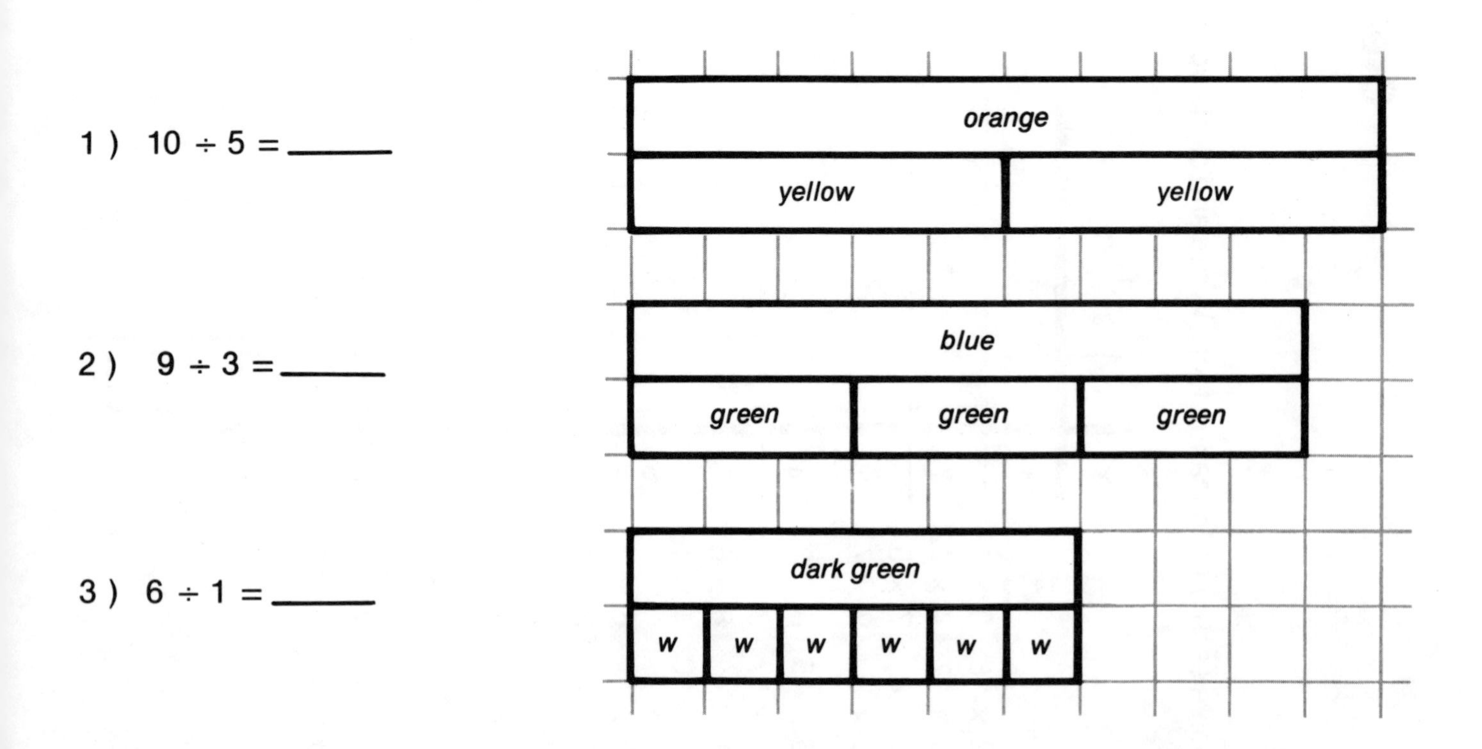

Introducing Division

(Let white = 1)

12 ÷ 4 = ?

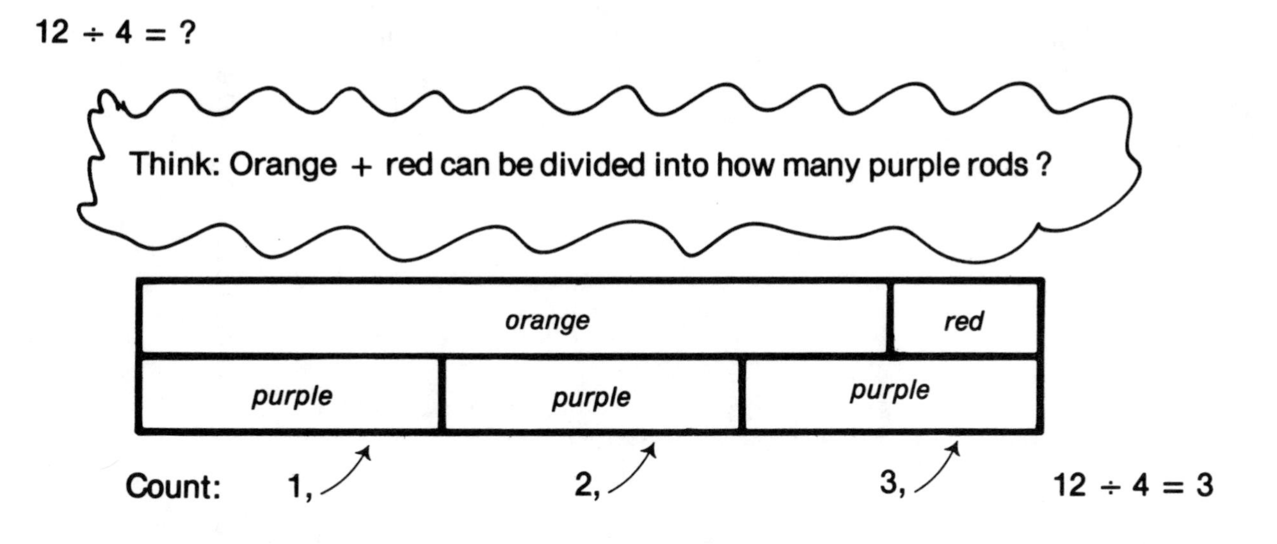

Draw the rod picture to show the division. Fill in each answer (quotient).

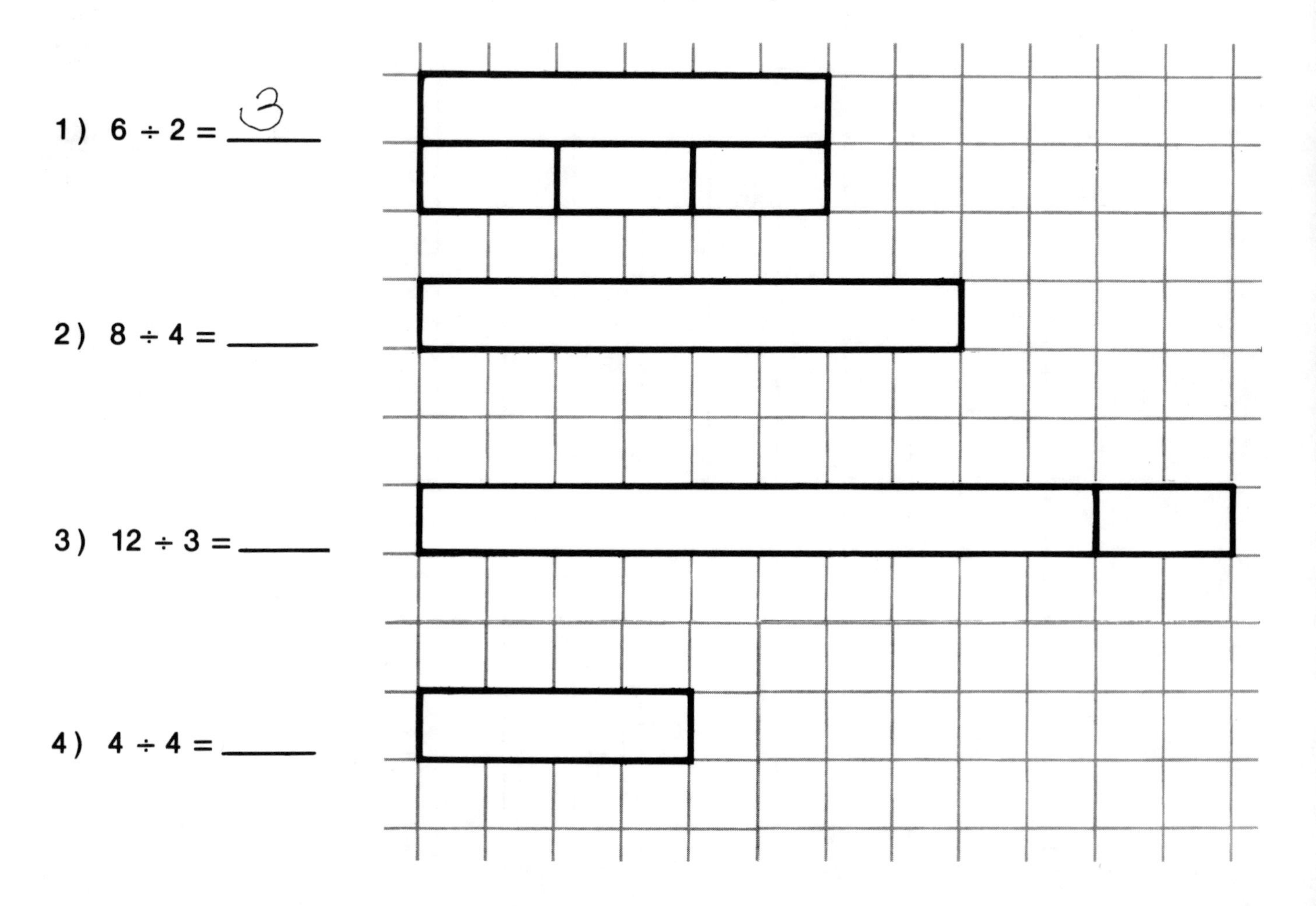

Practicing Division

(Let white = 1)

Draw the rod picture to show the division. Fill in the answer (quotient).

1) $15 \div 3 =$ ______

2) $10 \div 2 =$ ______

3) $14 \div 7 =$ ______

4) $16 \div 4 =$ ______

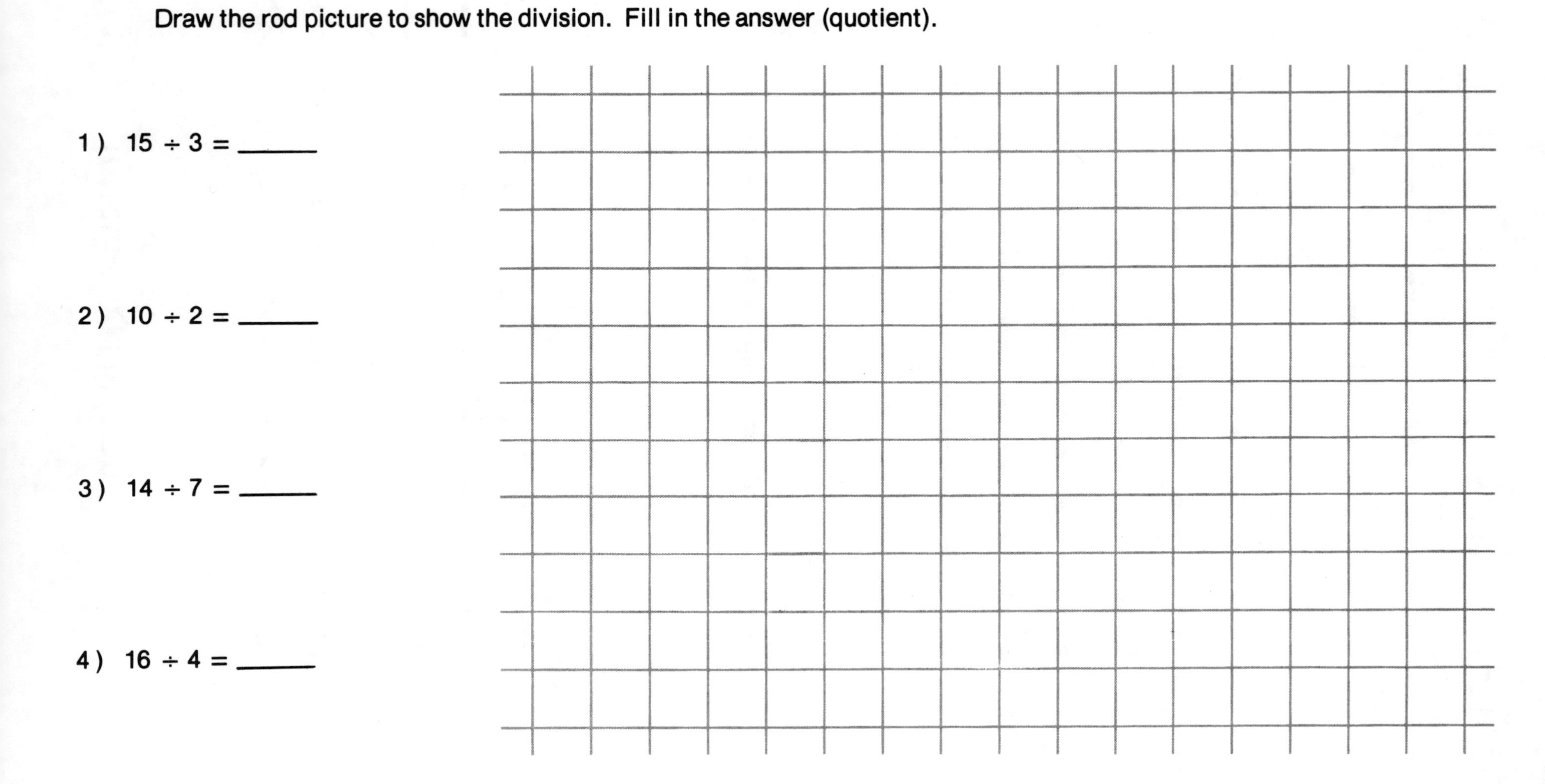

Practicing Division

(Let white = 1)

Draw the rod picture to show the division. Fill in each answer (quotient).

1) $18 \div 3 =$ ______

2) $15 \div 5 =$ ______

3) $10 \div 10 =$ ______

4) $16 \div 2 =$ ______

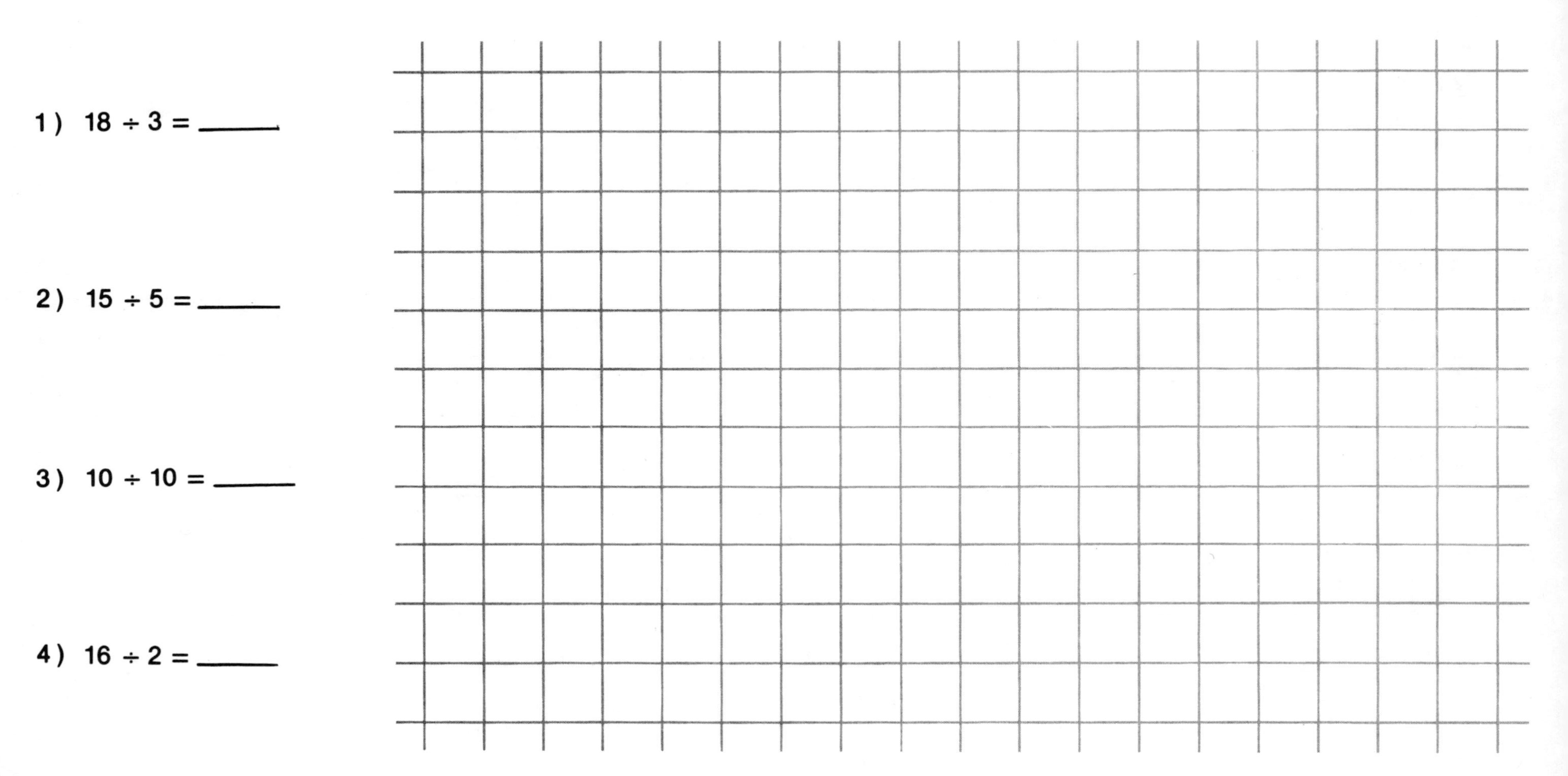

Dividing Larger Numbers

(Let white = 1)

Draw the rod picture to show the division. Fill in each answer (quotient).

1) $24 \div 6 =$ ______

2) $21 \div 7 =$ ______

3) $25 \div 5 =$ ______

4) $18 \div 2 =$ ______

Dividing Still Larger Numbers

(Let white = 1)

Draw the rod picture for each division. Fill in each answer (quotient). Make rod trains beside the worksheet to check your answers.

1) 42 ÷ 6 = ______

2) 32 ÷ 4 = ______

3) 36 ÷ 9 = ______

4) 40 ÷ 5 = ______

Writing Division

(Let white = 1)

Write the division problem that matches the rod picture. Make rod trains beside the worksheet to check your answers.

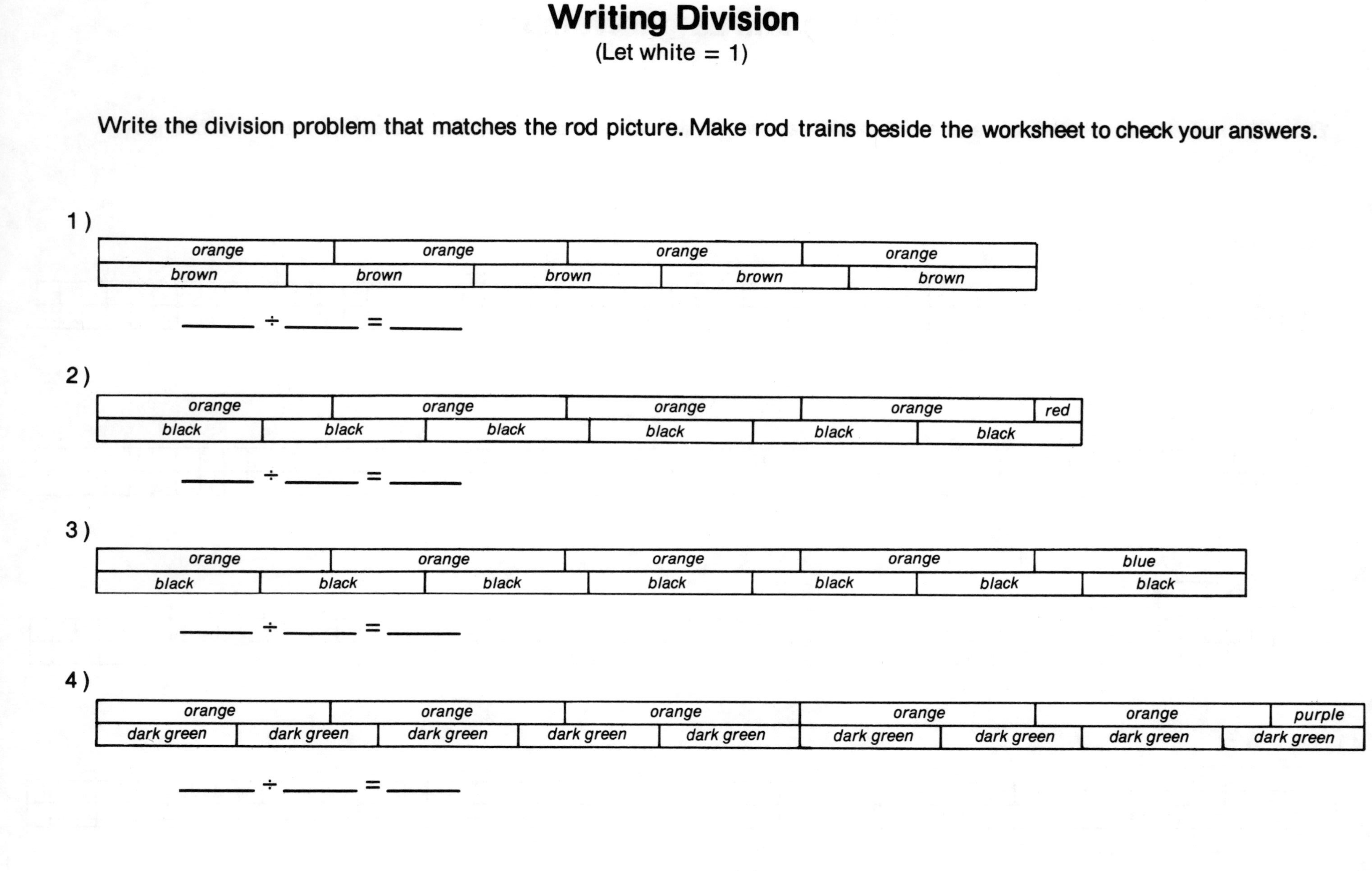

Writing Division

(Let white = 1)

Write the division problem that matches the rod picture. Make rod trains beside the worksheet to check your answers.

1)

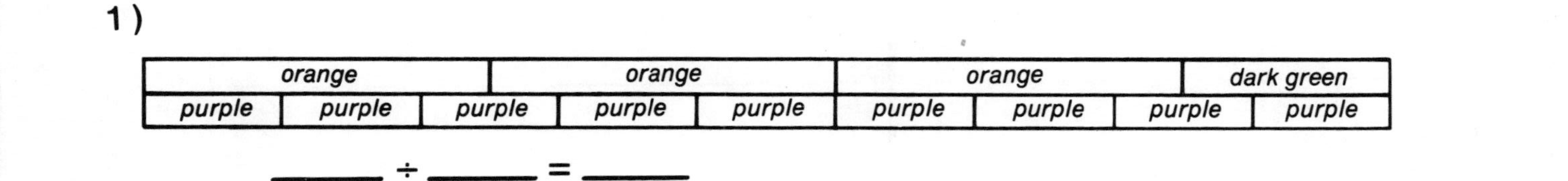

______ ÷ ______ = ______

2)

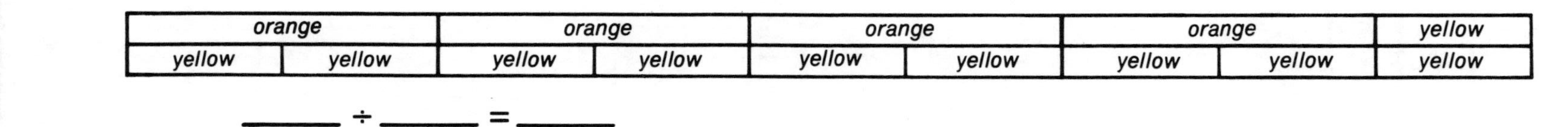

______ ÷ ______ = ______

3)

orange		orange		orange		orange		brown
dark green	dark green	dark green	dark green	dark green	dark green	dark green	dark green	

______ ÷ ______ = ______

4)

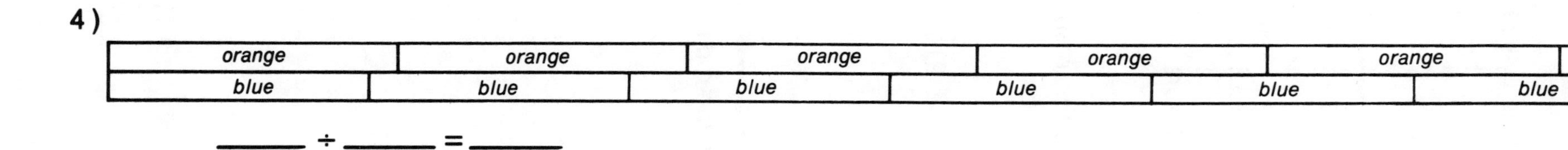

______ ÷ ______ = ______

Writing Division Another Way

(Let white = 1)

Another way of writing 14 ÷ 2 is $2\overline{)14}$.

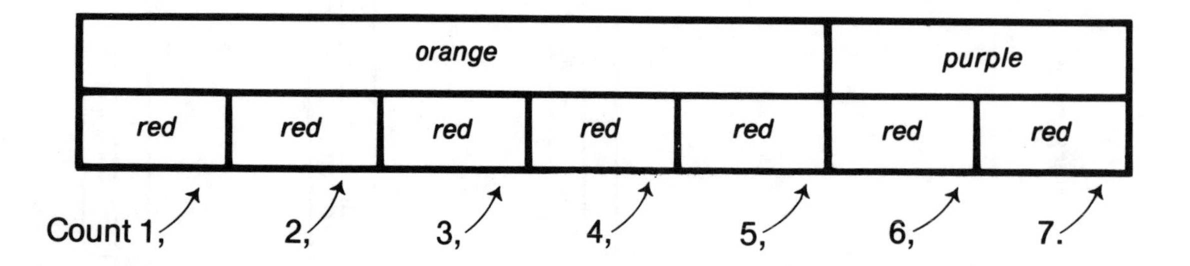

Write this result as:

$$2\overline{)14}^{\;7}$$

(Read as "14 divided by 2 equals 7.")

Draw the rod picture to show the division. Fill in each answer (quotient).

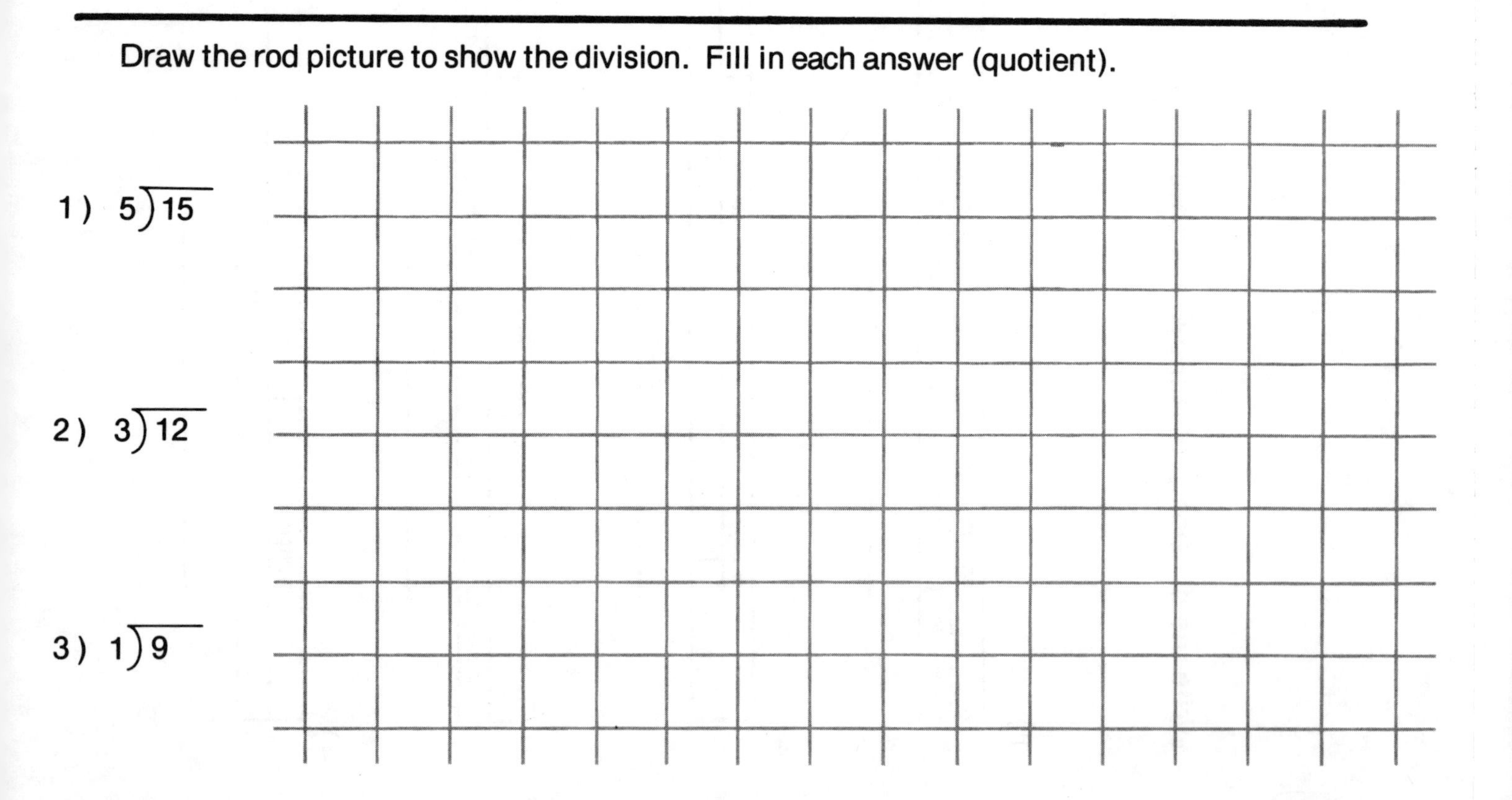

Finding Remainders

(Let white = 1)

Another way of writing 9 ÷ 4 is $4\overline{)9}$.

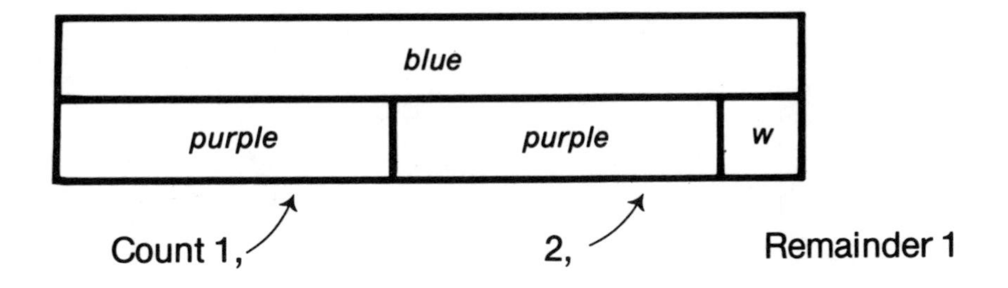

Write the result as:

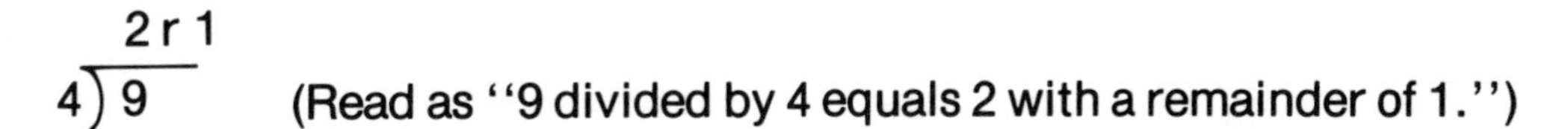

(Read as "9 divided by 4 equals 2 with a remainder of 1.")

Draw the rod picture to show the division. Fill in each answer (quotient and remainder).

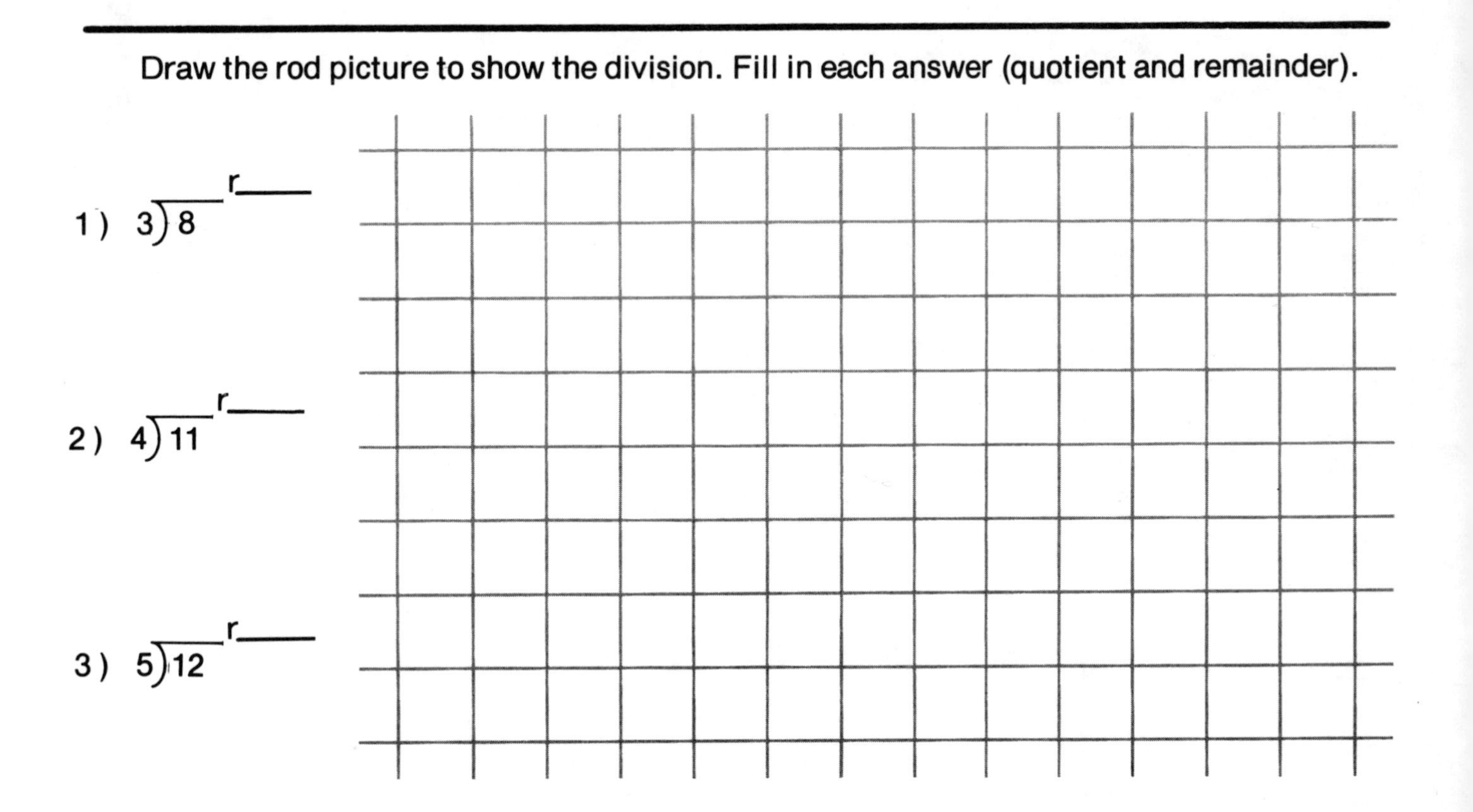

Dividing With Remainders

(Let white = 1)

Draw the rod picture to show the division. Fill in each answer (quotient and remainder).

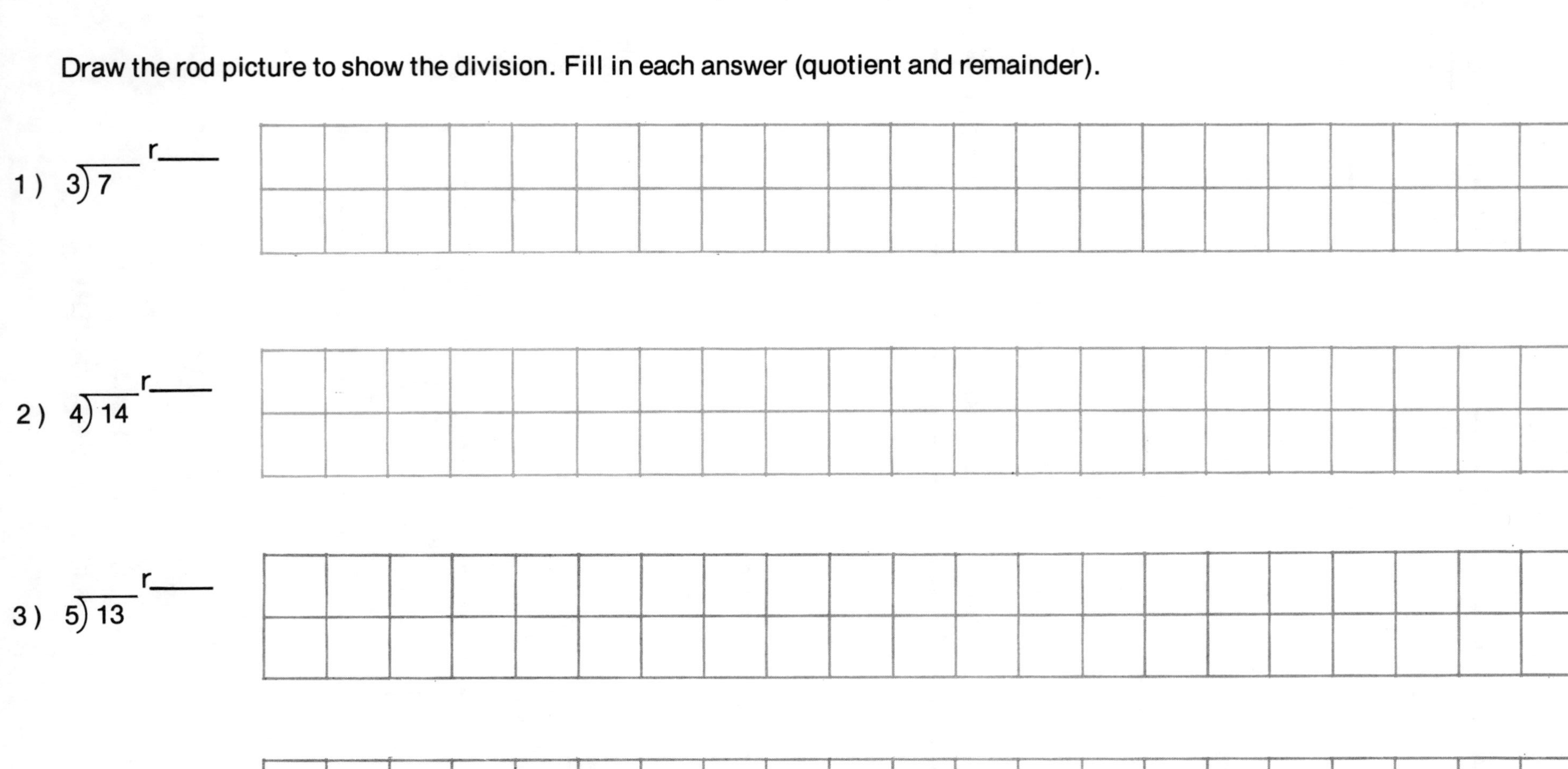

1) $3\overline{)7}$ r____

2) $4\overline{)14}$ r____

3) $5\overline{)13}$ r____

4) $7\overline{)12}$ r____

Dividing With Remainders

(Let white = 1)

Draw the rod picture to show the division. Fill in each answer (quotient and remainder).

Dividing Larger Numbers

(Let white = 1)

Show the rod picture for each division. Fill in each answer (quotient and remainder). Make rod trains beside the worksheet to check your answers.

1) $8\overline{)29}$ r____

2) $9\overline{)30}$ r____

3) $7\overline{)28}$ r____

4) $6\overline{)33}$ r____

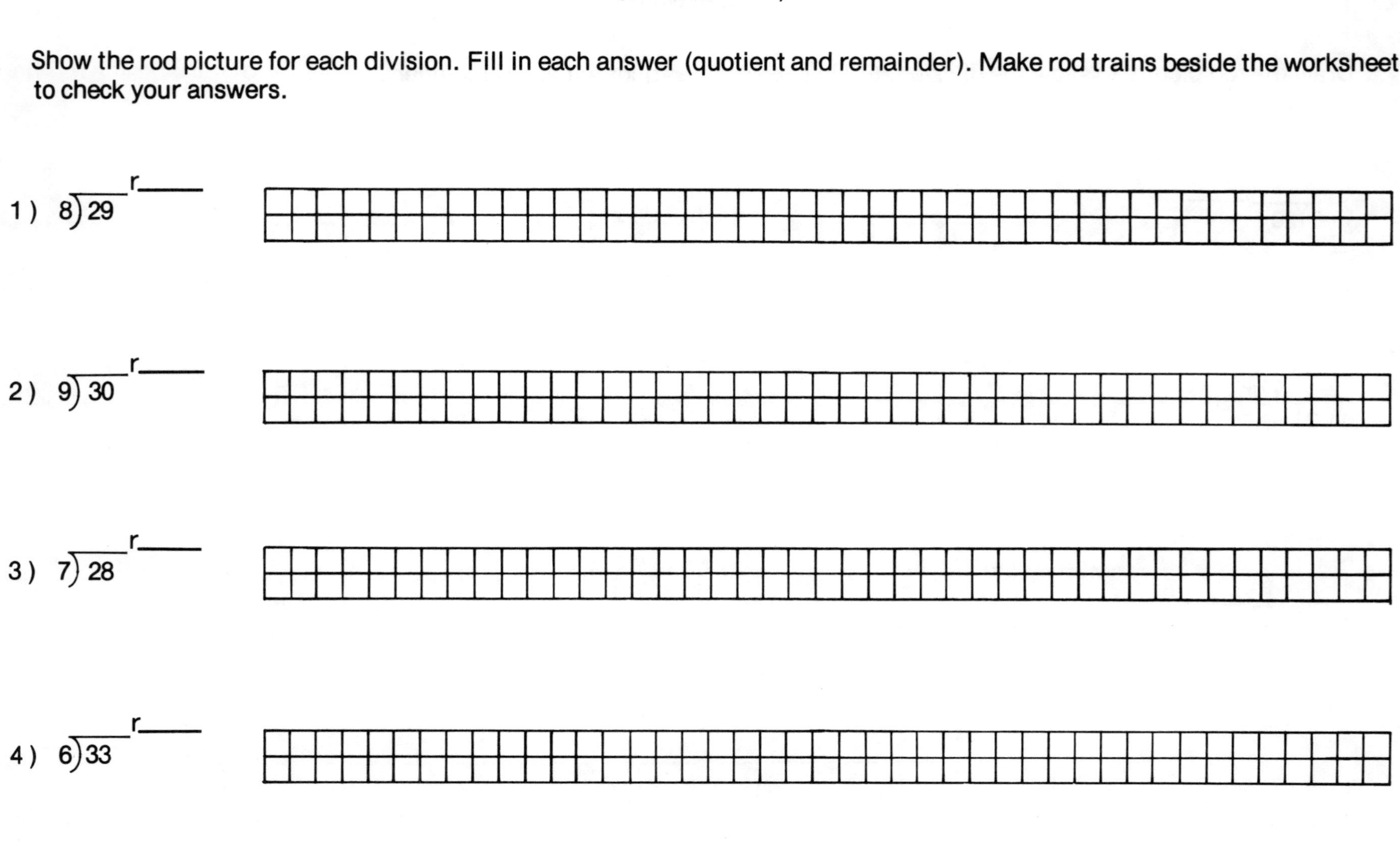

Dividing Larger Numbers

(Let white = 1)

Show the rod picture for each division. Fill in each answer (quotient and remainder). Make rod trains beside the worksheet to check your answers.

1) $6\overline{)40}$ r____

2) $8\overline{)36}$ r____

3) $7\overline{)41}$ r____

4) $8\overline{)39}$ r____

Writing Division With Remainders

(Let white = 1)

Write the division problem that matches the rod picture. Make rod trains beside the worksheet to check your answers.

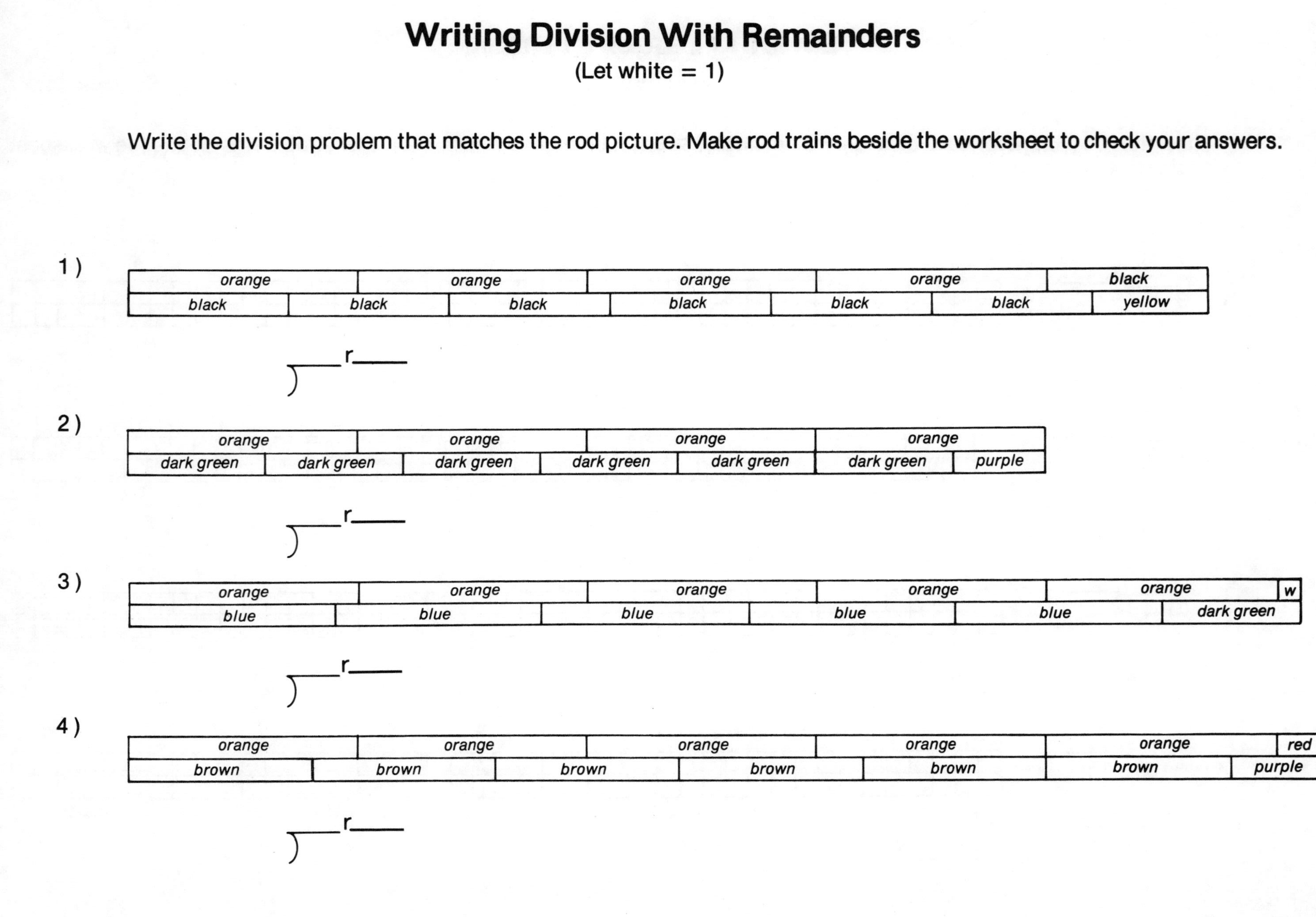

Finding Larger Quotients

(Let white = 1)

Write the division problem that matches the rod picture. Make rod trains beside the worksheet to check your answers.

1)

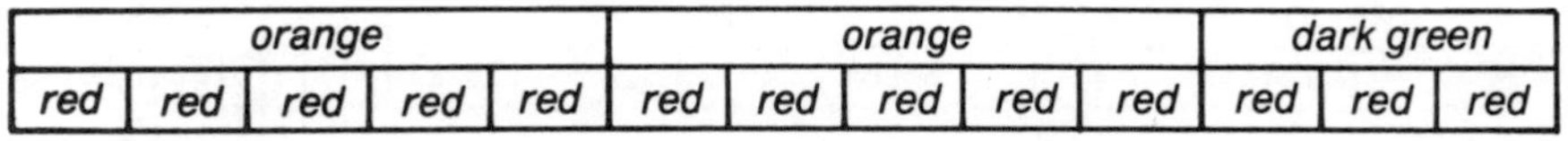

orange | orange | dark green

red | red | red | red | red | red | red | red | red | red | red | red | red

) r____

2)

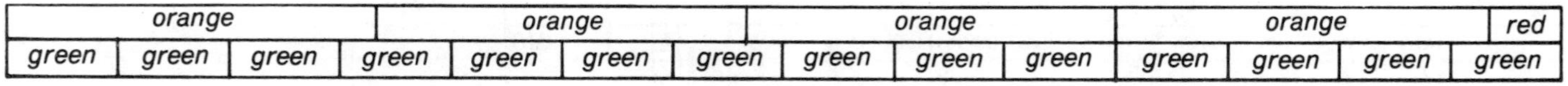

orange | orange | orange | orange | red

green | green | green | green | green | green | green | green | green | green | green | green | green | green

) r____

3)

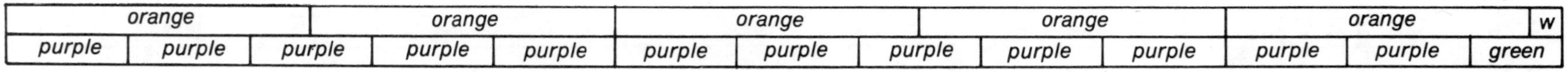

orange | orange | orange | orange | orange | w

purple | purple | purple | purple | purple | purple | purple | purple | purple | purple | purple | purple | green

) r____

4)

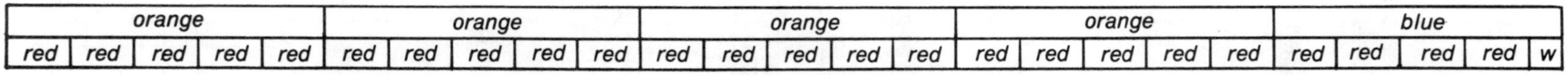

orange | orange | orange | orange | blue

red | red | red | red | red | red | red | red | red | red | red | red | red | red | red | red | red | red | red | red | red | red | red | red | w

) r____

Selected Answers and Comments

The goal of this workbook is to give visual meaning to the operations of multiplication and division, both through counting and measuring techniques. These worksheets are helpful to students learning these operations in grades 3-6 and to older students requiring review.

Pages 1-14: General Comments

The first fourteen worksheets introduce multiplication through the repeated addition model. One-color trains are made with the rods. For example:

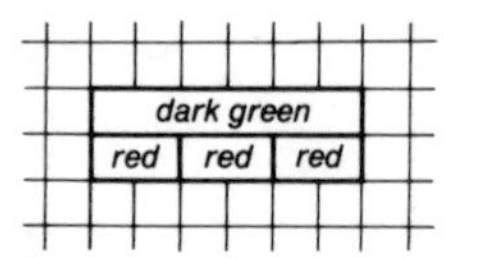

If white = 1, these rods show:
$3 \times 2 = 2 + 2 + 2 = 6$

The graph paper helps the transition from the concrete stage (rods) to the abstract stage (number). Students may wish to color their rod pictures with crayons matching the rod colors.

Pages 1-14: Selected Answers

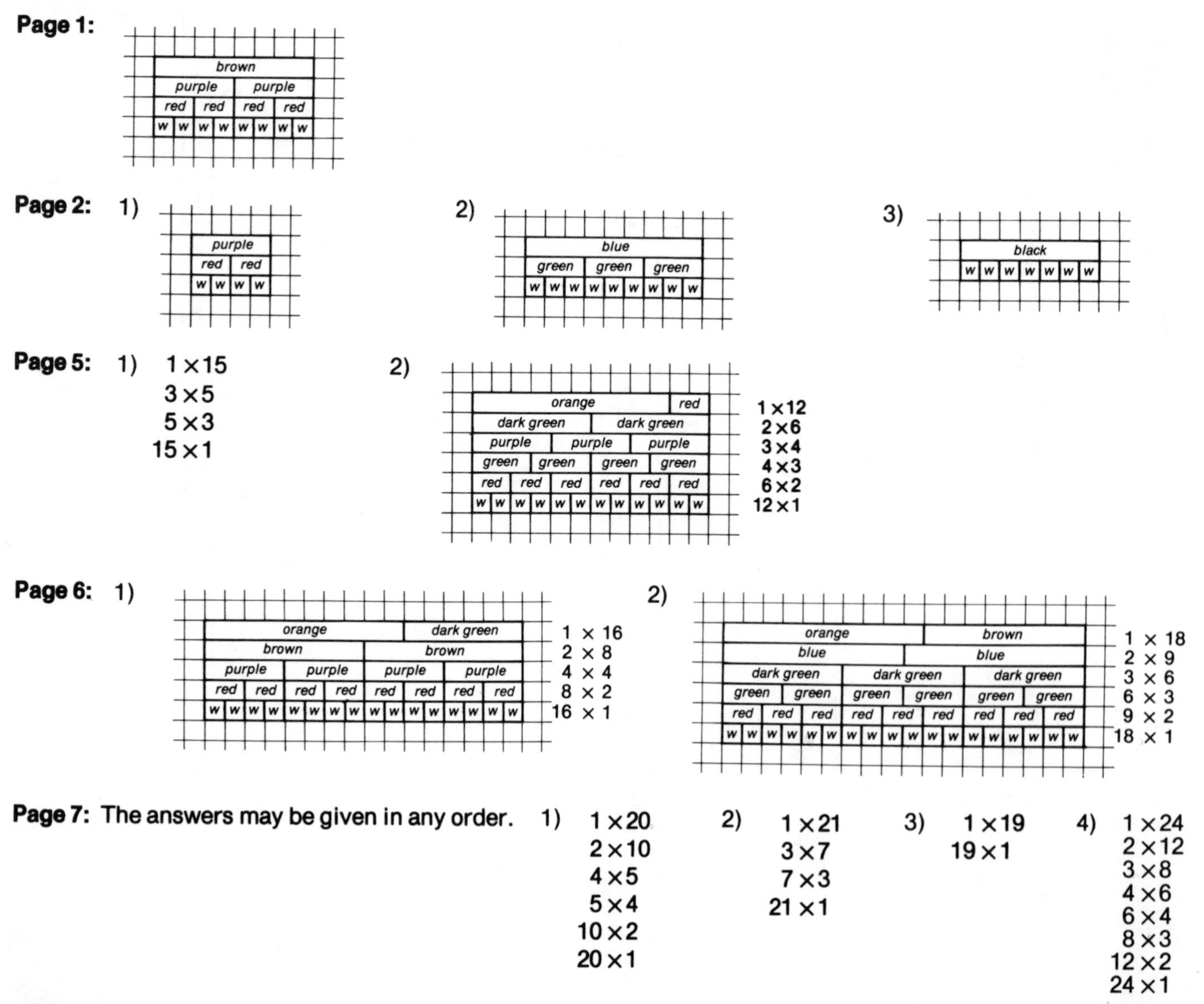

Page 8: The answers may be given in any order.

1)	2)	3)	4)
1 × 14	1 × 25	1 × 22	1 × 18
2 × 7	5 × 5	2 × 11	2 × 9
7 × 2	25 × 1	11 × 2	9 × 2
14 × 1		22 × 1	3 × 6
			6 × 3
			18 × 1

Page 11: The missing fact and product: $4 \times 6 = 24$

Page 12: The missing fact and product: $4 \times 8 = 32$

Page 13: The missing fact and product: $6 \times 8 = 48$

Page 14: The missing fact and product: $5 \times 6 = 30$

Pages 15-18: General Comments

Each one-color train can be made into a rectangle by placing the rods side-by-side. The rectangular model

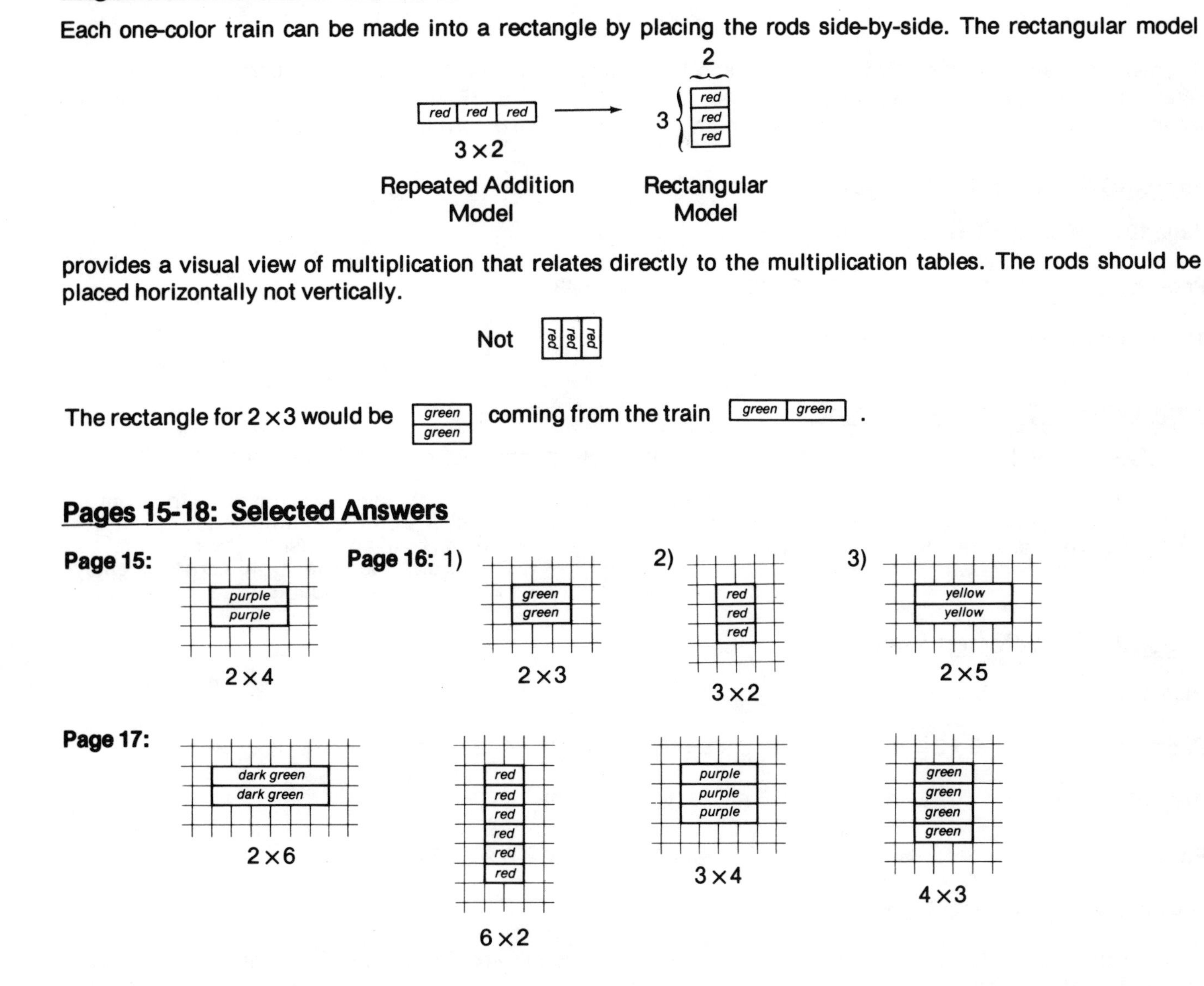

provides a visual view of multiplication that relates directly to the multiplication tables. The rods should be placed horizontally not vertically.

Not

The rectangle for 2×3 would be [green / green] coming from the train [green green].

Pages 15-18: Selected Answers

Page 15: 2×4

Page 16: 1) 2×3 2) 3×2 3) 2×5

Page 17: 2×6, 6×2, 3×4, 4×3

Page 18:

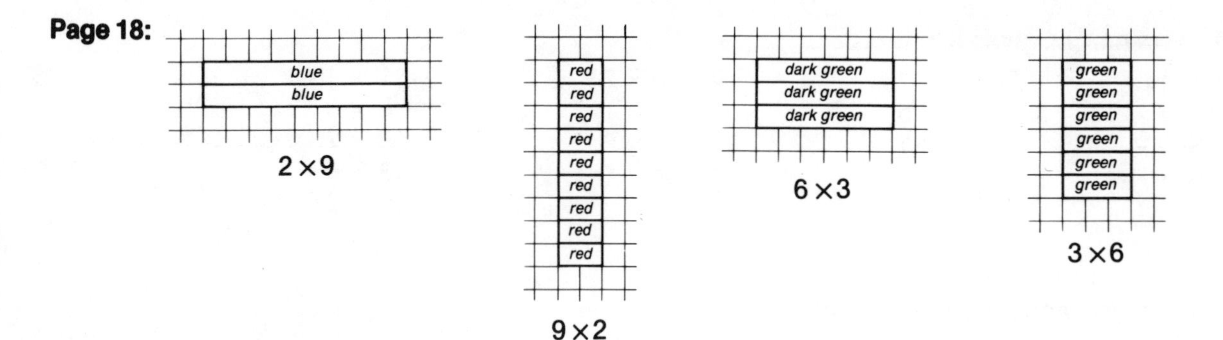

Pages 19-20: General Comments

There are some cases in multiplication when the two factors are the same numbers. For example:

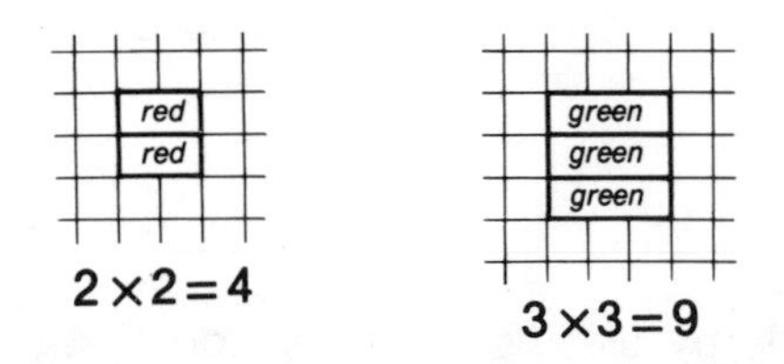

The geometric shape formed by the rods is a special rectangle known as a square. Numbers like 4 and 9 are called **square numbers.** The set of square numbers is 1, 4, 9, 16, 25, 36, 49, 81, 100, etc. Many students will use the familiar multiplication facts for square numbers to help them determine other less familiar facts.

Pages 19-20: Selected Answers

Page 19: $3 \times 3 = 9$ $4 \times 4 = 16$

Page 20: $5 \times 5 = 25$ $6 \times 6 = 36$ $7 \times 7 = 49$

The next three square numbers are 64, 81, and 100.

Pages 21-28: General Comments

The placement of the rod rectangles on the multiplication table gives visual meaning to the table:

This box where 12 is written is not only the intersection of the 3 row and the 4 column but also is the lower right corner of a rectangle with dimensions 3×4.

The rectangle 3×4 (made with 3 purple rods) fits on the multiplication table. The analogy of an artist "signing" the rod picture in the lower right corner is enjoyable for many students.

Pages 21-28: Selected Answers

Page 21: $2 \times 5 = 10$

Page 22: 1) $5 \times 3 = 15$ 2) $3 \times 5 = 15$ 3) $4 \times 5 = 20$ 4) $2 \times 7 = 14$

Page 27: 1) $7 \times 5 = 35$ 2) $5 \times 7 = 35$ 3) $9 \times 8 = 72$

Page 28: 1) $9 \times 4 = 36$ 2) $7 \times 9 = 63$ 3) $8 \times 7 = 56$

Pages 29-42: General Comments

These pages provide fun drill and practice with the multiplication tables. They are self-checking, as the correct answers create pictures of words or shapes or answers to riddles. The use of rods to help students find answers to the multiplication facts is encouraged as needed. A student can simply fill in "known" facts and use the rods to solve any "unknown" facts. Pages 37-42 extend the rod model to multiplication involving one-digit times two-digit numbers.

Pages 29-42: Selected Answers

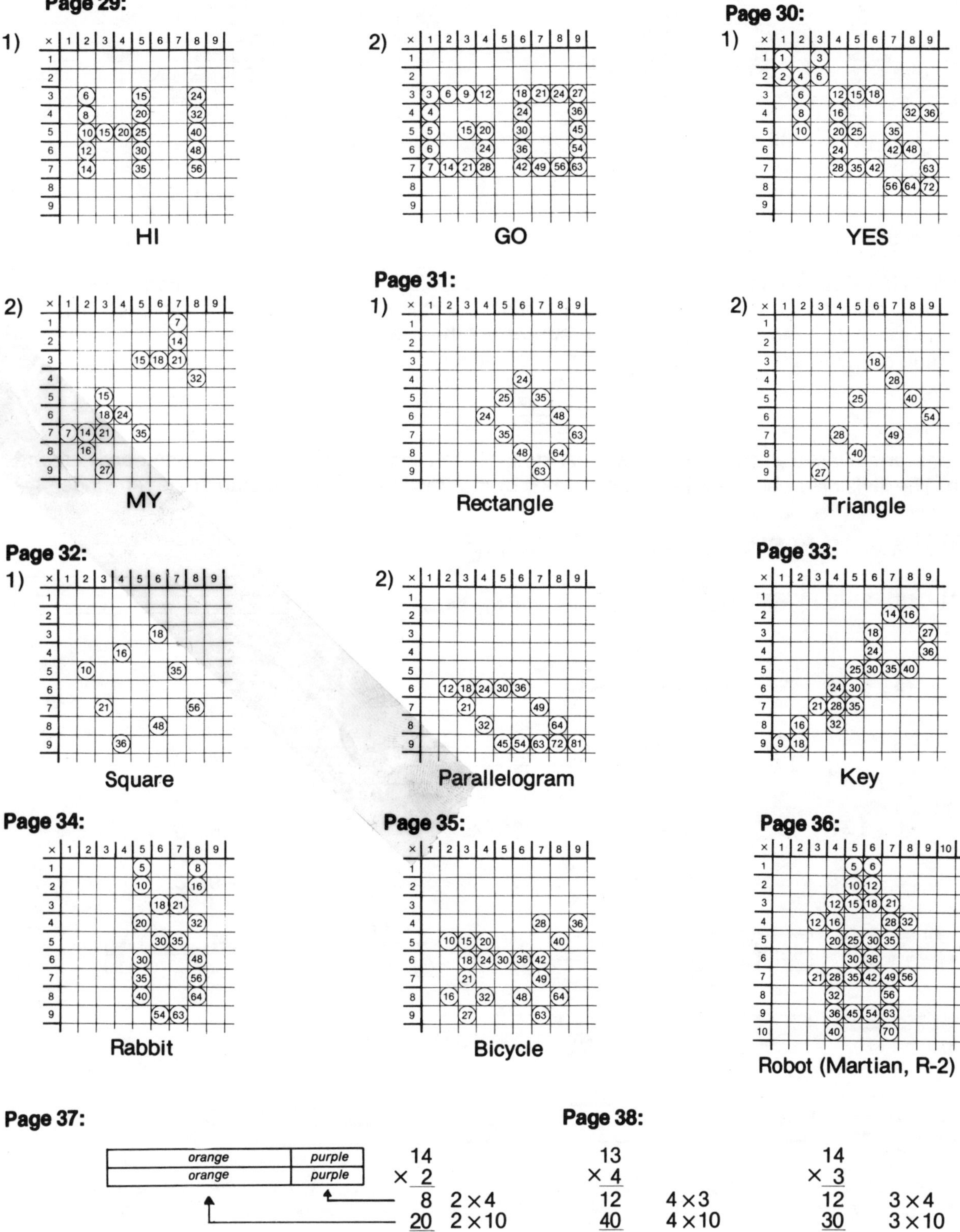

Page 37:

orange	purple
orange	purple

$$\begin{array}{rl} 14 & \\ \times\ \underline{2} & \\ 8 & 2\times4 \\ \underline{20} & 2\times10 \\ 28 & \end{array}$$

Page 38:

$$\begin{array}{rl} 13 & \\ \times\ \underline{4} & \\ 12 & 4\times3 \\ \underline{40} & 4\times10 \\ 52 & \end{array}$$

$$\begin{array}{rl} 14 & \\ \times\ \underline{3} & \\ 12 & 3\times4 \\ \underline{30} & 3\times10 \\ 42 & \end{array}$$

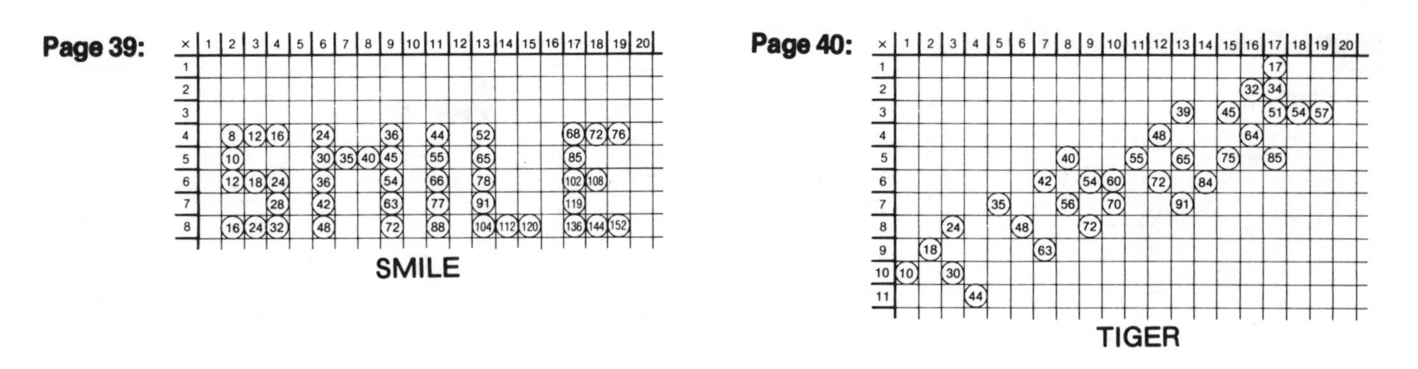

Pages 41-56: General Comments

Division with rods is viewed as finding how many rods of the same color fit into a given rod. Pages 41-48 give the students many opportunities to use rods to solve division problems, written with the notation ÷. In preparation for division with remainders, page 49 introduces another way of writing division, $\overline{)\quad}$
The rods give tangible meaning to the concept of a remainder in division on pages 49-56.

orange		w
purple	purple	green

$4\overline{)11}$ = 2 r3

Even though not formally shown on these worksheets, students can see the relationship between division and multiplication. The division problem above can be checked by multiplying 2 × 4 and adding in the remainder 3.

Pages 41-56: Selected Answers (Sample for Each Page)

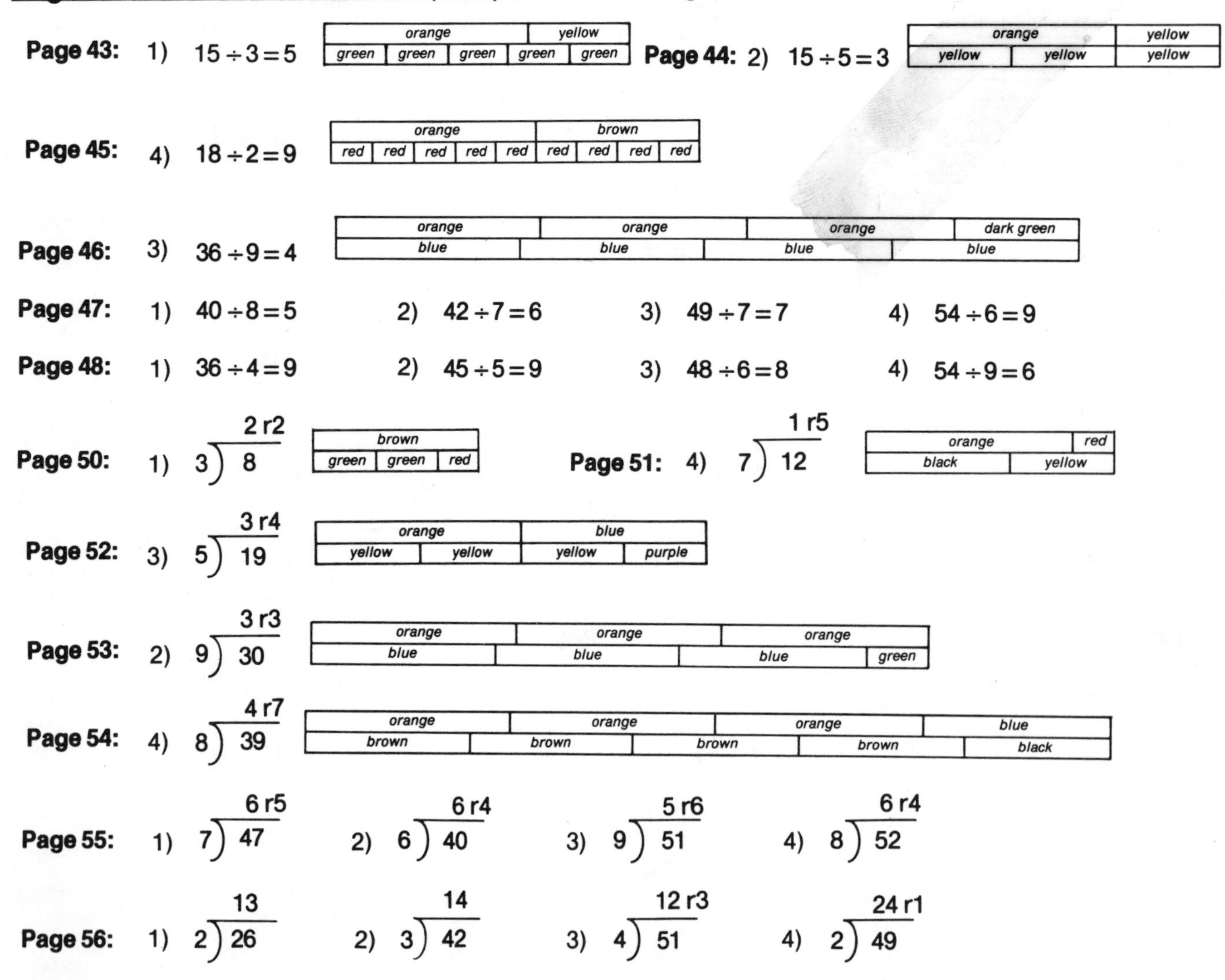

Page 43: 1) $15 \div 3 = 5$

Page 44: 2) $15 \div 5 = 3$

Page 45: 4) $18 \div 2 = 9$

Page 46: 3) $36 \div 9 = 4$

Page 47: 1) $40 \div 8 = 5$ 2) $42 \div 7 = 6$ 3) $49 \div 7 = 7$ 4) $54 \div 6 = 9$

Page 48: 1) $36 \div 4 = 9$ 2) $45 \div 5 = 9$ 3) $48 \div 6 = 8$ 4) $54 \div 9 = 6$

Page 50: 1) $3\overline{)8}$ 2 r2

Page 51: 4) $7\overline{)12}$ 1 r5

Page 52: 3) $5\overline{)19}$ 3 r4

Page 53: 2) $9\overline{)30}$ 3 r3

Page 54: 4) $8\overline{)39}$ 4 r7

Page 55: 1) $7\overline{)47}$ 6 r5 2) $6\overline{)40}$ 6 r4 3) $9\overline{)51}$ 5 r6 4) $8\overline{)52}$ 6 r4

Page 56: 1) $2\overline{)26}$ 13 2) $3\overline{)42}$ 14 3) $4\overline{)51}$ 12 r3 4) $2\overline{)49}$ 24 r1

THE COMPLETE GUIDE TO OUTDOOR WOOD PROJECTS

Step-By-Step Instructions for 50 Projects

CHANHASSEN, MINNESOTA
www.creativepub.com

Contents

18705 Lake Drive East
Chanhassen, Minnesota 55317
1-800-328-3895
www.creativepub.com

Printed in China
10 9 8 7 6 5 4 3 2 1

President: Ken Fund
Vice President/Publisher: Linda Ball
Vice President/Retail Sales & Marketing: Kevin Haas

Executive Editor: Bryan Trandem
Creative Director: Tim Himsel
Managing Editor: Michelle Skudlarek
Editorial Director: Jerri Farris

Lead Editor: Karen Ruth
Copy Editor: Nancy Baldrica
Proofreader: Tracy Stanley
Senior Art Director: David Schelitzche
Mac Designer: Jon Simpson
Illustrators: Jan-Willem Boer, David Schelitzche, Jon Simpson, Earl Slack
Production Manager: Helga Thielen

THE COMPLETE GUIDE TO OUTDOOR WOOD PROJECTS
Created by: The Editors of Creative Publishing international, Inc., in cooperation with Black & Decker.

ISBN 1-58923-202-X (softcover)

Library of Congress
Cataloging-in-Publication Data

The complete guide to outdoor wood projects : step-by-step instructions for over 50 projects / [created by the editors of Creative Publishing International, Inc., in cooperation with Black & Decker].
p. cm.
Includes index.
ISBN 1-58923-202-X (pbk. : perfect bound)
1. Woodwork--Amateurs' manuals. I. Title: At head of title: Black & Decker. II. Creative Publishing International. III. Black & Decker Corporation (Towson, Md.)

TT185.C67 2004
684'.08--dc22

2004023007

Portions of The Complete Guide to Outdoor Wood Projects are taken from the Black & Decker® books: *Sheds, Gazebos & Outbuildings*, and *The Backyard Playground*. Other titles from Creative Publishing international include:
Complete Guide to Painting & Decorating; Complete Guide to Home Plumbing; Complete Guide to Home Wiring; Complete Guide to Building Decks; Complete Guide to Creative Landscapes; Complete Guide to Home Masonry; Complete Guide to Home Carpentry; Complete Guide to Home Storage; Complete Guide to Windows & Doors; Complete Guide to Bathrooms; Complete Guide to Ceramic & Stone Tile; Complete Guide to Flooring; Complete Guide to Kitchens; Complete Guide to Roofing & Siding; Complete Photo Guide to Home Repair; Complete Photo Guide to Home Improvement; Complete Photo Guide to Outdoor Home Improvement.

Introduction

Looking for ways to enhance your yard and garden? Wood furnishings and accessories quickly add attractive and functional features to your yard, deck, or patio. The natural beauty of wood is perfectly at home in the outdoors, and the easy-to-build projects in this *Complete Guide to Outdoor Wood Projects* could soon make your outdoor home more liveable.

The projects contained in this book make life a little easier for anyone who enjoys the outdoors. There are plans for a variety of benches and tables, planters and garden accessories, birdhouses and feeders, children's play projects and storage options. Best of all, they are designed for simplicity—you don't have to be an accomplished woodworker to get impressive results.

The project plans for the 50 projects in *The Complete Guide to Outdoor Wood Projects* are complete, detailed and easy to follow. Each project has a cutting list, shopping list and construction drawing. Step-by-step directions are accompanied by photographs to guide you through each project. Throughout the book, tip boxes give pointers you can add to your woodworking knowledge.

Every project in this book is designed to be built with basic, affordable hand tools and portable power tools. You don't have to be an expert with a shop full of expensive tools to successfully create these useful, attractive projects. If you don't have a lot of experience with portable power tools, making outdoor projects is a great place to start. Compared to indoor woodworking projects, the lumber is less expensive and the construction techniques are less complex.

In additon to the complete project plans, the opening Materials & Techniques section covers the background information you'll need to be most successful in building your projects. Tools & Materials (pages 8 to 11) introduces the tools and materials needed for the projects. Woodworking Techniques (pages 12 to 15) introduces you to some simple, time-tested methods. Finishing (pages 16 to 19) gives you detailed how-to instructions for painting, sealing, staining and maintaining your new outdoor projects.

132"-3/8"

Materials & Techniques

Here is information on tools and materials and some basic woodworking techniques and tips to get you started—and to help you finish your project well. You'll find information on the wood and materials used in the projects, and how best to finish your projects to protect them from the weather.

Tools & Materials

Tools You Will Use

At the start of each project, a set of symbols shows which power tools are used to complete the project as it is shown. In some cases, optional tools, such as a power miter saw or a table router, may be suggested for speedier work. You will also need a set of basic hand tools: hammer, screwdrivers, tape measure, level, combination square, framing square, compass, wood chisels, nail set, putty knife, utility knife, straightedge, C-clamps and pipe or bar clamps. Where required, specialty hand tools are listed for each project.

Circular saw to make straight cuts. For long cuts, use a straightedge guide. Install a carbide-tipped combination blade for most projects.

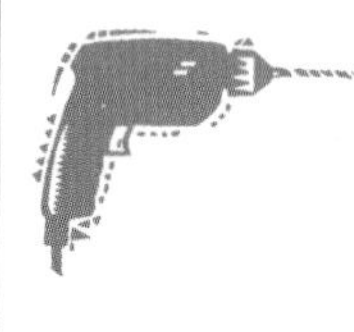

Drill for drilling holes and driving screws. Accessories help with sanding and grinding tasks. Use a corded or cordless drill with variable speed.

Jig saw for making contoured and internal cuts and for short straight cuts. Use the recommended blade for each type of wood or cutting task.

Power sander to prepare wood for a finish and to smooth sharp edges. Owning several types of power sanders is helpful.

Belt sander for resurfacing rough wood. Can also be used as a stationary sander when mounted on its side on a flat worksurface.

Router to cut structural grooves (rabbets) in wood. Also ideal for making a variety of decorative edges and roundover cuts.

Power miter saw for making angled cuts in narrow stock. Miter scales and a guide fence make it easy to set the saw quickly for precise angles.

Organizing Your Worksite

Working safely and comfortably is important to successfully completing your woodworking projects. Taking the time to set up your worksite before you begin will make your progress from step to step much smoother.

You will need a solid work surface, usually at waist level, to help you maintain a comfortable work angle. A portable work bench is sufficient for many of the smaller projects in this book. For larger projects, a sturdy sheet of plywood clamped to sawhorses will work well. In some cases you will need to use the floor for layout or assembly space.

Portable power tools and hand tools offer a level of convenience that is a great advantage over stationary power tools, but using them safely and conveniently requires some basic housekeeping. Whether you are working in a garage, a basement, or outdoors, it is important to establish a flat, dry holding area where you can store tools. Dedicate an area of your worksite for tool storage, and be sure to return tools to that area once you are finished with them.

It is also important that all waste, including lumber scraps and sawdust, be disposed of in a timely fashion. Check with your local waste disposal department before throwing away any large scraps of building materials or finishing material containers.

If you are using corded power tools outdoors, always use grounded extension cords connected to a ground fault circuit interrupter (GFCI) receptacle. Keep cords neat and out of traffic lanes at all times. Remember that most of the materials you will be working with are flammable and should be stored away from furnaces and water heaters.

Safety Tips

- *Always wear eye and hearing protection when operating power tools.*
- *Choose a well-ventilated work area when cutting or shaping wood and when using finishing products.*
- *Some wood sawdust is toxic—cedar especially—so wear an appropriate dust mask when sawing, routing and sanding.*

Materials Used in This Book

Sheet goods:
Sheet goods are manufactured products generally sold in 4 ft. × 8 ft. sheets of various thicknesses. Some retailers sell half sheets.
PLYWOOD: Wood product made of layers of thin veneer glued together. The two outer veneer faces are graded from A to D, with A being smooth and free of defects and D having knots, discoloration, splits and repairs. Grade A is best for painting.
EXTERIOR PLYWOOD: Plywood made with 100% waterproof glue. Available in the same grades as interior plywood.
CEMENTBOARD: A rigid material with a cement core faced on both sides with fiberglass mesh. Used as a backing material for tile.
LATTICE: Panels woven from ¼" or ⅜" strips of cedar or treated lumber.
PLYWOOD SIDING: Decorative exterior plywood available faced with Douglas fir, cedar or redwood, in various textures such as smooth and rough sawn, and a variety of groove patterns like kerfed, channel and reverse board and batten.
SHEET ACRYLIC: Clear plastic product available in thicknesses from 1/16" to 1".

Dimension lumber: The "nominal" size of lumber is usually larger than the actual size. For example, a 1 × 4 board measures ¾" × 3½".
CEDAR: Excellent for outdoor use, cedar is a lightweight aromatic softwood with a natural resistance to moisture and insects. Smooth cedar is best for furniture. Cedar can be painted, stained or sealed with a clear sealer.
PINE: A basic softwood used for interior projects. For exterior use it must be painted. "Select" grade is finish-quality, mostly free of knots and other imperfections; #2-or-better is a grade lower than select but more commonly available.
REDWOOD: Like cedar, it is lightweight, moisture, rot and insect resistant. It takes paint well, but its beautiful color and grain are better suited to stains and clear finishes.
TREATED LUMBER: Construction grade pine that has been treated with chemical preservatives and insecticides. *See box on page 11 for more information.*

Other wood products:
CEDAR LAP SIDING: Cedar boards cut with a bevel so the boards can overlap.
DOWELS: Round wooden rods available in a variety of diameters.
WOOD MOLDINGS: Available in a vast range of styles and sizes. Most types of molding are available in a variety of woods.
WOOD PLUGS: ⅜"-dia. × ¼"-thick disks with a slightly conical shape used to plug screw holes.

Fasteners and adhesives:
All fasteners used outdoors must be corrosion-resistent, such as galvanized, aluminum, stainless steel or coated. Galvanized fasteners are the cheapest, but can cause staining on cedar. Aluminum and stainless steel will not stain, but may be more difficult to find and are more expensive. Coated deck screws are guaranteed not to corrode or stain.

DECK SCREWS: Similar to wallboard screws, these have a light shank and coarse threads, making them ideal for fastening soft woods.
NAILS: Finish nails have a thin-shank and a cup-shaped head. They are driven below the surface with a nail set. Common (box) nails have wide, flat heads.
MOISTURE-RESISTANT GLUE: Any exterior wood glue, such as plastic resin glue.
EPOXY: A two-part glue that bonds powerfully and quickly.
CERAMIC TILE ADHESIVE: Multipurpose thin-set mortar applied with a V-notch trowel.
MISCELLANEOUS HARDWARE: 10" utility wheels; brass butt hinges; steel axle rod; eye hooks; cotter pins; hitch pins; friction hinges; chain; lag eye screws; specialty hardware as noted.

Other materials:
HARDWARE CLOTH: Galvanized, welded, and woven sceening in a variety of mesh sizes.
CERAMIC FLOOR TILE: Sturdy tile suitable in situations where durability is required. Available in sizes from 1 × 1" to 12 × 18".
MISCELLANEOUS MATERIALS: 1" 22-gauge copper strips; galvanized metal flashing.

Materials

These outdoor wood projects vary considerably in size and style, but they can be constructed with materials readily available at any home improvement center.

Lumber: Cedar (A, B, C, D, and F) and redwood (E) are warm-colored softwoods that are insect and rot resistant. Both are ideal for outdoor furnishings. Because of their attractive color and grain, they usually are left unfinished or coated with a clear finish. Pine (G) is an easy-to-cut softwood often used for projects that will be painted. Cedar is available as dimensional lumber and timbers with either smooth (C) or rough sawn surfaces (A).

Sheet goods: Plywood siding (A and B) is available in a number of textures and patterns. Exterior plywood (C) is made with exterior grade glue. Plywood products should be stained or painted to increase their longevity. Cedar lattice (D) is available in ¼" or ⅜" thicknesses. NOTE: Most sheet goods are sold in 4 ft. × 8 ft. sheets, in ¼", ½", or ¾" thicknesses; some types also are sold in 2 ft. × 4 ft. and 4 ft. × 4 ft. sheets.

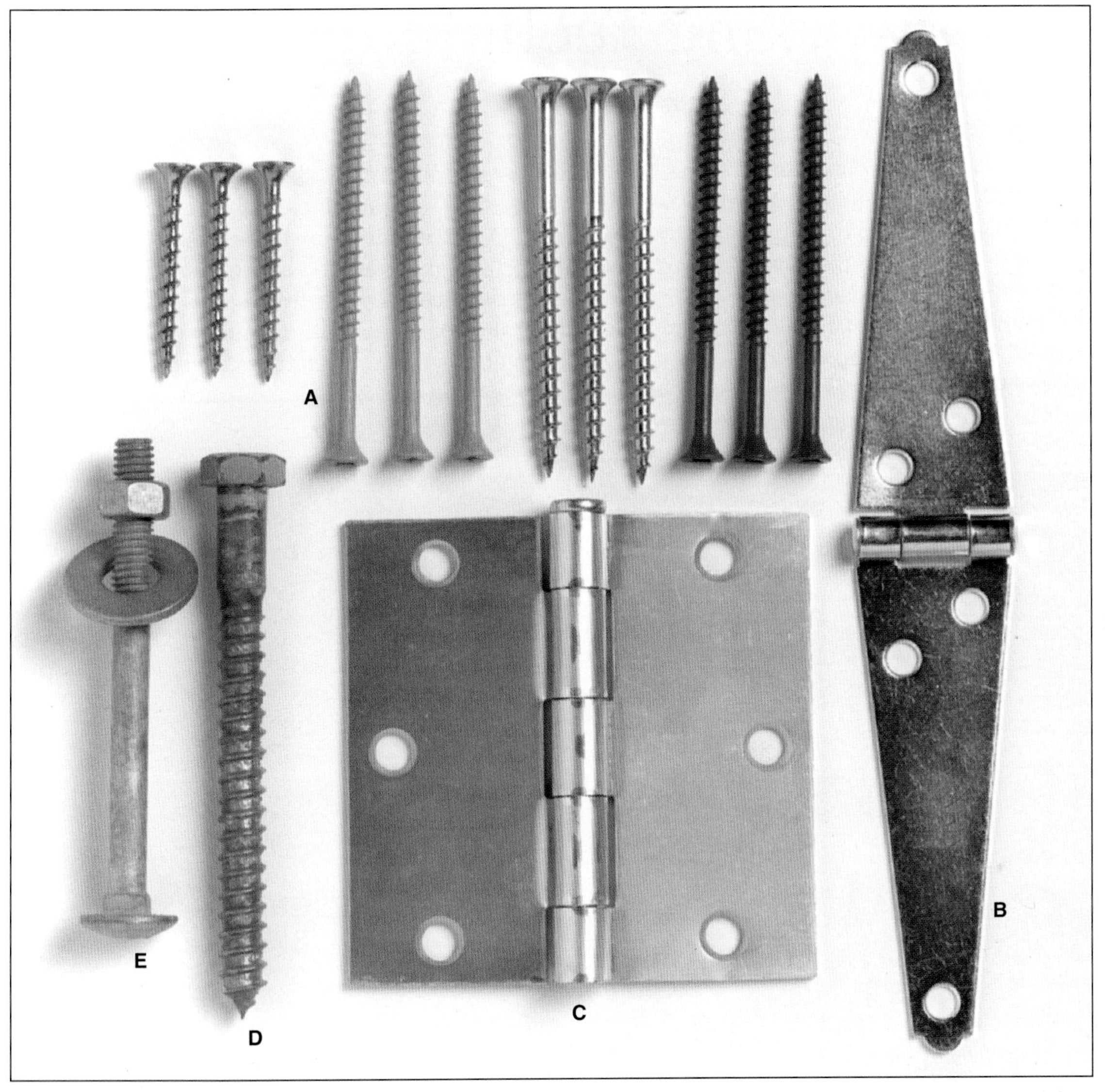

Hardware: Corrosion resistant hardware is necessary for your outdoor wood projects. Deck screws (A) are available as zinc coated, coated, or stainless steel. Zinc coated and galvanized hardware will stain unsealed cedar and redwood. Galvanized strap hinges (B) are readily available in many sizes. Brass butt hinges (C) add a nice touch to any project and are weather resistant as well. Galvanized lag screws (D) are useful for securing larger lumber. Galvanized carriage bolts (E) are used for parts that need to rotate.

TREATED LUMBER

Lumber treated with CCA (green treated), a long-time favorite for building decks, play structures and landscaping features, has been banned due to its arsenic content. Arsenic is a proven poison and a carcinogen and children are highly sensitive to it because they are still developing. Children ingest arsenic from treated lumber when they place their hands in their mouths. Do not use CCA treated lumber for any projects that children will be using. New treatment chemicals do not contain arsenic, but do contain other chemicals that are dangerous if you inhale sawdust or if you burn the wood. Always wear gloves when handling treated lumber and always wear an OSHA approved particle mask when sawing, sanding or routing. Never burn treated lumber scraps or sawdust.

Woodworking Techniques

Cutting

Circular saws and jig saws cut wood as the blade passes up through the material, which can cause splintering or chipping on the top face of the wood. For this reason, always cut with your workpiece facedown.

To ensure a straight cut with a circular saw, clamp a straightedge to your workpiece to guide the foot of the saw as you cut **(photo A).**

To make an internal cutout in your workpiece, drill starter holes near cutting lines and use a jig saw to complete the cut **(photo B).**

A power miter saw is the best tool for making straight or angled cuts on narrow boards and trim pieces **(photo C).** This saw is especially helpful for cutting hardwood. An alternative is to use an inexpensive hand miter box fitted with a backsaw **(photo D).**

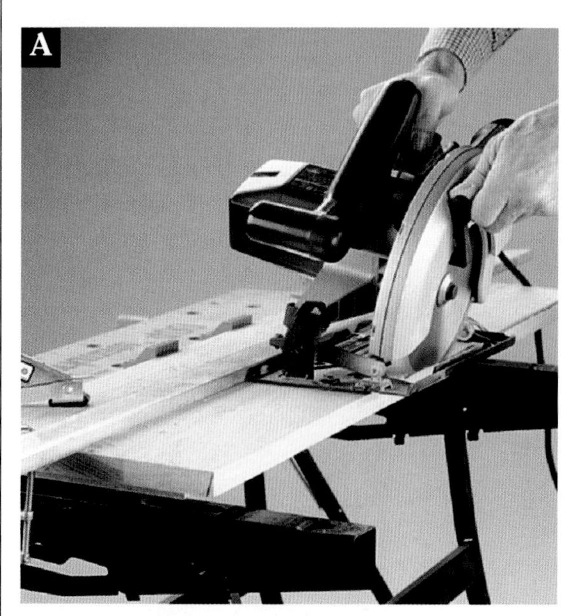

The foot of the circular saw rides along the straightedge to make straight, smooth cuts.

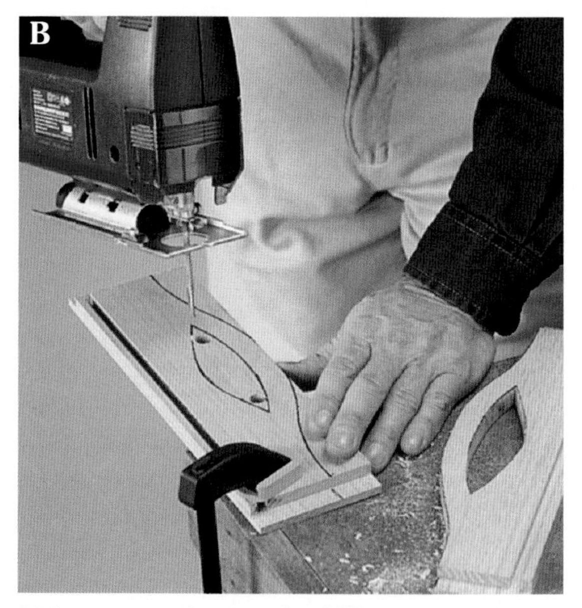

Make contoured cutouts by drilling starter holes and cutting with a jig saw.

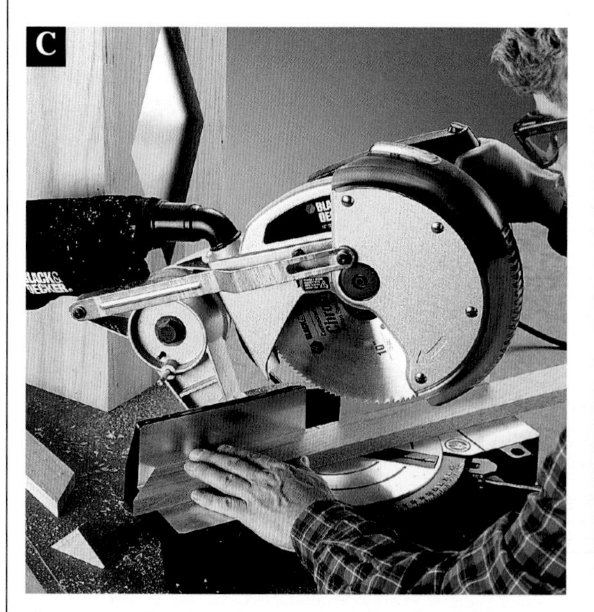

A power miter saw is easy to use and quickly makes clean, accurate angle cuts in any wood.

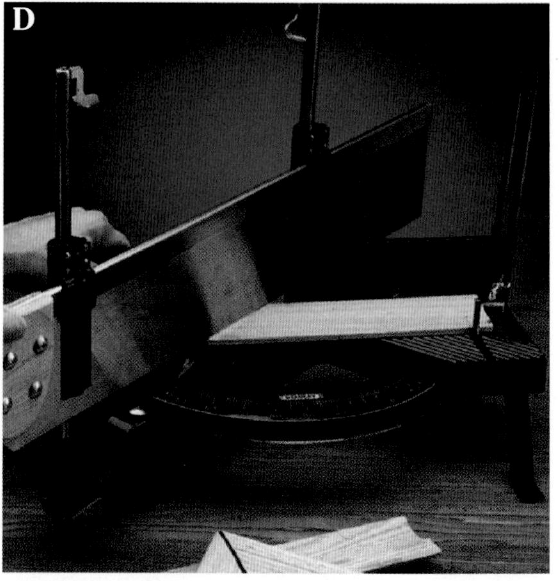

A hand miter box keeps your backsaw in line for making a full range of angle cuts.

Shaping

Create detailed shapes by drawing a grid pattern on your workpiece. Use the grid to mark accurate centers and endpoints for the shapes you will cut. Make smooth roundovers and curves using a standard compass **(photo E).**

You can also create shapes by enlarging a drawing detail, using a photocopier and transferring the pattern to the workpiece.

A belt sander makes short work of sanding tasks and is also a powerful shaping tool. Mounting a belt sander to your workbench allows you to move and shape the workpiece freely—using both hands **(photo F).** Secure the sander by clamping the tool casing in a benchtop vise or with large handscrew or C-clamps. Clamp a scrap board to your bench to use as a platform, keeping the workpiece square and level with the sanding belt.

To ensure that matching pieces have an identical shape, clamp them together before shaping **(photo G).** This technique is known as gang-sanding.

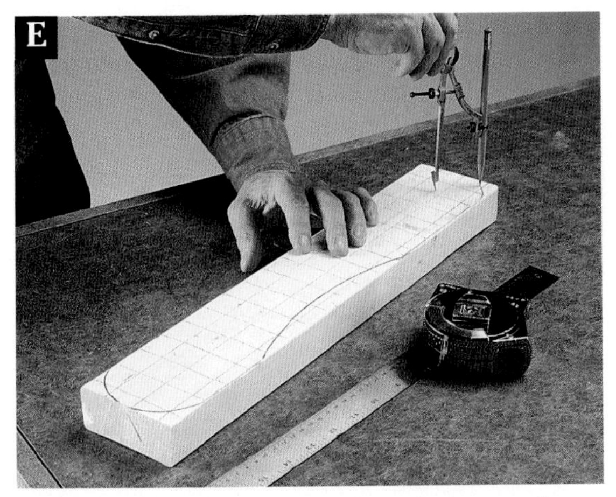

Use a square grid pattern and a compass to draw patterns on your workpiece.

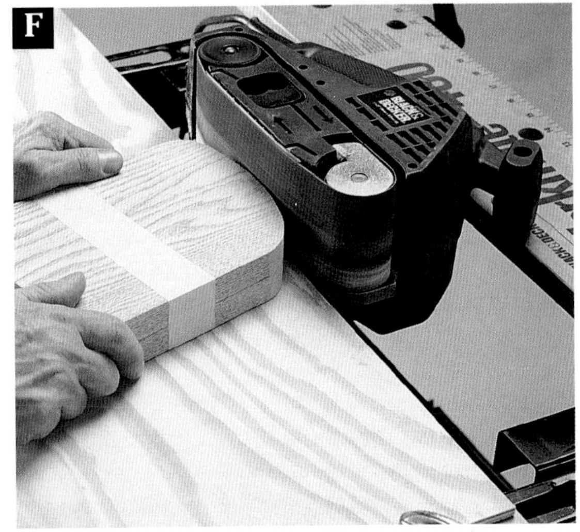

Clamp a belt sander and a scrap board to the workbench to create a stationary shaping tool.

Gang-sanding is an easy method for creating two or more identical parts.

Squaring a Frame

Squaring is an important technique in furniture construction. A frame or assembly that is not square will result in a piece that teeters on two legs or won't stand up straight. Always check an assembly for square before fastening the parts together.

To square a frame, measure diagonally from corner to corner **(photo H).** When the measurements are equal, the frame is square. Adjust the frame by applying inward pressure to diagonally opposite corners. A framing square or a combination square can also be used to see if two pieces form a right angle.

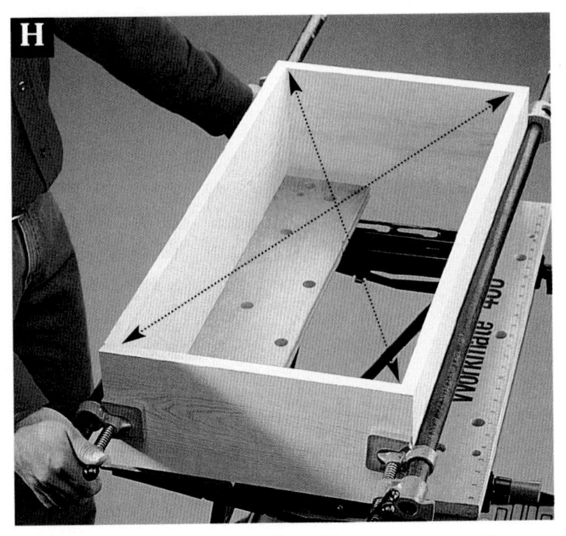

Clamp frame parts together. Then, measure the diagonals to check for square before fastening.

Piloting and Drilling

Pilot holes make it easier to drive screws or nails into a workpiece, and they also remove some material which keeps the fastener from splitting the wood. If you find that your screws are still difficult to drive or that the wood splits, switch to a larger piloting bit. If the screws are not holding well or are stripping the pilot holes, use a smaller bit to pilot subsequent holes. When drilling pilot holes for finish nails, use a standard straight bit.

A combination pilot bit drills pilot holes for the threaded and unthreaded sections of the screw shank, as well as a counterbore recess that allows the screw to seat below the surface of the workpiece **(photo A).** The counterbore portion of the bit drills a ⅜"-dia. hole to accept a standard wood plug. A bit stop with a setscrew allows you to adjust the drilling depth.

When drilling a hole through a workpiece, clamp a scrap board to the piece on the side where the drill bit will exit **(photo B).** This "backer board" will prevent the bit from splintering the wood and is especially important when drilling large holes with a spade bit.

To make perfectly straight or uniform holes, mount your drill to a portable drill stand **(photo C).** The stand can be adjusted for drilling to a specific depth and angle.

A combination pilot bit drills pilot holes and counterbores for wood screws in one step.

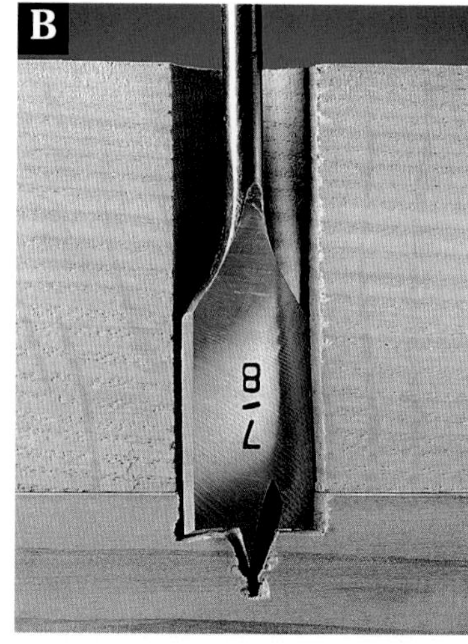

Use a scrap backer board to prevent tearout when drilling through a workpiece.

A portable drill stand helps you drill straight or angled holes.

Gluing

A gluing surface should be smooth and free of dust but not sanded. Glue and fasten boards soon after they are cut—machined surfaces, which dry out over time, bond best when they are freshly cut.

Before gluing, test-fit the pieces to ensure a proper fit. Then, clean the mating edges with a clean, dry cloth to remove dust **(photo D).**

Apply glue to both surfaces and spread it evenly, using a stick or your finger **(photo E).** Use enough glue to cover the area, with a small amount of excess.

Promptly assemble and clamp the pieces with enough clamps to apply even pressure to the joint. Watch the glue oozing from the joint to gauge the distribution of pressure. Excessive "squeeze-out" indicates that the clamps are too tight or that there is too much glue. Wipe away excess glue with a damp—not wet—cloth.

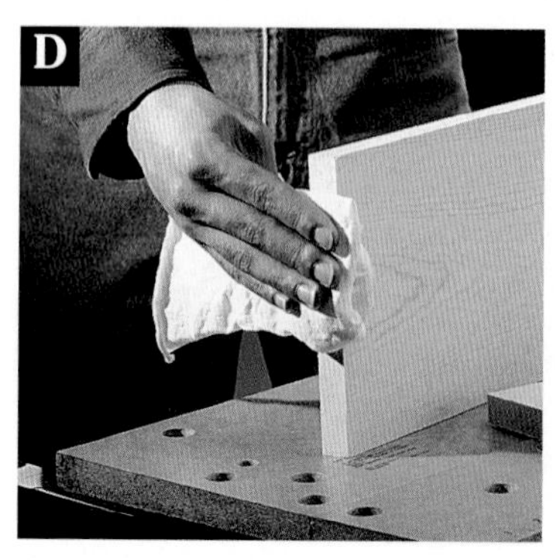

Clean the mating surfaces with a cloth to remove dust.

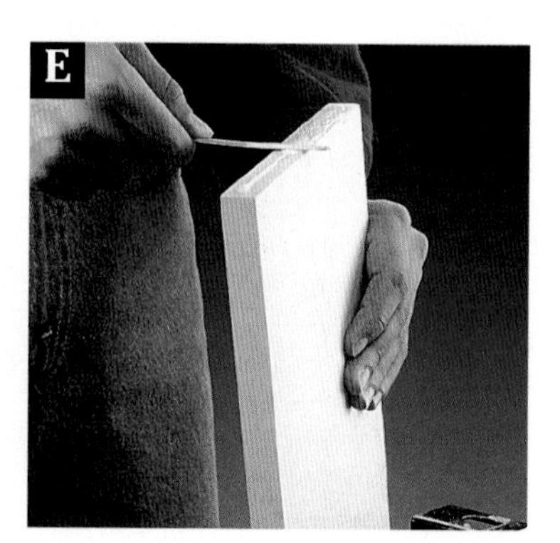

Spread glue evenly over the entire mating surface of both pieces.

Prepping Wood for Finishing Touches

Most projects require that nail heads be set below the surface of the wood, using a nail set **(photo F).** Choose a nail set with a point slightly smaller than the nail head.

Screws that have been driven well below the surface (about ¼") can be hidden by filling the counterbores with glued wood plugs **(photo G).** Tap the plug into place with a wood mallet or a hammer and scrap block, leaving the plug just above the surface. Then, sand the plug smooth with the surrounding surface.

Fill nail holes and small defects with wood putty **(photo H).** When applying a stain or clear finish to a project, use a tinted putty to match the wood, and avoid smearing it outside the nail holes. Use putty to fill screw holes on painted projects.

A power drill with a sanding drum attachment helps you sand contoured surfaces smooth **(photo I).**

Use a palm sander to finish-sand flat surfaces. To avoid sanding through thin veneers, draw light pencil marks on the surface and sand just until the marks disappear **(photo J).**

To finish-sand your projects, start with medium sandpaper (100- or 120-grit) and switch to increasingly finer papers (150- to 220-grit).

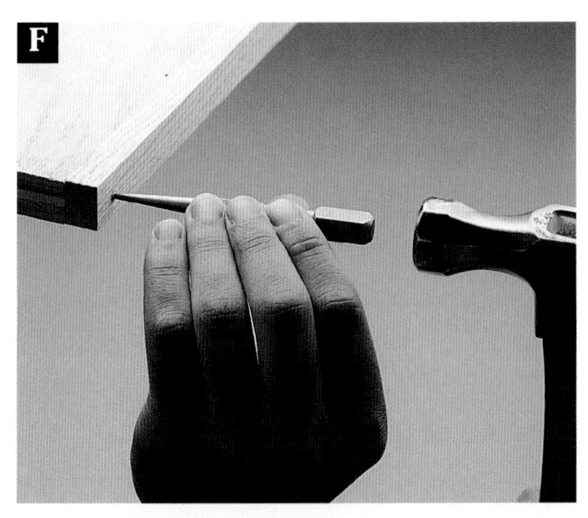

Set finish nails below the surface, using a nail set slightly smaller than the head of the nail.

Apply glue to wood plugs and insert them into screw counterbores to hide the screws.

Fill holes and wood defects with plain or tinted wood putty.

Smooth curves and hard-to-reach surfaces with a drum attachment on your power drill.

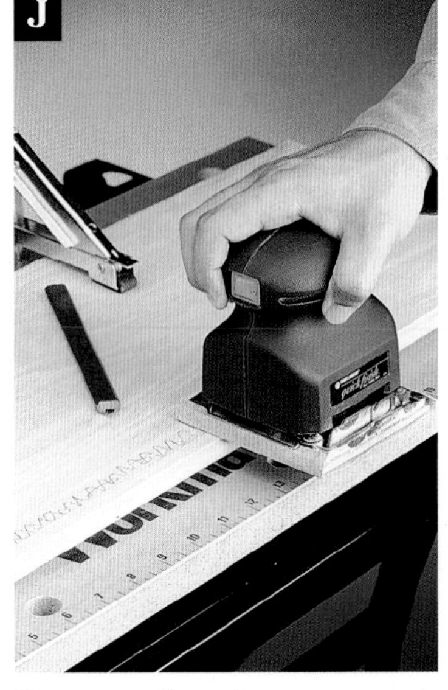

Draw pencil marks on veneered surfaces to prevent oversanding.

Finishing & Staining

Applying Paint

Painting wood is very much like painting walls and other common do-it-yourself painting projects. Whenever you paint anything, preparation is critical. For wood, that means sanding the surface until it is flat and smooth, then sealing with primer so the paint absorbs evenly.

Although it is a different product, primer is applied using the same techniques as paint. In addition to sealing the wood, it keeps resins in the wood from bleeding through the paint layer. If the wood you're painting has highly resinous knots, like cedar, redwood or pine, you may need to use a special stain killing primer.

Cleanup solvents, thinning agents, drying time, and coverage vary widely from one enamel paint to another. Read the manufacturer's directions carefully. Most paints will be dry to the touch quite quickly, and are ready for a second coat in a short time period. Since you will be sanding in between paint coats, you will get the best results if you allow each coat to dry for 12 to 24 hours.

For best results, designate a clean, dust-free area for painting. Ideally you should sand in one area and paint in another.

Finish-sand the wood. Vacuum the surfaces or wipe with a tack cloth after you sand to remove

Finish-sand the wood.

Prime the wood with an even coat of primer.

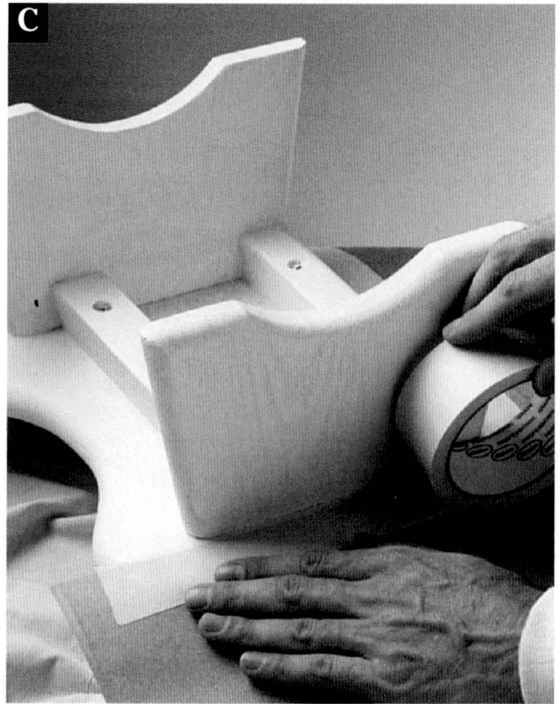

Mask any adjacent areas that will be painted a different color, using masking tape.

all traces of sanding dust from the workpiece **(photo A).**

Prime the wood with an even coat of primer **(photo B).** Use water-based primer with water-based paint, and oil-based primer with oil-based paint. Smooth out brush marks as you work, and sand with 220-grit sandpaper when dry.

Mask any adjacent areas that will be painted a different color, using masking tape. Press the edges of the tape firmly against the wood **(photo C).**

Apply a thin coat of paint, brushing with the grain **(photo D).** Heavy layers of paint will tend to sag and form an uneven surface. Loading your brush with too much paint will also cause drips to form on edges and around joints.

For a smooth surface that's free from lap marks, hold your paint brush at a 45° angle, and apply just enough pressure to flex the bristles slightly **(photo E).**

When dry, sand with 400-grit sandpaper, then wipe with a tack cloth. Apply at least one more heavier coat, sanding and wiping with a tack cloth between additional coats. Darker colors may require more coats than lighter colors. Do not sand the last coat.

Option: Apply clear polyurethane topcoat to surfaces that will get heavy wear. Before applying, wet-sand the paint with 600-grit wet/dry sandpaper, then wipe with a tack cloth. Use water-based polyurethane over latex paint, and oil-based over oil-based paint **(photo F).**

Apply a thin coat of paint, brushing with the grain.

For a smooth surface free from lap marks, hold your paint brush at a 45° angle, and apply just enough pressure to flex the bristles slightly.

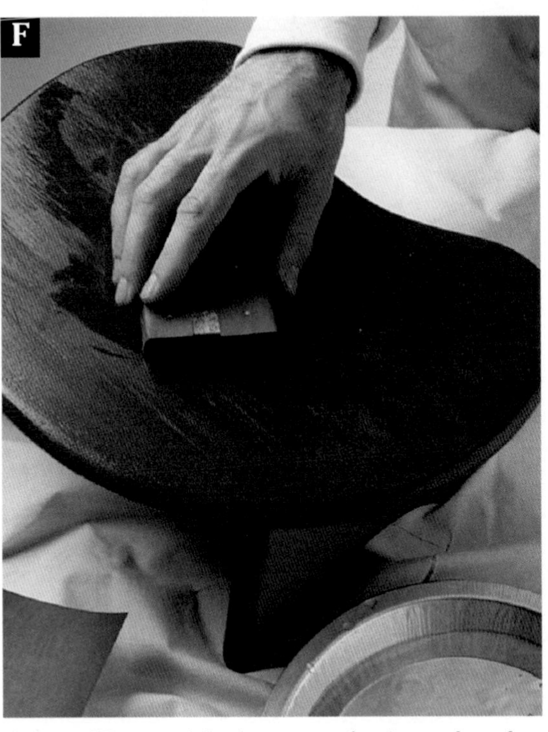

Option: Wet sand the last coat of paint and apply clear polyurethane topcoat to surfaces that will get heavy wear.

Applying Sealers and Stains

You can leave your solid cedar and redwood outdoor wood projects unfinished, or seal or stain them. Leaving cedar unfinished allows it to weather and age to a mellow gray. Sealing protects the grain from raising but allows for a slow progression to gray, unless you use a product containing ultraviolet blockers. Staining protects the wood grain and adds color. Plywood, plywood siding and pine boards need to be sealed, stained or painted because they do not have the natural rot resistance of solid cedar and redwood.

Sealers and stains differ from paint. Paints require primers to bond properly to wood and usually require multiple coats for coverage. Because of the layers of paint, all but the roughest wood textures are obscured or hidden. Paint bonds to the wood surface, but does not penetrate. It also forms a moisture resistant layer. If moisture does penetrate the wood, the paint will blister and peel when the water evaporates. Exterior oil-based polyurethane finishes are similar to paint in this way. Properly applied stain penetrates the wood surface, rather than sitting on top of it and allows water vapor to evaporate from within the wood. Depending on the type, wood sealers may create a waterproof barrier somewhat like paint, or may create a semi-permeable barrier like stain.

Wood sealers partially or totally block the wood's pores, preventing water penetration. When wood absorbs water and dries repeatedly, the grain becomes raised and rough. While most wood sealers do not prevent the wood from graying, some sealers now come with ultra-violet blockers to prevent graying. Because sealers are clear, the wood's grain and color are still visible after treatment. Sealers and waterproofers must be reapplied annually for best results.

Exterior staining products are available in a number of water- or oil-based options. The options are based on how much pigment is in the product. Wood-toned stains have only a small amount of pigment. This stain does not provide much protection from ultra-violet rays unless it also contains a UV blocking agent. Semi-transparent stains have more pigment than wood-toned stains, but still allow the grain and texture of the wood to show through. Solid color stain is the most durable of the wood stains. It allows the texture of the wood to show, but obscures the color and most of the grain **(photo A).**

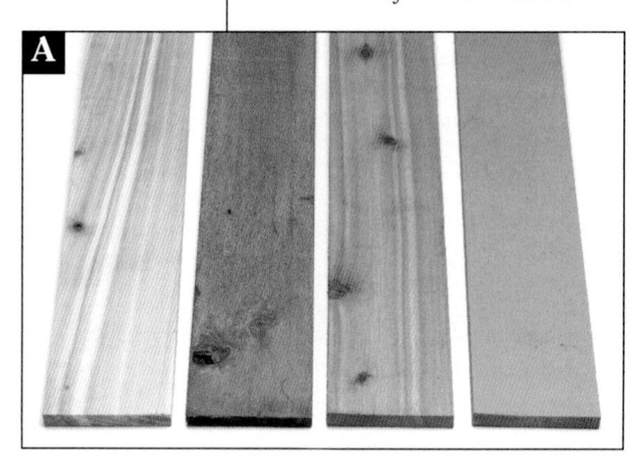

New, raw cedar varies in color from a light tan to a deep red. Without a finish, cedar, redwood and teak will weather to a light gray. Semi-transparent stains are available in wood tones and offer a small amount of protection from graying. Solid color stains are the most durable and obscure the natural colors of the wood, but allow the wood's texture to be visible.

To seal or stain new wood, the wood must be clean and dry. Sanding is not necessary when applying sealers and stains because the products penetrate the wood. Mix the products thoroughly, especially stains. The pigments in stains settle very quickly, so to ensure consistent color it is important to stir frequently. Apply stain in a well-ventilated area or wear a respirator.

Use a natural bristle brush for applying oil based products and a synthetic brush for latex products. Special staining brushes with shorter bristles drip less than a standard paint brush but can be difficult to find. Do not apply sealers or stains in direct sun. Stain absorbs and dries more quickly than paint, and hot surfaces will further speed this process, making it difficult to create a consistent finish. Remember to protect surrounding surfaces with drop cloths (it is difficult or impossible to remove wood stains from cement).

To apply stain using a brush, load the brush and apply the stain in the direction of the wood grain. To prevent lap marks, brush backward from dry to wet for subsequent brush strokes **(photo B).** Always work on one board at a time. For sheet goods, move from left to right, covering an entire top to bottom swath at a time.

A brush is the best applicator for exterior stain. To prevent uneven coverage, brush from dry to wet.

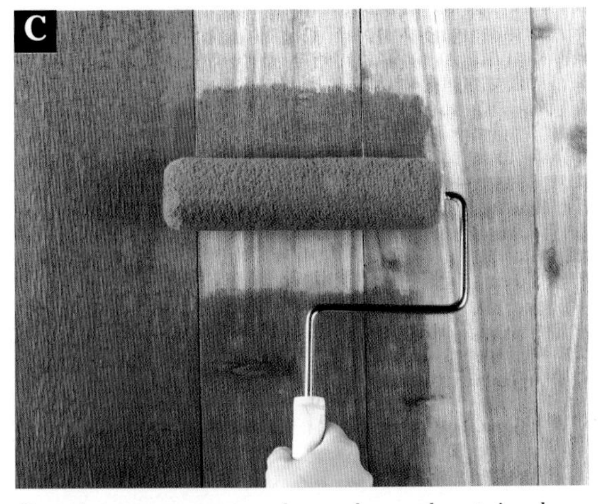

Rough-sawn or textured woods can be stained with a ⅜" nap roller. Cover one section at a time and roll back from dry to wet to prevent uneven color distribution.

Before applying stain or sealer to previously finished wood, test the wood's absorbency by sprinkling it with drops of water. If they are not absorbed after 5 minutes, the wood will need to be stripped before refinishing.

Maintain a wet edge as you go. Do not stop in the middle of a board or panel. Check the manufacturer's instructions concerning application of a second coat.

Stain can be applied with a roller, but it is recommended only for rough surfaces. Use a roller with a ⅜" nap. Stir the stain thoroughly and pour a small amount into a roller tray. Apply the stain to one board at a time in the direction of the grain. For sheet goods, cover top-to-bottom swath at a time. Do not stop in the middle of a board or panel. Roll from dry to wet (backrolling) to prevent lap marks **(photo C).**

To apply sealers and preservatives, stir the product thoroughly. Saturate the surface until the wood absorbs no more sealer and a wet sheen forms. Do not overapply (puddling of the product indicates overapplication) and do not apply a second coat.

Wood sealers may need to be reapplied every year or two years. Stains usually last 3 to 7 years depending on the pigment content and exposure to weather. Before reapplying sealer or stain, it is important to do a "splash test" to determine if the wood is porous enough to absorb a new application of sealer or stain. Drip or spray large water droplets onto the wood surface. If the droplets are not absorbed within 5 minutes, the old finish must be removed **(photo D).** Use an exterior deck stripper and wood cleaners to prepare the surface for refinishing.

Maintain your outdoor wood projects by regularly cleaning or refinishing them. A power washer can make this a quick job **(photo E).** Make sure you follow directions carefully.

In harsh climates, move items to a sheltered area in the winter, or cover with a tarp. This will prevent excess weathering and help your projects last longer.

A pressure washer can be used to clean and brighten unfinished outdoor projects, or to strip and clean finished surfaces for refinishing. Remember that high water pressure can destroy the soft wood fibers of cedar and redwood.

BLACK & DECKER
PLUNGE ROUTER
2 HP

Ornaments & Decorations

Outdoor ornaments and decorations provide the special touches that make your outdoor home a reflection of your personal style. Whether it's an arbor that leads to your perennial garden, a planter that displays showy flowers, or an accent piece for your yard, the simple projects shown here will help you distinguish your landscape with eye-catching wood structures.

Freestanding Arbor

Create a shady retreat on a sunny patio or deck with this striking arbor.

PROJECT POWER TOOLS

This freestanding arbor combines the beauty and durability of natural cedar with an Oriental-inspired design. Set it up on your patio or deck, or in a quiet corner of your backyard—it adds just the right finishing touch to turn your outdoor living space into a showplace geared for relaxation and quiet contemplation. The arbor has a long history as a focal point in gardens and other outdoor areas throughout the world. And if privacy and shade are concerns, you can enhance the sheltering quality by adding climbing vines that weave their way in and out of the trellis. Or simply set a few potted plants around the base to help the arbor blend in with the outdoor environment. Another way to integrate plant life into your arbor is to hang decorative potted plants from the top beams.

This arbor is freestanding, so it easily can be moved to a new site whenever you desire. Or, you can anchor it permanently to a deck or to the ground and equip it with a built-in seat.

Sturdy posts made from 2 × 4 cedar serve as the base of the arbor, forming a framework for a 1 × 2 trellis system that scales the sides and top. The curved cutouts that give the arbor its Oriental appeal are made with a jig saw, then smoothed out with a drill and drum sander for a more finished appearance.

Construction Materials

Quantity	Lumber
2	1 × 2" × 8' cedar
5	2 × 2" × 8' cedar
9	2 × 4" × 8' cedar
3	2 × 6" × 8' cedar

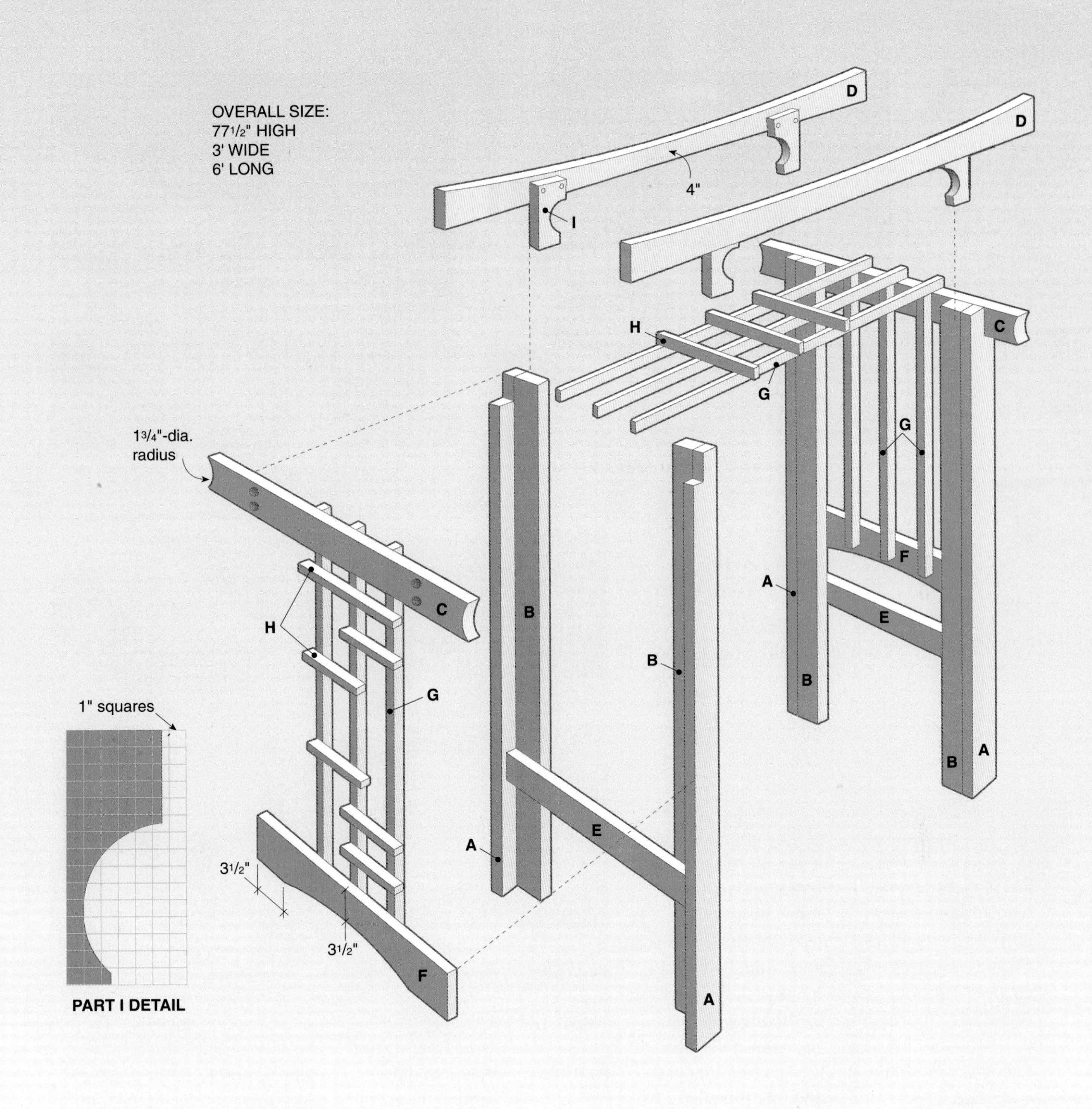

Cutting List

Key	Part	Dimension	Pcs.	Material
A	Leg front	1½ × 3½ × 72"	4	Cedar
B	Leg side	1½ × 3½ × 72"	4	Cedar
C	Cross beam	1½ × 3½ × 36"	2	Cedar
D	Top beam	1½ × 5½ × 72"	2	Cedar
E	Side rail	1½ × 3½ × 21"	2	Cedar

Cutting List

Key	Part	Dimension	Pcs.	Material
F	Side spreader	1½ × 5½ × 21"	2	Cedar
G	Trellis strip	⅞ × 1½ × 48"	9	Cedar
H	Cross strip	⅞ × 1½ × *	15	Cedar
I	Brace	1½ × 5½ × 15"	4	Cedar

Materials: Wood glue, wood sealer or stain, #10 × 2½" wood screws, ⅜"-dia. × 2½" lag screws (8), 6" lag screws (4), 2½" and 3" deck screws, finishing materials.

Note: Measurements reflect the actual size of dimension lumber.

*Cut to fit

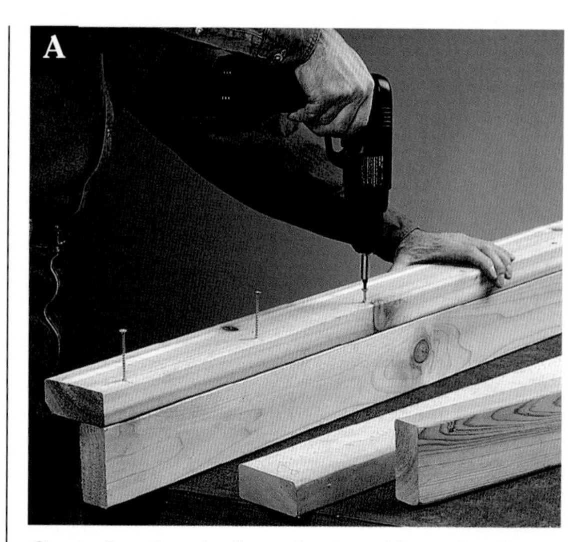

Create four legs by fastening leg sides to leg fronts at right angles.

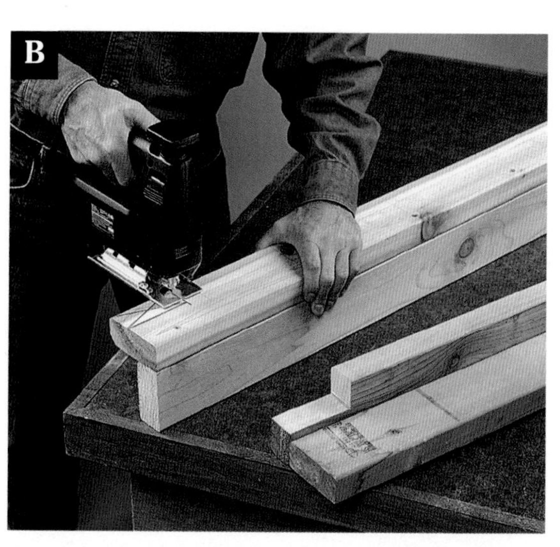

Cut a notch in the top of each of the four legs to hold the cross beams.

Directions: Freestanding Arbor

MAKE THE LEGS.
Each of the four arbor legs is made from two 6'-long pieces of 2 × 4 cedar, fastened at right angles with 3" deck screws.

1. Cut the leg fronts (A) and leg sides (B) to length. Position the leg sides at right angles to the leg fronts, with top and bottom edges flush. Apply moisture-resistant glue to the joint. Attach the leg fronts to the leg sides by driving evenly spaced screws through the faces of the fronts and into the edges of the sides **(photo A).**

2. Use a jig saw to cut a 3½"-long × 2"-wide notch at the top outside corner of each leg front **(photo B).** These notches cradle the cross beams when the arbor is assembled.

TIP

Climbing vines help any man-made structure blend into a natural environment. Common climbing vines include ivy, clematis, morning glory, and wild rose. Check with the professionals at your local gardening center for planting suggestions and other information.

MAKE THE CROSS BEAMS, RAILS & SPREADERS.

1. Cut cross beams (C) to length. Cut a small arc at both ends of each cross beam. Start by using a compass to draw a 3½"-diameter semicircle at the edge of a strip of cardboard. Cut out the semicircle, and use the strip as a template for marking the arcs **(photo C).** Cut out the arcs with a jig saw. Sand the cuts smooth with a drill and drum sander.

A piece of cardboard acts as a template when you trace the outline for the arc on the cross beams.

2. Cut two spreaders (F) to length. The spreaders fit just above the rails on each side. Mark a curved cutting line on the bottom of each spreader (see *Diagram,* page 23). To mark the cutting lines, draw starting points 3½" in from each end of a spreader. Make a reference line 2" up from the bottom of the spreader board. Tack a casing nail on the reference line, centered between the ends of the spreader. With the spreader clamped to the work surface, also tack nails into the work surface next to the starting lines on the spreader. Slip a thin strip of metal or plastic between the casing nails so the strip bows

Lag-screw the cross beams to the legs, and fasten the spreaders and rails with deck screws to assemble the side frames.

Attach trellis strips to the cross brace and spreader with deck screws.

out to create a smooth arc. Trace the arc onto the spreader, then cut along the line with a jig saw. Smooth with a drum sander. Use the first spreader as a template for marking and cutting the second spreader.

3. Cut the rails (E) to length. They are fitted between pairs of legs on each side of the arbor, near the bottom, to keep the arbor square.

ASSEMBLE THE SIDE FRAMES.

Each side frame consists of a front and back leg, joined together by a rail, spreader and cross beam.

1. Lay two leg assemblies parallel on a work surface, with the notched board in each leg facing up. Space the legs so the inside faces of the notched boards are 21" apart. Set a cross beam into the notches, overhanging each leg by 6". Also set a spreader and a rail between the legs for spacing.

2. Drill ⅜" pilot holes in the cross beam. Counterbore the holes to a ¼" depth, using a counterbore bit. Attach the cross beam to each leg with glue. Drive two ⅜"-dia. × 2½" lag screws through the cross beam and into the legs **(photo D).**

3. Position the spreader between the legs so the top is 29½" up from the bottoms of the legs. Position the rail 18" up from the leg bottoms. Drill ⅛" pilot holes in the spreader and rail. Counterbore the holes. Keeping the legs parallel, attach the pieces with glue and drive 3" deck screws through the outside faces of the legs and into the rail and spreader.

ATTACH THE SIDE TRELLIS PIECES.

Each side trellis is made from vertical strips of cedar 2 × 2 that are fastened to the side frames. Horizontal cross strips will be added later to create a decorative cross-hatching effect.

1. Cut three vertical trellis strips (G) to length for each side frame. Space them so they are 2⅜" apart, with the ends flush with the top of the cross beam **(photo E).**

2. Drill pilot holes to attach the trellis strips to the cross beam and spreader. Counterbore the holes and drive 2½" deck screws. Repeat the procedure for the other side frame.

TIP

Drill counterbores for lag screws in two stages: first, drill a pilot hole for the shank of the screw; then, use the pilot hole as a center to drill a counterbore for the washer and screw head.

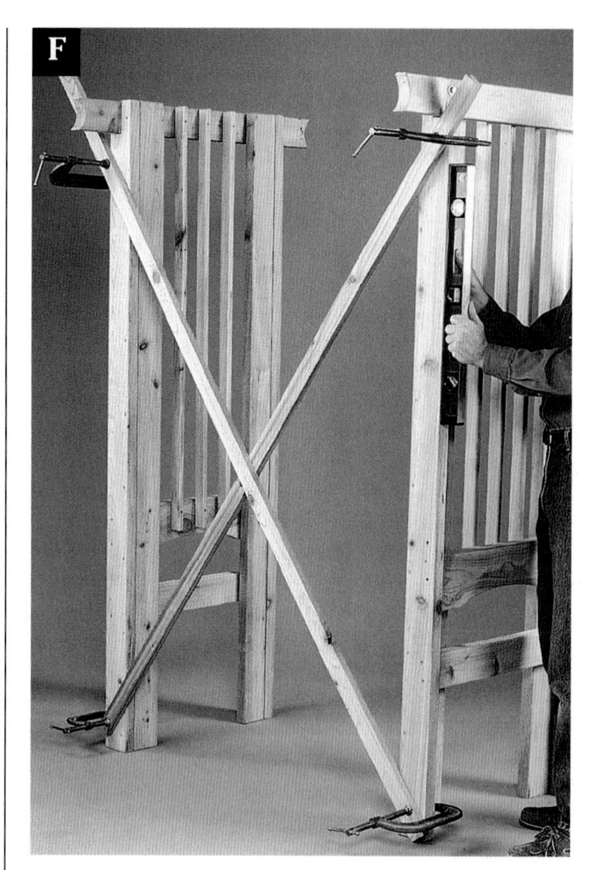

Use long pieces of 1×4 to brace the side frames in an upright, level position while you attach the top beams.

Lock the legs in a square position after assembling the arbor by tacking strips of wood between the front legs and between the back legs.

CUT AND SHAPE TOP BEAMS.

1. Cut two top beams (D) to length. Draw 1½"-deep arcs at the top edges of the top beams, starting at the ends of each of the boards.

2. Cut the arcs into the top beams with a jig saw. Sand smooth with a drum sander.

TIP

There are no firm rules about arbor placement. It can be positioned to provide a focal point for a porch, patio or deck. Placed against a wall or at the end of a plain surface, arbors improve the general look of the area. With some thick, climbing vines and vegetation added to the arbor, you can also disguise a utility area, such as a trash collection space.

ASSEMBLE TOP AND SIDES.

1. Because the side frames are fairly heavy and bulky, you will need to brace them in an upright position to fasten the top beams between them. A simple way to do this is to use a pair of 1 × 4 braces to connect the tops and bottoms of the side frames **(photo F).** Clamp the ends of the braces to the side frames so the side frames are 4' apart, and use a level to make sure the side frames are plumb.

2. Mark a centerpoint for a lag bolt 12¾" from each end of each top beam. Drill a ¼" pilot hole through the top edge at the centerpoint. Set the top beams on top of the cross braces of the side frames. Mark the pilot hole locations onto the cross beams. Remove the top beams and drill pilot holes into the cross beams. Secure the top beams to the cross beams with 6" lag screws.

3. Cut four braces (I) to length, and transfer the brace cutout pattern from the *Diagram* on page 23 to each board. Cut the patterns with a jig saw. Attach the braces at the joints where the leg fronts meet the top beams, using 2½" deck screws. To make sure the arbor assembly stays in position while you complete the project, attach 1 × 2 scraps between the front legs and between the back legs **(photo G).**

4. Cut and attach three trellis strips (G) between the top beams.

Attach the trellis cross strips to spice up the design and assist climbing plants.

ADD TRELLIS CROSS STRIPS.

1. Cut the cross strips (H) to 7" and 10" lengths. Use wood screws to attach them at 3" intervals in a staggered pattern on the side trellis pieces **(photo H).** You can adjust the sizes and placement of the cross strips but, for best appearance, retain some symmetry of placement.

2. Fasten cross strips to the top trellis in the same manner. Make sure the cross strips that fit across the top trellis are arranged in similar fashion to the side strips.

APPLY FINISHING TOUCHES.

1. To protect the arbor, coat the cedar wood with clear wood sealer. After the finish dries, the arbor is ready to be placed onto your deck or patio or in a quiet corner of your yard.

2. Because of its sturdy construction, the arbor can simply be set onto a hard, flat surface. If you plan to install a permanent seat in the arbor, you should anchor it to the ground. For decks, try to position the arbor so you can screw the legs to the rim of the deck or toenail the legs into the deck boards. You can buy fabricated metal post stakes, available at most building centers, to use when anchoring the arbor to the ground.

TIP

Create an arbor seat by resting two 2 × 10 cedar boards on the rails in each side frame. Overhang the rails by 6" or so, and drive a few 3" deck screws through the boards and into the rails to secure the seat.

Trellis Planter

Two traditional yard furnishings are combined into one compact package.

PROJECT POWER TOOLS

The decorative trellis and the cedar planter are two staples found in many yards and gardens. By integrating the appealing shape and pattern of the trellis with the rustic, functional design of the cedar planter, this project showcases the best qualities of both furnishings.

Because the 2 × 2 lattice trellis is attached to the planter, not permanently fastened to a wall or railing, the trellis planter can be moved easily to follow changing sunlight patterns, or to occupy featured areas of your yard. It is also easy to move into storage during non-growing seasons. You may even want to consider installing wheels or casters on the base for greater mobility.

Building the trellis planter is a very simple job. The trellis portion is made entirely of strips of 2 × 2 cedar, fashioned together in a crosshatch pattern. The planter bin is a basic wood box, with panel sides and a two-board bottom with drainage holes, that rests on a scalloped base. The trellis is screwed permanently to the back of the planter bin.

Stocking the trellis planter with plantings is a matter of personal taste and growing conditions. In most areas, ivy, clematis and grapevines are good examples of climbing plants that can be trained up the trellis. Ask at your local gardening center for advice on plantings. Plants can be set into the bin in containers, or you can fill the bin with potting soil and plant directly in the bin.

CONSTRUCTION MATERIALS

Quantity	Lumber
1	2 × 6" × 8' cedar
1	2 × 4" × 6' cedar
4	2 × 2" × 8' cedar
3	1 × 6" × 8' cedar
1	1 × 2" × 6' cedar

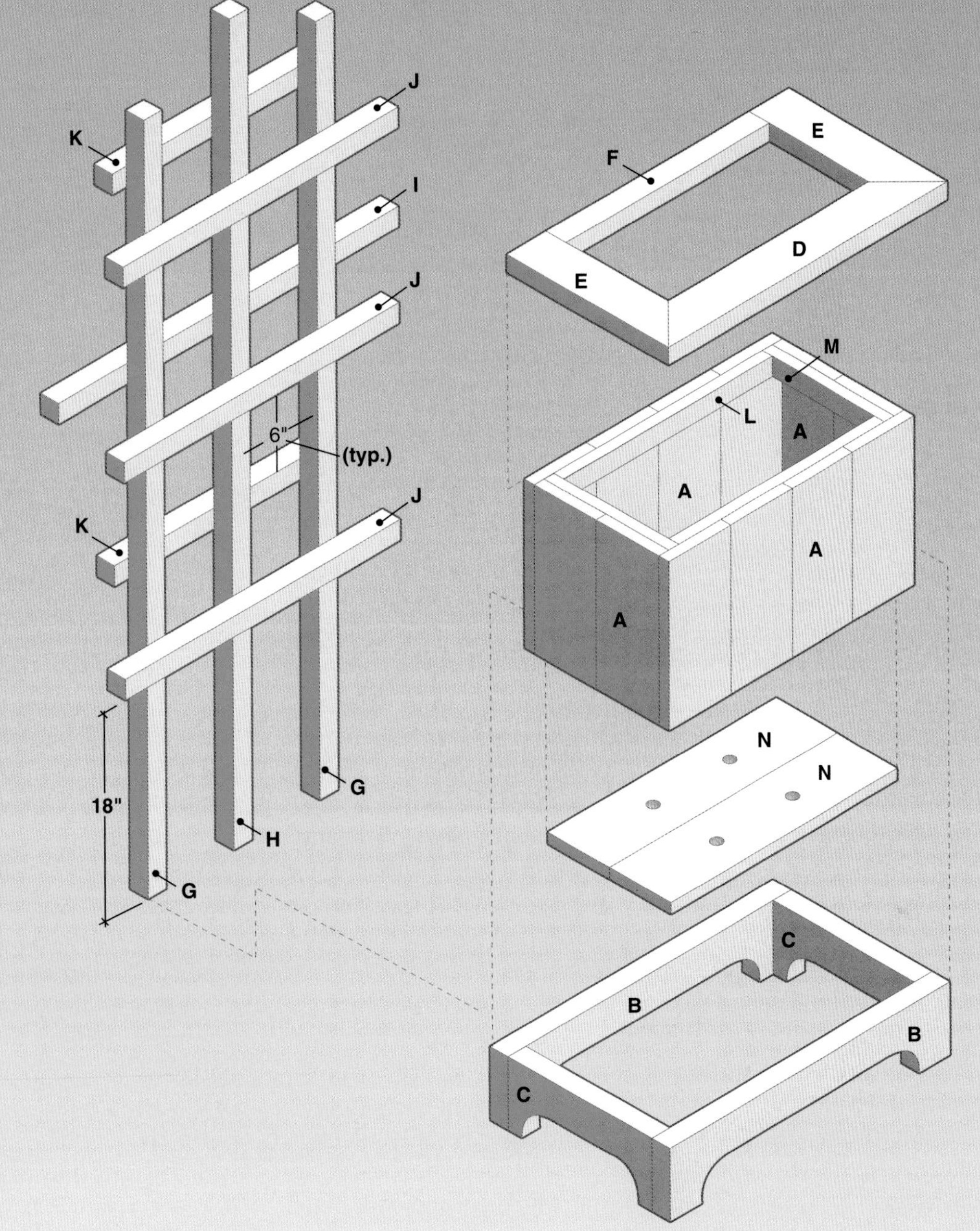

Key	Part	Dimension	Pcs.	Material
A	Box slat	⅞ × 5½ × 13"	12	Cedar
B	Base front/back	1½ × 5½ × 25"	2	Cedar
C	Base end	1½ × 5½ × 12¾"	2	Cedar
D	Cap front	1½ × 3½ × 25"	1	Cedar
E	Cap end	1½ × 3½ × 14¼"	2	Cedar
F	Cap back	1½ × 1½ × 18"	1	Cedar
G	End post	1½ × 1½ × 59½"	2	Cedar

Cutting List

Key	Part	Dimension	Pcs.	Material
H	Center post	1½ × 1½ × 63½"	1	Cedar
I	Long rail	1½ × 1½ × 30"	1	Cedar
J	Medium rail	1½ × 1½ × 24"	3	Cedar
K	Short rail	1½ × 1½ × 18"	2	Cedar
L	Long cleat	⅞ × 1½ × 18½"	2	Cedar
M	Short cleat	⅞ × 1½ × 11"	2	Cedar
N	Bottom board	⅞ × 5½ × 20¼"	2	Cedar

Materials: Moisture-resistant glue, #8 2" wood screws, 1⅝" and 2½" deck screws, finishing materials.

Note: Measurements reflect the actual size of dimension lumber.

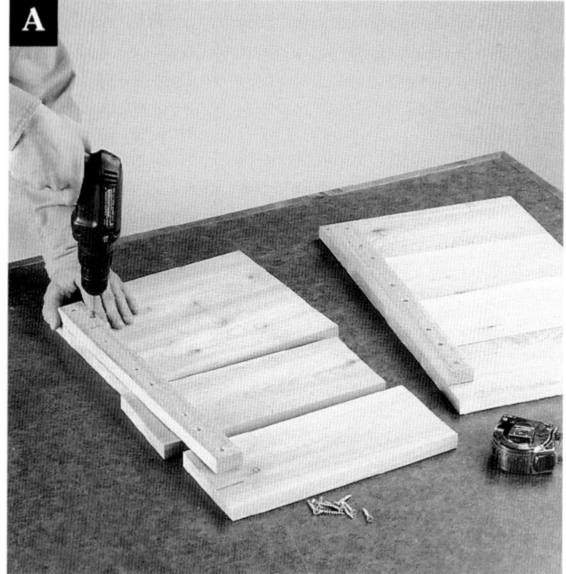

Attach the side cleats flush with the tops of the side boards.

Use a jig saw to make scalloped cutouts in all four base pieces—make sure the cutouts in matching pieces are the same.

Directions: Trellis Planter

BUILD THE PLANTER BIN.

1. Cut the box slats (A) and cleats (L, M) to length. Arrange the slats edge to edge in two groups of four and two groups of two, with tops and bottoms flush.

2. Center a long cleat (L) at the top of each set of four slats, so the distance from each end of the cleat to the end of the panel is the same. Attach the cleats to the four-slat panels by driving 1⅝" deck screws **(photo A)** through the cleats and into the slats.

3. Lay the short cleats (M) at the tops of the two-slat panels. Attach them to the slats the same way.

4. Arrange all four panels into a box shape and apply moisture-resistant wood glue to the joints. Attach the panels by driving 1⅝" deck screws through the four-slat panels and into the ends of the two-slat panels.

INSTALL THE BIN BOTTOM.

1. Cut the bottom boards (N) to length. Set the bin upside down on your work surface, and mark reference lines on the inside faces of the panels, ⅞" in from the bottom of the bin. Insert the bottom boards into the bin, aligned with the reference lines to create a ⅞" recess. Scraps of 1× cedar can be slipped beneath the bottom boards as spacers.

2. Drill ⅛" pilot holes through the panels. Counterbore the holes slightly with a counterbore bit. Fasten the bottom boards by driving 1⅝" deck screws through the panels, and into the edges and ends of the bottom boards.

BUILD THE PLANTER BASE.

The planter base is scalloped to create feet at the corners.

1. Cut the base front and back (B) and the base ends (C) to length. To draw the contours for the scallops on the front and back boards, set the point of a compass at the bottom edge of the base front, 5" in from one end. Set the compass to a 2½" radius, and draw a curve to mark the curved end of the cutout (see *Diagram*, page 29). Draw a straight line to connect the tops of the curves, 2½" up from the bottom of the board, to complete the scalloped cutout.

2. Make the cutout with a jig saw, then sand any rough spots in the cut. Use the board as a template for marking a matching cutout on the base back.

3. Draw a similar cutout on one base end, except with the point of the compass 3½" in from the ends. Cut out both end pieces with a jig saw **(photo B).**

4. Draw reference lines for wood screws, ¾" from the ends of the base front and back. Drill three evenly spaced pilot holes through the lines. Counterbore the holes. Fasten the base ends between the base front and back by driving three evenly spaced deck screws at each joint.

ATTACH THE BIN TO THE BASE.

1. Set the base frame and planter bin on their backs. Position the planter bin inside

The recess beneath the bottom boards in the planter bin provides access for driving screws.

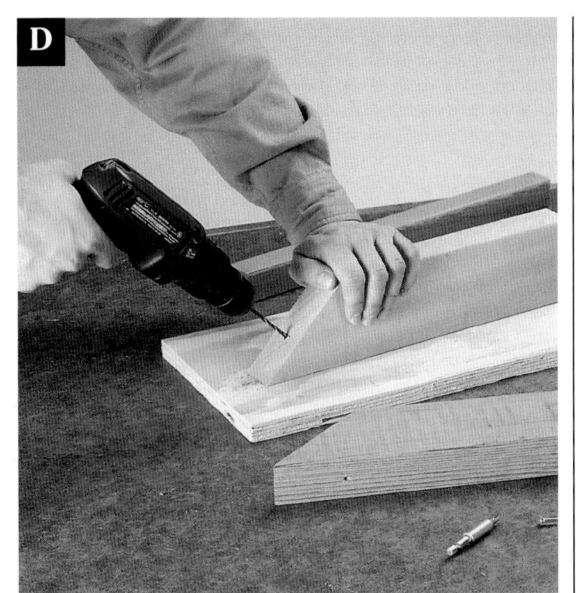

Before attaching the cap ends, drill pilot holes through the mitered ends of the cap front ends.

the base so it extends ⅞" past the top of the base.
2. Drive 1⅝" deck screws through the planter bin and into the base to secure the parts **(photo C).**

MAKE THE CAP FRAME.
1. Cut the cap front (D), cap ends (E) and cap back (F) to length. Cut 45° miters at one end of each cap end, and at both ends of the cap front.
2. Join the mitered corners by drilling pilot holes through the joints **(photo D).** Counterbore the holes. Fasten the pieces with glue and 2½" deck screws. Clamp the cap front and cap ends to the front of your worktable to hold them while you drive the screws.
3. Fasten the cap back between the cap ends with wood screws, making sure the back edges are flush. Set the cap frame on the planter bin so the back edges are flush. Drill pilot holes. Counterbore them. Drive 2½" deck screws through the cap frame and into the side and end cleats.

MAKE THE TRELLIS.
The trellis is made from pieces in a crosshatch pattern. The exact number and placement of the pieces is up to you—use the same spacing we used (see *Diagram),* or create your own.
1. Cut the end posts (G), center post (H) and rails (I, J, K) to length. Lay the end posts and center post together side by side with their bottom edges flush, so you can gang-mark the rail positions.
2. Use a square as a guide for drawing lines across all three posts, 18" up from the bottom. Draw the next line 7½" up from the first. Draw additional lines across the posts, spaced 7½" apart.
3. Cut two 7"-wide scrap blocks, and use them to separate the posts as you assemble the trellis. Attach the rails to the posts in the sequence shown in the *Diagram,* using 2½" screws **(photo E).** Alternate from the fronts to the backs of the posts when installing the rails.

APPLY FINISHING TOUCHES.
Fasten the trellis to the back of the planter bin so the bottoms of the posts rest on the top edge of the base. Drill pilot holes in the posts. Counterbore the holes. Drive 2½" deck screws through the posts and into the cap frame. With a 1"-dia. spade bit, drill a pair of drainage holes in each bottom board. Stain the project with an exterior wood stain.

Temporary spacers hold the posts in position while the trellis crossrails are attached.

PROJECT
POWER TOOLS

Planters

These cedar planters are simple projects that can transform a plain plant container into an attractive outdoor accessory.

CONSTRUCTION MATERIALS

Quantity	Lumber
1	1 × 10" × 6' cedar
1	¼ × 20 × 20" hardboard or plywood

Add a decorative touch to your deck, porch or patio with these stylish cedar planters. Created using square pieces of cedar fashioned together in different design patterns, the styles shown above feature circular cutouts that are sized to hold a standard 24-ounce coffee can. To build them, simply cut 1 × 10 cedar to 9¼" lengths, then make 7¼"-diameter cutouts in the components as necessary. We used a router and template to make the cutouts with production speed. Follow the assembly instructions (see page 35 and the diagrams on page 33) to create the three designs above. Or, you can create your own designs by rearranging the components or altering the cutout size.

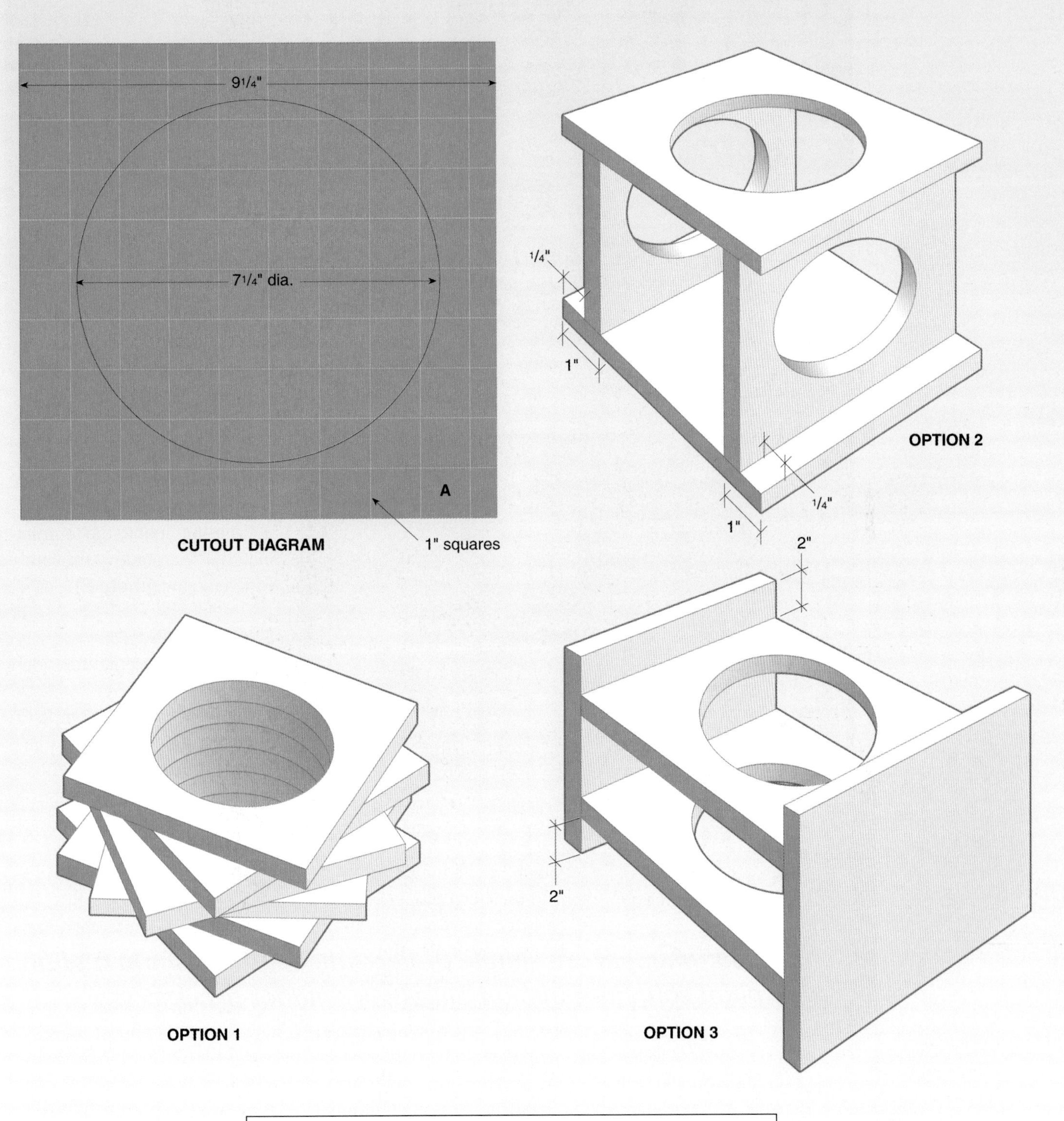

Cutting List

Key	Part	Dimension	Pcs.	Material
A	Component	¾ × 9¼ × 9¼"	*	Cedar

Materials: Moisture-resistant glue, 2" deck screws, 24-ounce coffee can, finishing materials.

* Number of pieces varies according to planter style.

Note: Measurements reflect the actual size of dimension lumber.

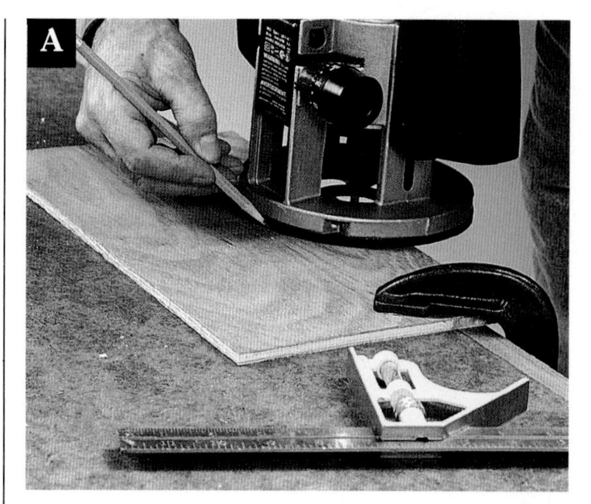

Outline the router base onto scrap material to help determine the router-base radius.

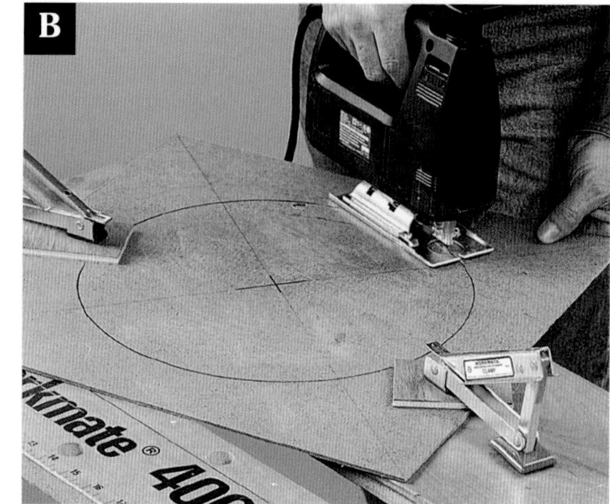

Cut out the router template using a jig saw.

Drill a starter hole for the router bit in the centers of the components.

Directions: Planters

MAKE THE ROUTER TEMPLATE.

Using a router and a router template is an excellent method for doing production-style work with uniform results. To create the cutout components for the planters, make a circular template to use as a cutting guide for the router. To determine the size of the template circle, add the radius of your router base to the radius of your finished cutout (3⅝" in the project as shown).

1. Begin by finding the radius of your router base. First, install a 1"-long straight bit in your router. (For fast cutting, use a ¾"-diameter bit, but make sure you use the same bit for making the template and cutting the components.) Make a shallow cut into the edge of a piece of scrap wood. With the router bit stopped, trace around the outside edge of the router base with a pencil **(photo A).**

2. Measure from the perimeter of the router cut to the router-base outline to find the radius. Add 3⅝" for the radius of a 24-ounce coffee can. Using a compass, draw a circle with this measurement onto the template material. Cut out the template with a jig saw **(photo B).**

> TIP
>
> *Plunge routers are routers with a bit chuck that can be raised or lowered to start internal cuts. If you own a plunge router, use it to cut parts for this project. Otherwise, drill a starter hole as shown, for a standard router.*

MAKE THE PARTS.

The planters are built from identical components (A).

1. Cut the number of components required for your design. Make circular cutouts on those components that require them. To do that, draw diagonal lines connecting the corners of the component. The point of intersection is the center of the square board. Center the template on the component and clamp it in place.

2. Use a drill to bore a 1"-diameter starter hole for the router bit (unless you are using a plunge router—see *Tip,* left) at the center **(photo C).** Position the router bit inside the hole.

3. Turn the router on and move it away from the starter hole until the router base contacts the template. Pull the router in a counterclockwise direction around the inside of the template to make the cutout. Sand sharp edges with sandpaper.

Assembly Options

Option 1. *Attach the pieces on the stacked planter from top to bottom, ending with a solid base. To make this stacked planter, you need six pieces of 1 × 10 cedar. Cut them to length, and rout circular shapes in five of them. The solid piece will be the base. Stack the pieces on top of the base component. Place a painted coffee can in the center and arrange the sections to achieve a spiraling effect (see* Diagram, *page 33). Use a pencil to mark the locations of the pieces. Remove the can and fasten the pieces together using glue and deck screws. Attach the pieces by driving the deck screws through the lower pieces into the upper pieces, fastening the base last.*

Option 2. *Use four components on this option to create a planter with three cutout components and a solid base. Measure and mark lines 1" from each side edge on the solid component and one of the cutout components. Attach the inner components with their inside faces flush with these lines. Fasten the solid component to the sides with moisture-resistant glue and deck screws. Attach the remaining cutout component to finish the planter. Insert a painted coffee can.*

Option 3. *Attach two components with circular cutouts to the inside faces of two solid components to make this planter. Measure and mark guidelines 2" from the top and bottom edges on the two solid components. Fasten the two cutout components between the others with moisture-resistant glue and deck screws, making sure their outside edges are flush with the drawn guidelines. Insert a painted coffee can.*

Sundial

This throwback to ancient times casts a shadow of classical elegance in your yard or garden—and it tells time, too.

PROJECT POWER TOOLS

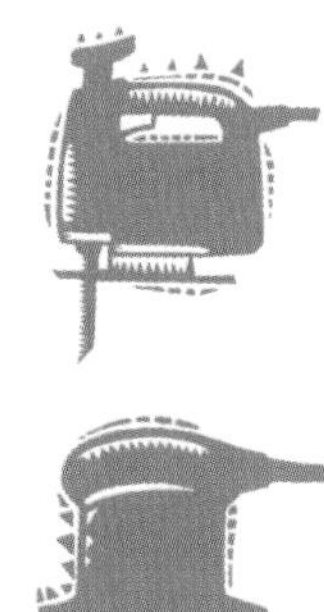

Sundials have been popular garden accessories for just about as long as there have been formal gardens. While their time-telling function is not as critical as it was around the time of the Roman Empire, sundials today continue to dot formal gardens, and even suburban flower beds, throughout the world.

The design for the cedar sundial shown here contains a few elements that harken back to ancient times. The fluted pillar suggests the famous architectural columns of Old World cathedrals and amphitheaters. The plates at the top and base are trimmed with a Roman Ogee router bit for a Classical touch. Making these cuts requires a router and several different types of router bits. If you don't mind a little plainer look, you can bypass the router work and simply round over the parts with a power sander.

This sundial is not just another stylish accent for your yard or garden. You can actually use it to tell time. Simply mount the triangular shadow-caster (called a "gnomon") to the face of the sundial, and orient it so it points north. Then calibrate the face at twelve hourly intervals. That way you know that the positioning of the numbers is accurate.

This sundial is secured to the ground with a post anchor that fits around a mounting block on the underside of the sundial. The anchor is driven into the ground, making it easy to move the sundial (you'll appreciate this the first time Daylight Saving Time comes around).

CONSTRUCTION MATERIALS

Quantity	Lumber
1	1 × 12" × 4' cedar
1	6 × 6" × 4' cedar
1	4 × 4" × 4' cedar

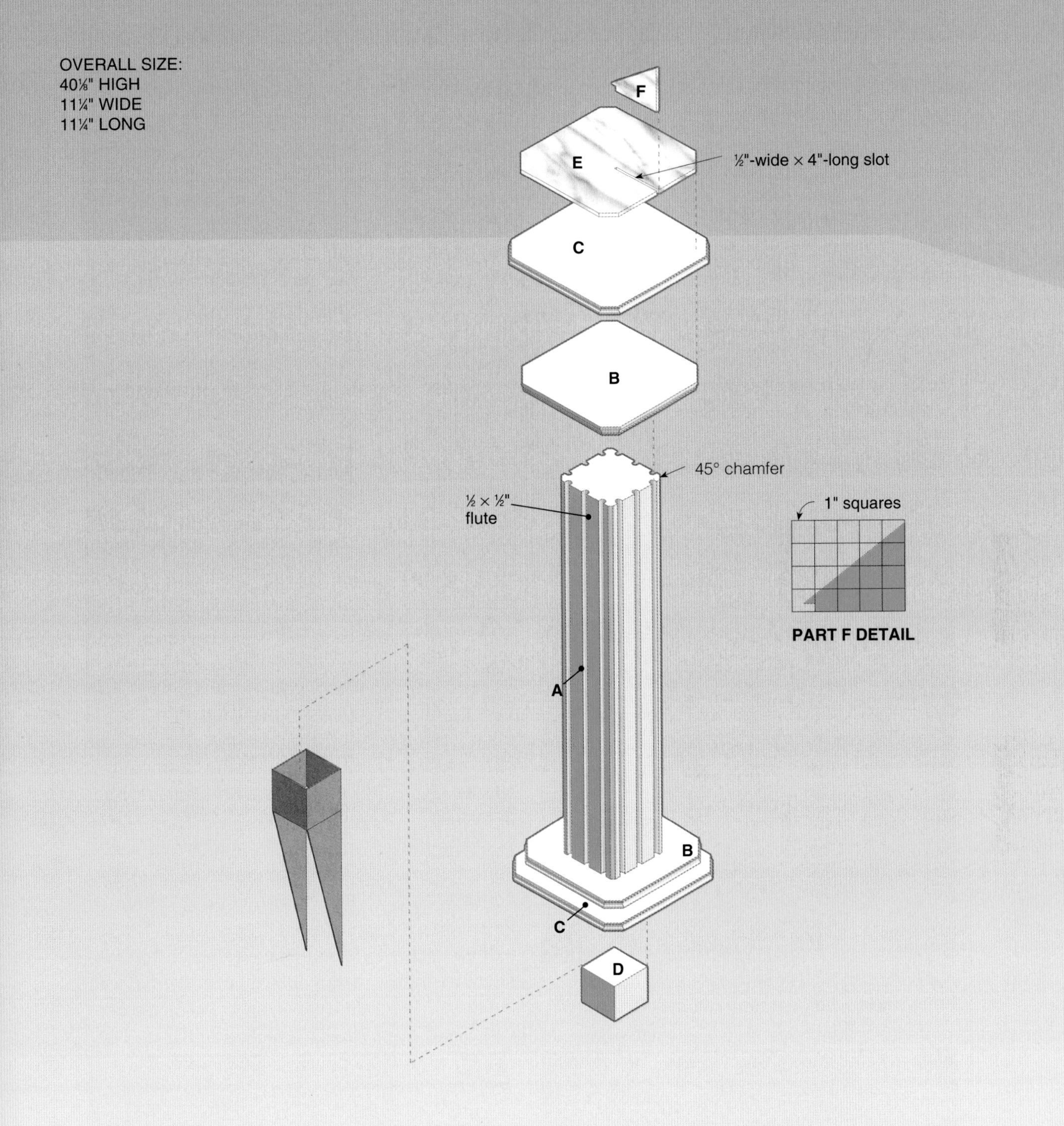

Cutting List

Key	Part	Dimension	Pcs.	Material
A	Column	5½ × 5½ × 32¾"	1	Cedar
B	Inner plate	⅞ × 10 × 10"	2	Cedar
C	Outer plate	⅞ × 11¼ × 11¼"	2	Cedar

Cutting List

Key	Part	Dimension	Pcs.	Material
D	Mounting block	3½ × 3½ × 3"	1	Cedar
E	Dial face	⅜ × 10 × 10"	1	Ceramic tile
F	Gnomon	¼ × 3½ × 16½"	1	Plexiglass

Materials: 1¼" and 2" galvanized deck screws, construction adhesive, silicone caulk, clock-face numbers, 4 × 4 metal post anchor.

Note: Measurements reflect the actual size of dimension lumber.

Directions: Sundial

MAKE THE COLUMN.

1. Cut the column (A) to length from a 4'-long 6 × 6" cedar post.

2. Use a combination square as a marking gauge to lay out three pairs of parallel lines lengthwise on each face of the column—the lines in each pair should be ½" apart **(photo A).** These lines form the outlines for the fluted grooves that will be cut into the post. The two outer-flute outlines should start ¾" from the edge, and the middle flute should be 1¼" from each outer outline. Install a ½" core box bit in your router—a core box bit is a straight bit with a rounded bottom.

3. Hook the edge guide on the foot of your router over the edge of the post (or, use a straightedge cutting guide to guide the router), and cut each ½"-deep flute in two passes.

4. After all 12 flutes are cut, install a 45° chamfering bit in your router and trim off all four edges of the column **(photo B).**

CUT THE FLAT PARTS.

Two flat, square boards are sandwiched together and attached at the top and bottom of the column.

1. Cut the inner plates (B) and outer plates (C) to size from 1 × 12 cedar.

2. After the plates are cut to final size, trim off a corner with 1" legs from all four corners of each plate, using a jig saw **(photo C).**

3. Install a piloted bit in your router to cut edge contours (we used a double ogee fillet bit), then cut the roundovers on all edges of the plates **(photo D).**

MAKE THE SUNDIAL FACE.

We used a piece of octagonal marble floor tile for the face (E) of our sundial. If you prefer, you can use inexpensive ceramic floor tile instead of marble, but either way you should purchase a piece of tile that is already cut to the correct size and shape for the project (cutting floor tile is very difficult).

1. Lay out the ¼"-wide, 4"-long slot for the gnomon, centered on one edge of the sundial face. Have the slot cut at a tile shop (if this is a problem, you are probably better off eliminating the slot than trying to cut it yourself).

2. Next, mark a 1"-square grid pattern on a small piece of ¼"-thick white plexiglass.

3. Lay out the shape and dimensions of the gnomon, following the *Grid Pattern* (F) on page 37.

4. Mount a wood-cutting blade with medium-sized teeth in your jig saw, and cut the gnomon shape.

5. Sand the edges with 100-grit sandpaper, then finish-sand with fine paper, up to 180-or 220-grit.

Lay out three ½"-wide flutes on each column face using a combination square as a marking gauge.

Trim off the corners of the column with a router and a 45° chamfering bit.

Cut triangular cutoffs with 1" legs at each corner of each plate.

Cut decorative profiles along the edges of the inner and outer plates.

Make sure screws driven through the outer plate are at least 2" in from the edges.

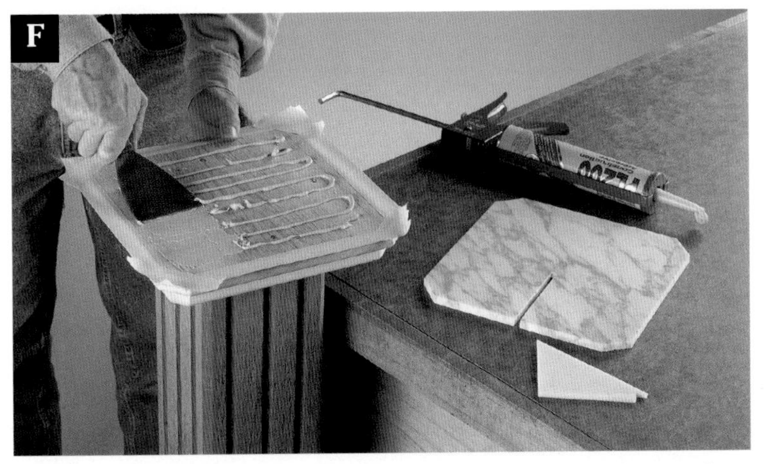

Frame the top of the top outer plate with masking tape, then apply a layer of construction adhesive to attach the sundial face.

6. Fit the gnomon into the slot on the sundial face.

ASSEMBLE THE SUNDIAL.

1. Attach an inner plate, centered, to each end of the column, using 2½" deck screws.
2. Cut the mounting block (D) to length from 4 × 4 cedar (or two pieces of cedar 2 × 4).
3. Attach the block to one face of the bottom outer plate, centered, using deck screws driven through the plate and into the block. Do not attach the base plate to the column assembly yet.
4. Attach the top outer plate to the top inner plate, making sure the overhang is equal on all four sides and the corners align **(photo E).**
5. Set the post on its base, and attach the sundial face (E) to the top outer plate, using construction adhesive **(photo F).** Make sure the face is centered and in alignment.
6. Place construction adhesive into the gnomon slot, then insert the gnomon (F) into the slot and press firmly. Seal all joints around the edges of the gnomon and the edges of the face, using clear silicone caulk.
7. Attach the bottom outer plate by laying the column on its side and driving 1¼" deck screws up through the outer plate and into the inner plate.
8. Coat wood parts with exterior wood stain.

INSTALL & CALIBRATE THE SUNDIAL.

1. Choose a sunny spot to install your sundial. Lay a piece of scrap wood on the spot, pointing directly north (use a magnetic compass for reference).
2. Purchase a metal post anchor for a 4 × 4 post (most have an attached metal stake about 18" long).
3. Drive the post anchor (G) into the ground at the desired location, making sure one side of the box part of the anchor is perpendicular to the scrap piece facing north.
4. Insert the mounting block on the base of the post into the anchor, making sure the gnomon is facing north.
5. Calibrate the sundial by marking a point at the edge of the shadow from the gnomon at the top of every hour. Apply hour markers at those points. We used metal Roman numerals from a craft store, attached with clear silicone caulk.

PROJECT
POWER TOOLS

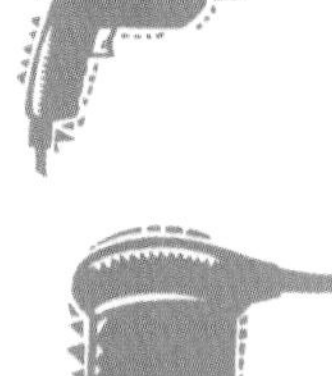

Driveway Marker

Build an inviting yard ornament that graces the entrance to your driveway or front walk and directs foot traffic where you want it to go.

CONSTRUCTION MATERIALS

Quantity	Lumber
1	2 × 4" × 8' cedar
1	1 × 6" × 6' cedar
1	1 × 6" × 8' cedar
4	1 × 2" × 8' cedar

Bestow a sense of order on your front yard by building this handsome cedar driveway marker. Position it on your lawn at the entry to your driveway to keep cars from wandering off the paved surface. Or, set a driveway marker on each side of your front walk to create a formal entry to your home.

This freestanding driveway marker has many benefits you'll appreciate. The fence-style slats slope away from the corner post to create a sense of flow. The broad corner post can be used to mount address numbers, making your home easier to find for visitors and in emergencies. And behind the front slats you'll find a spacious planter.

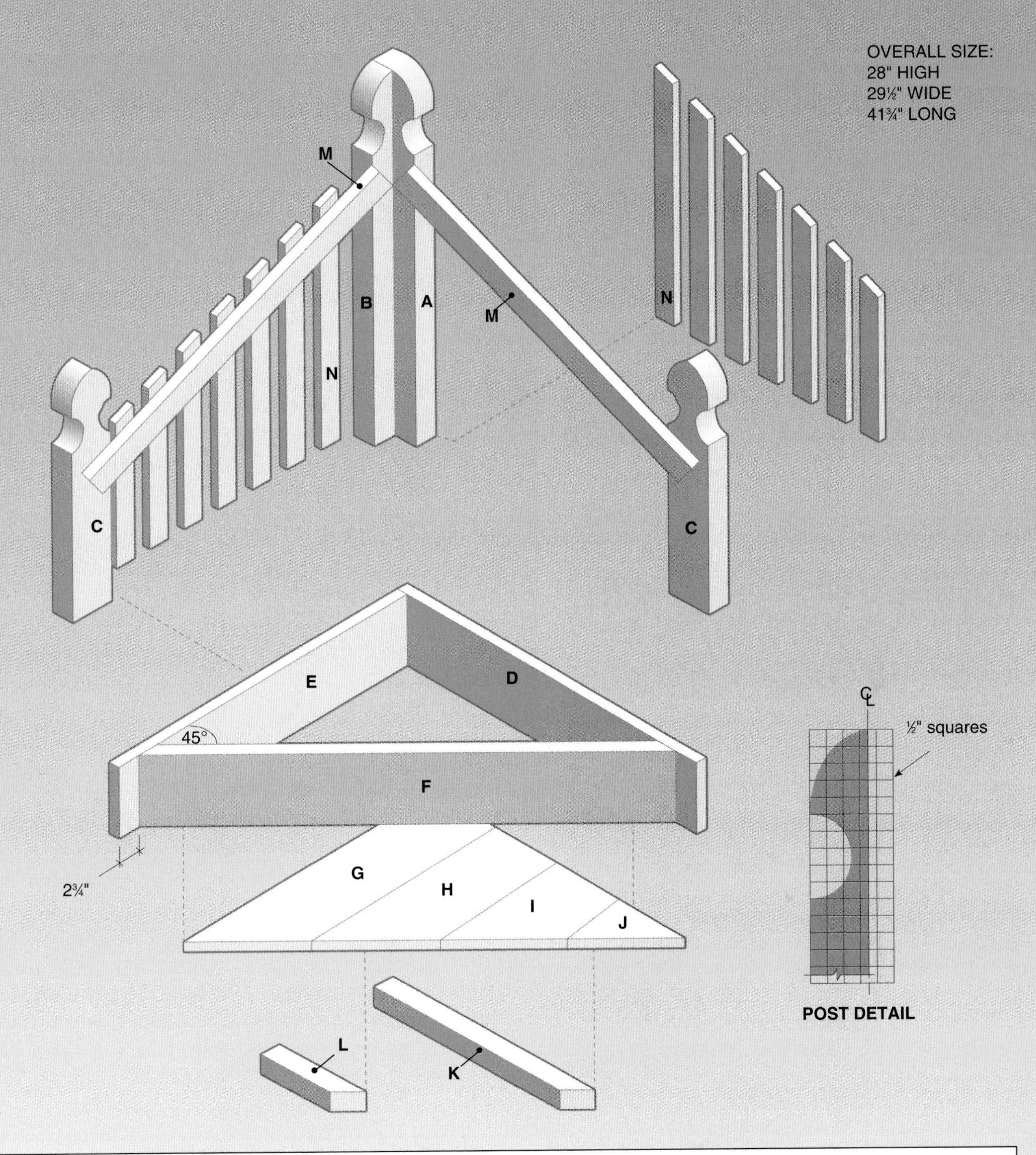

Cutting List

Key	Part	Dimension	Pcs.	Material
A	Corner post	1½ × 3½ × 28"	1	Cedar
B	Corner post	1½ × 1½ × 28"	1	Cedar
C	End post	1½ × 3½ × 18½"	2	Cedar
D	Planter side	⅞ × 5½ × 26½"	1	Cedar
E	Planter side	⅞ × 5½ × 25⅝"	1	Cedar
F	Planter back	⅞ × 5½ × 33"	1	Cedar
G	Bottom board	⅞ × 5½ × 23"	1	Cedar

Cutting List

Key	Part	Dimension	Pcs.	Material
H	Bottom board	⅞ × 5½ × 17"	1	Cedar
I	Bottom board	⅞ × 5½ × 11"	1	Cedar
J	Bottom board	⅞ × 5½ × 6"	1	Cedar
K	Long cleat	1½ × 1½ × 19"	1	Cedar
L	Short cleat	1½ × 1½ × 9"	1	Cedar
M	Stringer	⅞ × 1½ × 27"	2	Cedar
N	Slat	⅞ × 1½ × 20"	14	Cedar

Materials: Moisture-resistant glue, 1¼", 1½", 2" and 2½" deck screws, 2" brass numbers (optional), finishing materials.

Note: Measurements reflect the actual size of dimension lumber.

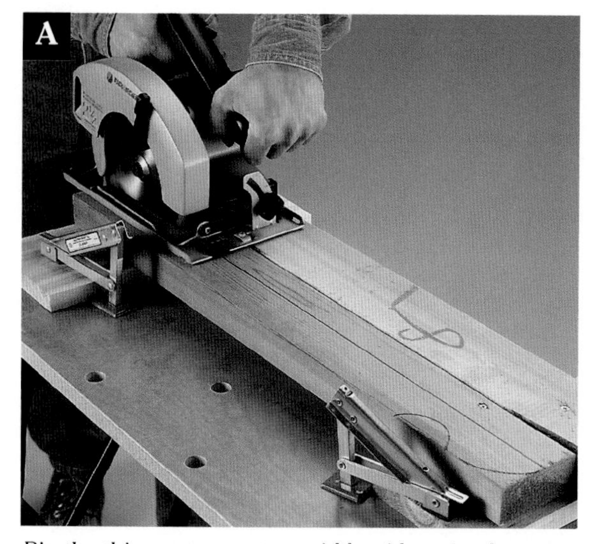

Rip the thin corner post to width with a circular saw.

Sand the top of the corner post assembly so the joint is smooth.

Lay the bin frame on the bottom boards and trace along the back inside edge to mark cutting lines.

Directions: Driveway Marker

CUT THE POSTS.
This driveway marker is a freestanding yard ornament supported by single 2 × 4 posts at each end and a doubled 2 × 4 post at the corner.
1. Cut the corner posts (A, B) and end posts (C) to length. Draw a ½"-square grid pattern at the top of one of the end posts, using the grid pattern on page 41 as a reference. Mark a centerpoint at the top of the post and draw the pattern as shown on one side. Reverse the pattern on the other side to create the finished shape. Use a jig saw to cut the end post to shape. Mount a drum sander attachment in your electric drill and use it to smooth out the cut.
2. Use the shaped end post as a template to mark the other end post. Cut and sand it.
3. To make the corner posts, mark centerpoints at the top of each corner post. Trace the contour of one end post on one side of the centerline.
4. On one corner post, draw a line down the length of the post, 2" in from the side with no contour cutout. This will be the narrower post (B). To rip this post to width, attach two pieces of scrap wood to your work surfaces. Screw the post, facedown, to the wood scrap (making sure to drive screws in the waste area of the post).
5. Butt a scrap of the same thickness as the post next to the post, to use as a guide for the circular saw. Attach the guide board to the wood scraps. Set the edge guide on the saw so it follows the outside edge of the scrap. Make the rip cut along the cutting line **(photo A).** Cut the contours at the tops of the corner posts and sand smooth.

BUILD THE CORNER POST.
1. Apply glue to the ripped edge of the narrower post board (B). Lay it on the face of the wider post board (A), so the joint at the corner is flush and the tops of the contours come together in a smooth line.
2. Drill ⅛" pilot holes in the wider board at 4" intervals. Counterbore the holes ¼" deep, using a counterbore bit. Drive 2½" deck screws through the wider board and into the edge of the narrower board. After the glue sets, sand the tops smooth **(photo B).**

MAKE THE PLANTER FRAME.
The triangular planter fits in the back of the driveway marker.
1. Cut the planter sides (D, E) to length, making square cuts at the ends. The ends of the planter back (F) are mitered so they fit flush against the sides when the bin is formed. Set your circular saw to make a 45° cut. Cut the planter back to length, making sure the bevels

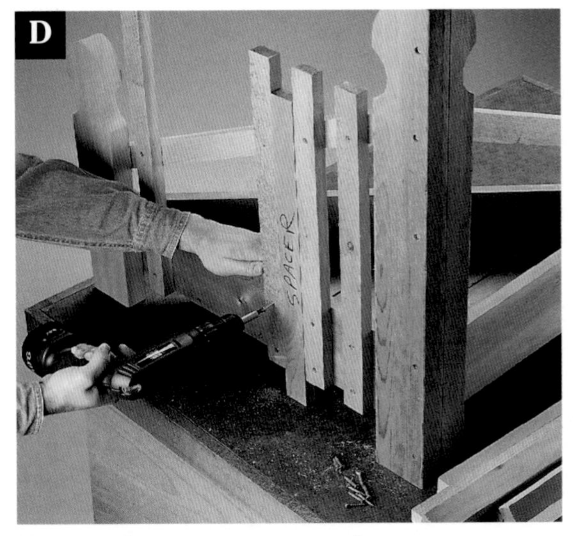

Use one slat as a spacer to set the correct gap as you fasten the slats to the bin and the stringers.

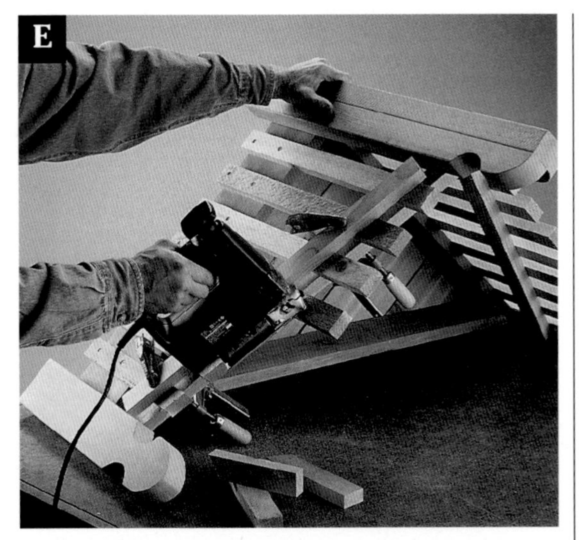

Use a cutting guide to trim the tops of the slats so they extend beyond the tops of the stringers.

both go inward from the same side (see *Diagram,* page 41).
2. Apply glue to the ends of the planter back. Assemble the back and the sides by drilling pilot holes in the outside faces of the sides. Counterbore the holes. Drive 2" deck screws through the sides and into the ends of the back. This will create a setback of about 2¾" from the joints to the ends of the sides.

ATTACH THE BIN BOTTOM.
1. Cut the bottom boards (G, H, I, J) to length. Lay the boards on your work surfaces, arranged from shortest to longest, and butted together edge to edge. Set the bin frame on top of the boards so the inside edges of the frame sides are flush with the outer edges of the boards, and the boards extend past the back edge of the frame. Trace along the inside of the frame back to mark cutting lines on the bottom boards **(photo C).** Cut them with a circular saw.
2. Cut the long cleat (K) and short cleat (L) to length, making a 45° miter cut at one end of each cleat. Turn the planter bin upside down. Attach the reinforcing cleats so one is 2½"-3" from Side D and the other is about 12" from the same side. Attach them by driving two 2" deck screws through each plant side and back into the end of each cleat.
3. Right the frame. Position the bottom boards flush with the bottom of the frame and attach them with glue and 2" deck screws, driven through the frame and into the ends of the bottom boards.

ATTACH THE
BIN AND POSTS.
1. Set the bin on 2"-tall spacers. Fit the corner post assembly over the front corner of the bin and attach with glue and 1½" deck screws.
2. Attach the end posts so each is 29½" away from the corner post assembly.

ATTACH THE
STRINGERS AND SLATS.
The stringers (M) are attached between the tops of the posts to support the tops of the slats.
1. Cut the stringers to length. Attach them to the insides of the posts so the top edges are 1½" below the bottom of the post contour at the point where the stringer meets each post.
2. Cut all slats (N) to length. (The tops will be trimmed after the slats are installed.) Attach them to the bin and the stringers, spaced at 1½" intervals, using 1¼" deck screws. Use a slat as a spacer **(photo D).** Install all 14 slats, making sure the bottoms are flush with the bin bottom.
3. Clamp a piece of 1 × 2 scrap against the outside faces of the slats for a cutting guide—the scrap should be directly opposite the stringer on the back side of the slats. Cut along the guide with a jig saw to trim the slats so the tops are slightly above the top of the stringer **(photo E).**

APPLY FINISHING TOUCHES.
Sand all exposed surfaces and apply two or more coats of exterior wood stain. If your marker will be visible from the curb, you may want to attach 2"-high brass numbers to the corner post to indicate your street address.

PROJECT
POWER TOOLS

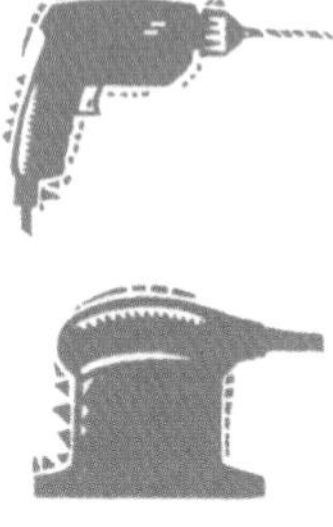

Plant Boxes

Build these simple plant boxes in whichever size or number best meets your needs.

CONSTRUCTION MATERIALS*

Quantity	Lumber
3	1 × 2" × 8' cedar
6	1 × 4" × 8' cedar
1	⅝" × 4 × 8' fir siding
1	¾" × 2 × 4' CDX plywood

*To build all three plant boxes as shown

Planters and plant boxes come in a vast array of sizes and styles, and there is a good reason for that. Everyone's needs are different. Rather than build just one planter that may or may not work for you, try this handy planter design. It can easily be changed to fit your space and planting demands.

This project provides measurements for planters in three sizes and shapes: short and broad for flowers or container plants; medium for spices and herbs or small trees and shrubs; and tall and narrow for vegetables or flowering vines that will cascade over the cedar surfaces. The three boxes are proportional to one another—build all three and arrange them in a variety of patterns, including the tiered effect shown above.

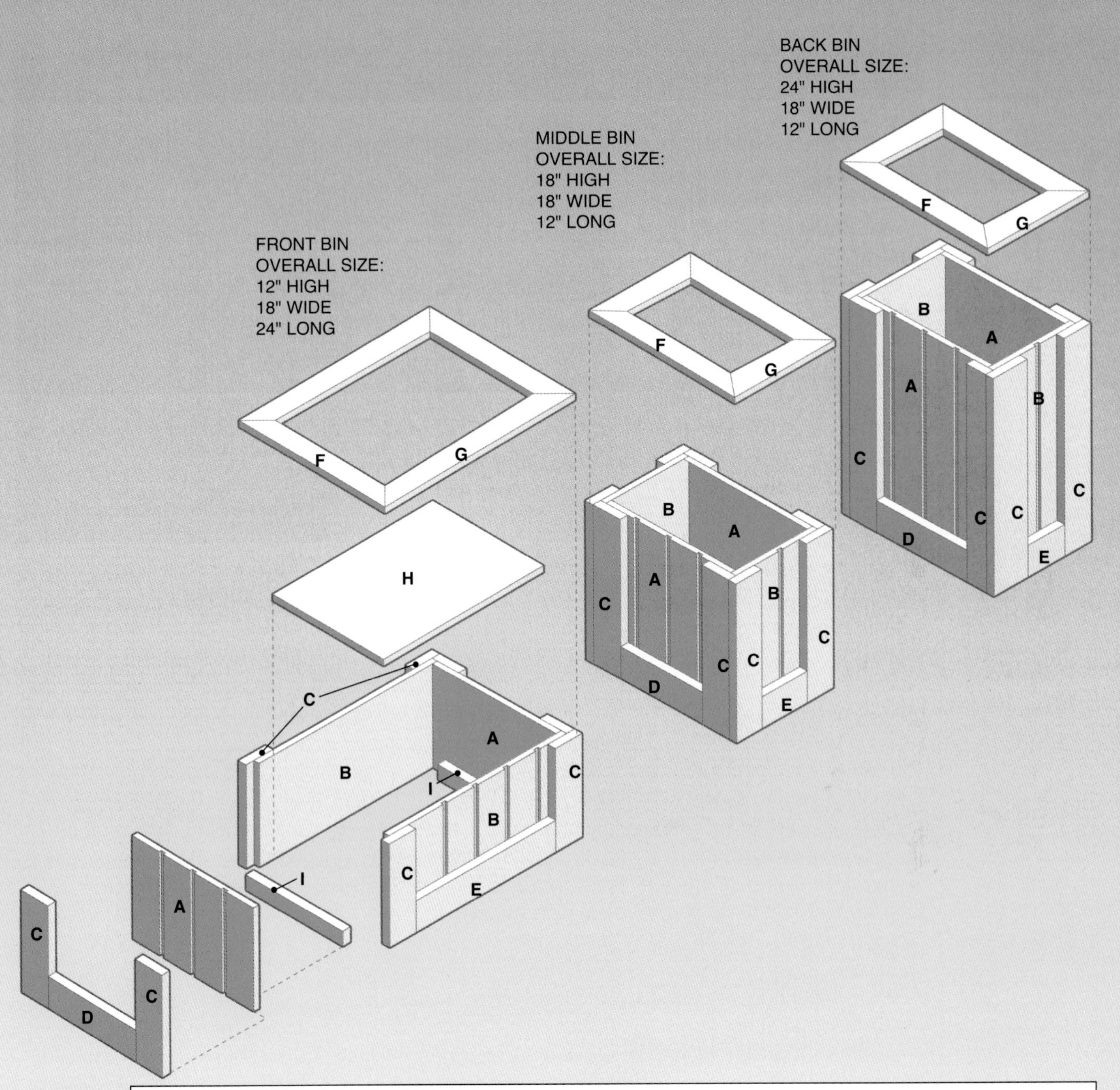

Cutting List

Key	Part	Front Bin Dimension	Pcs.	Middle Bin Dimension	Pcs.	Back Bin Dimension	Pcs.	Material
A	End panel	⅝ × 15 × 11⅛"	2	⅝ × 15 × 17⅛"	2	⅝ × 15 × 23⅛"	2	Siding
B	Side panel	⅝ × 22¼ × 11⅛"	2	⅝ × 10¼ × 17⅛"	2	⅝ × 10¼ × 23⅛"	2	Siding
C	Corner trim	⅞ × 3½ × 11⅛"	8	⅞ × 3½ × 17⅛"	8	⅞ × 3½ × 23⅛"	8	Cedar
D	Bottom trim	⅞ × 3½ × 9¼"	2	⅞ × 3½ × 9¼"	2	⅞ × 3½ × 9¼"	2	Cedar
E	Bottom trim	⅞ × 3½ × 17"	2	⅞ × 3½ × 5"	2	⅞ × 3½ × 5"	2	Cedar
F	Top cap	⅞ × 1½ × 18"	2	⅞ × 1½ × 18"	2	⅞ × 1½ × 18"	2	Cedar
G	Top cap	⅞ × 1½ × 24"	2	⅞ × 1½ × 12"	2	⅞ × 1½ × 12"	2	Cedar
H	Bottom panel	¾ × 14½ × 19½"	1	¾ × 14½ × 8½"	1	¾ × 14½ × 8½"	1	Plywood
I	Cleat	⅞ × 1½ × 12"	2	⅞ × 1½ × 12"	2	⅞ × 1½ × 12"	2	Cedar

Materials: 1¼", 1½" and 3" deck screws, 6d galvanized finish nails, finishing materials.

Note: Measurements reflect the actual size of dimension lumber.

Directions: Plant Boxes

Whatever the size of the plant box or boxes you are building, you'll use the same basic steps for construction. The only difference between the three boxes is the size of some components. If you need larger, smaller, broader or taller plant boxes than those shown, it's easy to create your own cutting list based on the *Diagram* and dimensions shown on page 45. If you are building several planters, do some planning and sketching to make efficient use of your wood and to save time by gang-cutting parts that are the same size and shape.

MAKE AND ASSEMBLE THE BOX PANELS.

The end and side panels are rectangular pieces of sheet siding fastened together with deck screws. You can use fir sheet siding with 4"-on-center grooves for a decorative look. Or, you can substitute any exterior-rated sheet goods (or even dimension lumber) to match the rest of your yard or home.

1. Cut the end panels (A) and side panels (B) to size, using a circular saw and straightedge cutting guide **(photo A).**

2. Lay an end panel facedown on a flat work surface and butt the side panel, face-side-out, up to the end of the end panel. Mark positions for pilot holes in the side panel. Drill ⅛" pilot holes. Counterbore the holes slightly so the heads are beneath the surface of the wood. Fasten the side panel to the end panel with 1½" deck screws.

3. Position the second side panel at the other end of the end panel and repeat the procedure.

4. Lay the remaining end panel facedown on the work surface. Position the side panel assembly over the end panel, placing the end panel between the side panels and keeping the edges of the side panels flush with the edges of the end panel. Drill pilot holes in the side panels. Counterbore the holes. Fasten the side panels to the end panel with deck screws.

Cut the end panels and side panels to size using a circular saw and a straightedge cutting guide.

ATTACH THE TRIM.

The cedar trim serves not only as a decorative element, but also as a structural reinforcement to the side panels. Most cedar has a rough texture on one side. For a rustic look, install your trim pieces with the rough side facing out. For a more finished appearance, install the pieces with the smooth side facing out.

1. Cut the corner trim (C) to length. Overlap the edges of the corner trim pieces at the corners to create a square butt joint. Fasten the corner trim pieces directly to the panels by driving 1¼" deck screws through the inside faces of the panels and into the corner trim pieces **(photo B).** For additional support, drive screws or galvanized finish nails through the overlapping corner trim pieces and into the edges of the adjacent trim piece (this is called "lock-nailing").

2. Cut the bottom trim pieces (D, E) to length. Fasten the pieces to the end and side panels, between the corner trim pieces. Drive 1¼" deck screws through the side and end panels and into the bottom trim pieces.

TIP

Make plant boxes portable by adding wheels or casters. If your yard or garden is partially shaded, attaching locking casters to the base of the plant boxes lets you move your plants to follow the sun, and can even be used to bring the plants indoors during colder weather. Use locking wheels or casters with brass or plastic housings.

Fasten the corner trim to the panels by driving deck screws through the panels into the trim.

Tip

Line plant boxes with landscape fabric before filling them with soil. This helps keep the boxes from discoloring, and also keeps the soil in the box, where it belongs. Simply cut a piece of fabric large enough to wrap the box (as if you were gift-wrapping it), then fold it so it fits inside the box. Staple it at the top, and trim off the excess. For better soil drainage, drill a few 1"-dia. holes in the bottom of the box and add a layer of small stones at the bottom of the box.

INSTALL THE TOP CAPS.

The top caps fit around the top of the plant box to create a thin ledge that keeps water from seeping into the end grain of the panels and trim pieces.

1. Cut the top caps (F, G) to length. Cut 45° miters at both ends of one cap piece, using a power miter saw or a miter box and backsaw. Tack the mitered cap piece to the top edge of the planter, keeping the outside edges flush with the outer edges of the corner trim pieces. For a proper fit, use this cap piece as a guide for marking and cutting the miters on the rest of the cap pieces.

2. Miter both ends of each piece and tack it to the box so it makes a square corner with the previously installed piece. If the corners do not work out exactly right, loosen the pieces and adjust the arrangement until everything is square. Permanently fasten all the cap pieces to the box with 6d galvanized finish nails.

INSTALL THE BOX BOTTOM.

The bottom of the planter box is supported by cleats (I) that are fastened inside the box, flush with the bottoms of the side and end panels.

1. Cut the cleats to length. Screw them to the end panels with 1½" deck screws **(photo C).** On the taller bins you may want to mount the cleats higher on the panels so the box won't need as much soil when filled. If you choose to do this, add cleats on the side panels for extra support.

2. Cut the bottom panel (H) to size from ¾"-thick exterior-rated plywood, such as CDX plywood. Set the bottom panel onto the cleats. You do not need to fasten it in place.

Attach 1 × 2 cleats to the inside faces of the box ends to support the bottom panel.

APPLY FINISHING TOUCHES.

After you've built all the boxes, sand all the edges and surfaces to remove rough spots and splinters. Apply two or three coats of exterior wood stain to all the surfaces to protect the wood. When the finish has dried, fill the boxes with potting soil. If you are using shorter boxes, you can simply place potted plants inside the planter box.

Prairie Windmill

With a mill section that turns and spins in the wind, this lively little windmill becomes the focal point of any garden.

PROJECT POWER TOOLS

Modeled loosely after the old turn-of-the-century windmills that dotted the prairie landscape, this fun, active garden accent may be just the thing to put some spice into your yard. With a solid, staked base firmly planted on the ground, this windmill spins and turns with the passing breezes.

We used cedar siding for the blades and tail of the mill section. The beveled cedar is the perfect shape and weight for catching the wind, and, because it's cedar, it will withstand the elements. The moving parts spin on lag screws and nylon washers, which perform better with moving parts than metal washers.

Overall, the most impressive part of this prairie windmill may be the geometrically striking tower section, which rises from the base to anchor the spinning mill above. Despite its size, the tower section is very easy to make. You can set the completed windmill in the heart of your garden, or position it in a corner of your yard to create a unique accent. Either way, you won't be disappointed. This is a fun project, and you'll get a glowing sense of satisfaction when you see it spinning and turning in the wind like a real windmill—just the way you built it.

Construction Materials

Quantity	Lumber
2	4 × 4" × 6' cedar
1	2 × 10" × 6' cedar
1	2×6" × 6' cedar
1	2×4" × 6' cedar
5	2 ×2"× 8' cedar
1	⅝ × 7" × 6' cedar siding
2	¾-dia. × 3' hardwood dowel

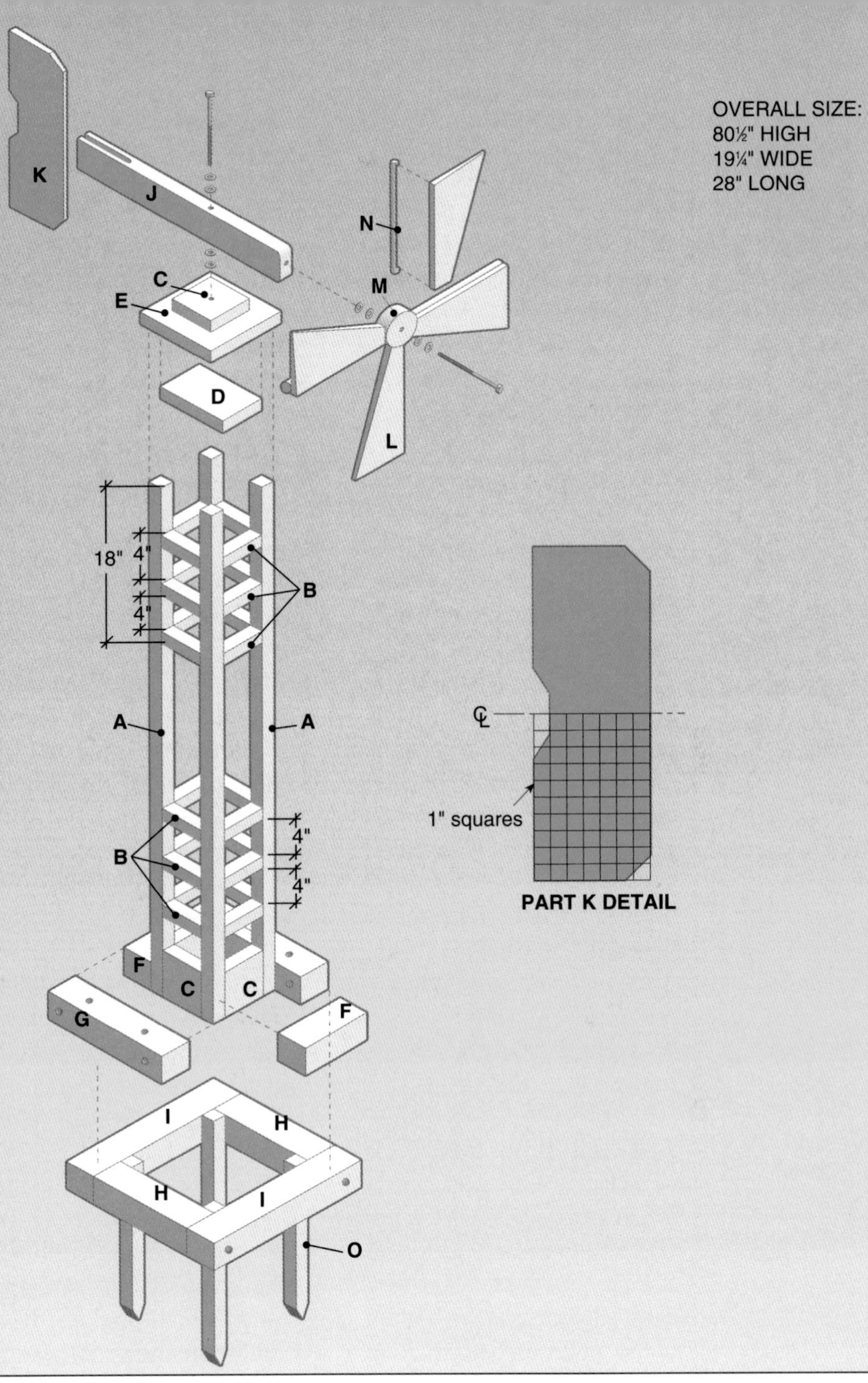

Cutting List

Key	Part	Dimension	Pcs.	Material
A	Post	1½ × 1½ × 60"	4	Cedar
B	Rail	1½ × 1½ × 5"	24	Cedar
C	Spacer	1½ × 5 × 5"	5	Cedar
D	Top insert	1½ × 5 × 8"	1	Cedar
E	Top	1½ × 9¼ × 9¼"	1	Cedar
F	Base end	3½× 3½ × 8"	2	Cedar
G	Base side	3½× 3½ × 15"	2	Cedar
H	Foot end	3½× 3½ × 12"	2	Cedar

Cutting List

Key	Part	Dimension	Pcs.	Material
I	Foot side	3½ × 3½ × 19"	2	Cedar
J	Shaft	1½ × 3½ × 26½"	1	Cedar
K	Tail	⅝ × 7 × 24	1	Cedar siding
L	Blade	⅝ × 7 × 12"	4	Cedar siding
M	Hub	1½ × 4 × 4"	1	Cedar
N	Backer rod	¾ × ¾ × 13"	4	Dowel
O	Stake	1½ × 1½ × 18"	4	Cedar

Materials: Moisture-resistant glue, epoxy glue, ⅜ × 6" and ⅜ × 8" lag screws, 1¼", 2½" and 3" deck screws, #4 × 1" panhead screws, 1"-dia. nylon washers, finishing materials.

Note: Measurements reflect the actual size of dimension lumber.

Directions: Prairie Windmill

BUILD THE TOWER FRAME. The main frame for the windmill tower is made up of four posts connected by a series of short rails.

1. Cut the posts (A) and rails (B).

2. Clamp all four posts in a row, and mark the rail locations on all the posts, starting 9" up from one end of the posts and following the spacing shown in the *Diagram* on page 49.

3. Unclamp the posts, and arrange them in pairs. Attach the rails between the posts at the location marks to create two ladderlike assemblies. Use moisture-resistant glue and a single countersunk 2½" deck screw, driven through each post and into each rail.

4. Once the two assemblies are completed, join them together with rails to create the tower frame.

ADD THE TOWER BOTTOM & TOP.

1. Cut the spacers (C), top insert (D), and tower top (E) to size.

2. Fit four of the spacers between the posts at the bottom of the tower, and attach them with glue and 2½" deck screws driven through countersunk pilot holes.

3. Draw diagonal lines between opposite corners on the fifth spacer—the point where the lines intersect is the center of the board. Center the spacer on the tower top (E), and attach it with glue and 1¼" deck screws. Do not drive screws within 1" of the centerpoint.

4. Drill a ¼"-dia. pilot hole for a ⅜"-dia. lag screw through the center, making sure to keep your drill perpendicular—the lag screw is driven later to secure the windmill to the tower.

5. Slip the top insert between the posts at the top of the tower, and fasten the insert with glue and screws.

6. Center the spacer and tower top over the top insert and fasten with glue and screws. After assembly is complete, use a power sander to smooth out all sharp edges **(photo A).**

The tower for the windmill is basically two ladder frames joined by rails. Sand sharp edges smooth.

ATTACH THE TOWER BASE. The base of the tower is a heavy frame made from 4 × 4 cedar. When the windmill is installed in your yard, the base is attached to a 4 × 4 frame that is staked into the ground.

1. Cut the base ends (F) and base sides (G) to size.

2. Attach the base ends to the tower with 3" deck screws. To attach the base sides, drill ¼"-dia. pilot holes for ⅜"-dia. lag screws in the base sides, then counterbore the pilot holes with a 1" spade bit. Drive ⅜ × 8" lag screws with metal washers through the base sides and into the base ends.

3. Cut the foot ends (H) and foot sides (I) to size.

4. Drill pilot holes for counterbored lag screws through the the foot sides, and secure each foot side to the foot ends with a ⅜ × 8" lag screw.

MAKE THE TAIL.

1. Cut the tail (K) to length from beveled cedar lap siding.

2. Draw a 1" grid pattern onto the board, then draw the shape shown in the *Part K Detail*, page 49. Make sure the notch is on the thick edge. Cut with a jig saw.

MAKE THE SHAFT.

1. Cut the shaft (J) to size from a cedar 2 × 4.

2. Draw a centerline on one long edge of the shaft.

3. Measure the thickness of the beveled siding at several points, including the thin edge and the thick edge. Using drill bits with the same diameters as the thicknesses of the siding, drill holes along the centerline at points that correspond with the width of the tail (make sure that you drill holes at each end of the slot outline).

4. Connect the holes with a pair of straight lines to create an outline for the tail notch. Cut along the outlines with a handsaw **(photo B).**

5. Next, drill a centered, 7⁄16"-dia. guide hole (for the lag screw that secures the shaft to the tower) through the top edge of the shaft, 9" from the front end. Also drill a ¼"-dia. pilot hole in the center of the front end of the shaft.

Drill holes of varying diameter to create an outline, then cut a slot for the tail into the shaft.

Drill ¾"-dia. × 1"-deep guide holes into the four sides of the hub to hold the backer rods.

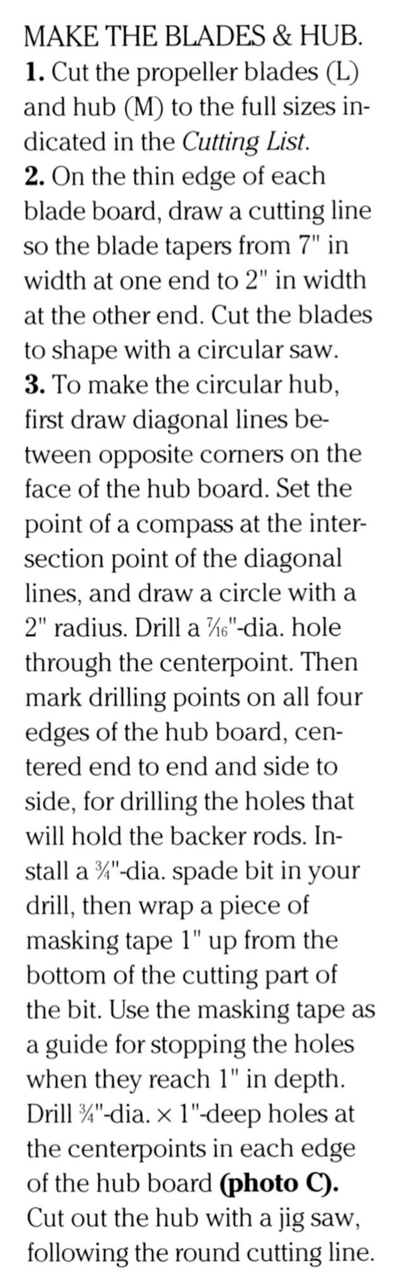

MAKE THE BLADES & HUB.
1. Cut the propeller blades (L) and hub (M) to the full sizes indicated in the *Cutting List*.
2. On the thin edge of each blade board, draw a cutting line so the blade tapers from 7" in width at one end to 2" in width at the other end. Cut the blades to shape with a circular saw.
3. To make the circular hub, first draw diagonal lines between opposite corners on the face of the hub board. Set the point of a compass at the intersection point of the diagonal lines, and draw a circle with a 2" radius. Drill a 7/16"-dia. hole through the centerpoint. Then mark drilling points on all four edges of the hub board, centered end to end and side to side, for drilling the holes that will hold the backer rods. Install a ¾"-dia. spade bit in your drill, then wrap a piece of masking tape 1" up from the bottom of the cutting part of the bit. Use the masking tape as a guide for stopping the holes when they reach 1" in depth. Drill ¾"-dia. × 1"-deep holes at the centerpoints in each edge of the hub board **(photo C).** Cut out the hub with a jig saw, following the round cutting line.
4. Cut the backer rods (N) from ¾" doweling, then sand a flat edge onto each rod, using a belt sander. Stop the flat edges 1" from the end of each dowel (this creates a flat mounting surface).

ASSEMBLE THE PROPELLER.
1. Attach the thick edge of each blade to the flat surface of a backer rod with three #4 × 1" panhead screws and epoxy glue **(photo D).**
2. Apply epoxy glue to the tail where it meets the shaft, and fasten it in the slot with 1¼" deck screws.
3. Before proceeding, apply exterior wood stain to all the wood parts, and apply paste wax in the guide holes in the hub and shaft.
4. Attach the blade assembly to the shaft with a ⅜ × 6" lag screw and pairs of 1"-dia. nylon washers inserted on each side of the hub.
5. Fasten the shaft to the tower with a ⅜ × 6" lag screw and pairs of nylon washers at the top and bottom edges of the shaft. Do not over-tighten the screws.

SET UP THE WINDMILL.
Position the 4 × 4" frame in the desired location in your yard or garden.
1. Cut the stakes (O), sharpening one end of each stake. Attach the stakes at the inside corners of the foot frame with screws, then drive the stakes into the ground. Attach the base frame to the foot frame with counterbored lag screws. Drive counterbored lag screws through the base sides and into the foot ends.

Glue the propeller blades into the holes to mount them on the hubs.

PROJECT
POWER TOOLS

Luminary

Dress up your yard or garden with these warm, decorative accents that look even better in bunches.

CONSTRUCTION MATERIALS*

Quantity	Lumber
1	2 × 8 × *" cedar
1	1 × 2" × 8' cedar
1	1"-wide × 5' copper strip

*The shortest length available at most lumber yards is 6'. Since this is much more than you need for a single luminary, ask a yardworker if they have any scraps that are at least 6" long.

Luminaries are decorative outdoor accents that hold and protect candles or glass lamp chimneys. Traditionally, they are arranged in groups around an entrance or along a garden pathway. The simple slat-built luminary design shown here is easy and inexpensive to make. All you need are a few pieces of 1 × 2" cedar, some 1"-thick strips of copper and a 6"-dia. cedar base. The copper trim, glass chimneys, and candles can be purchased at most craft stores. We used 22-gauge copper strips, which are thin enough to cut with scissors and will form easily around the luminary. Make sure to use copper nails to attach the strips.

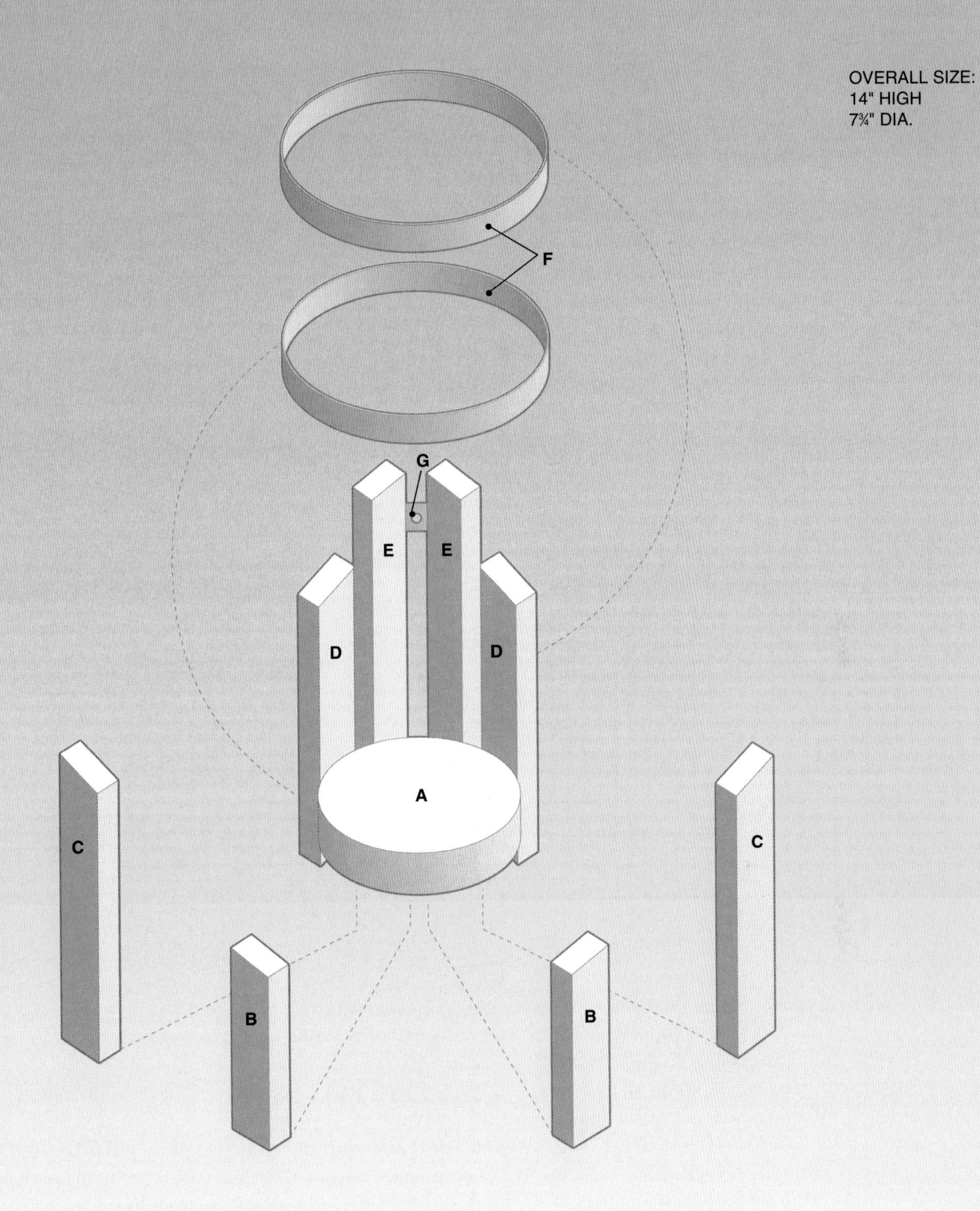

Cutting List

Key	Part	Dimension	Pcs.	Material
A	Base	1½ × 6 × 6"	1	Cedar
B	Front slat	⅞ × 1½ × 8"	2	Cedar
C	Short slat	⅞ × 1½ × 10"	2	Cedar
D	Middle slat	⅞ × 1½ × 12"	2	Cedar

Cutting List

Key	Part	Dimension	Pcs.	Material
E	Back slat	⅞ × 1½ × 14"	2	Cedar
F	Strap	1 × 25"	2	Copper
G	Hanger	1 × 3½"	1	Copper

Materials: 1⅝" deck screws, ¾" copper nails, candle and glass candle chimney.

Note: Measurements reflect the actual size of dimension lumber.

Directions: Luminary

MAKE THE BASE.

The base for the luminary is a round piece of cedar cut with a jig saw. Because the luminary slats are attached to the sides of the base, it is important that the base be as symmetrical and smooth as you can get it.

1. Start by cutting the base (A) to 7¼" in length from a piece of 2 × 8 cedar (this will result in a square workpiece).

2. Draw diagonal lines between opposite corners. The point where the lines intersect is the center of the board. Set the point of a compass at the centerpoint, and draw a 6"-dia., circular cutting line with the compass. Cut the base to shape along the cutting line, using a jig saw with a coarse-wood cutting blade (thicker blades are less likely to "wander" than thinner blades).

3. Clamp a belt sander to your worksurface on its side, making sure the sanding belt is perpendicular to the worksurface. Sand the edges of the base to smooth out any rough spots, using the belt sander as a stationary grinder **(photo A).** If you are making more than one luminary, cut and sand all the bases at once for greater efficiency.

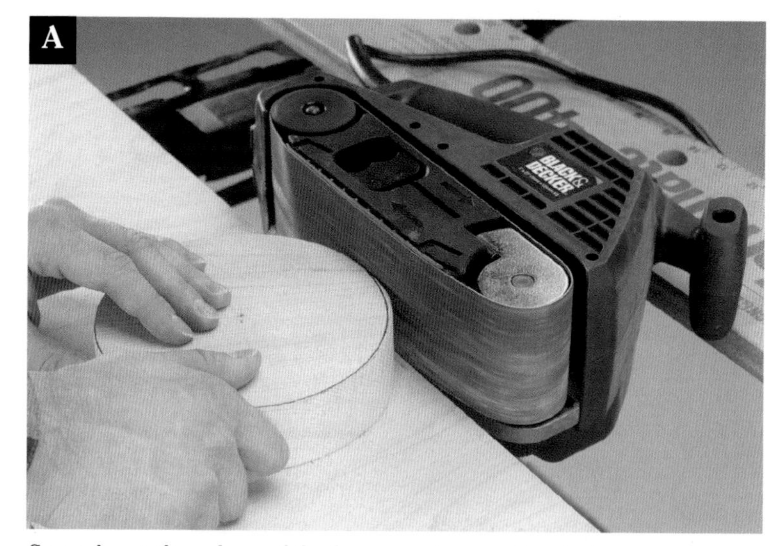

Smooth out the edges of the luminary base with a belt sander clamped to your worksurface.

Make a miter cut at the top of one slat, then use that slat as a guide for marking miter cuts on the rest of the slats.

MAKE THE SLATS.

The sides of each luminary are formed by four pairs of 1 × 2 cedar, cut to different lengths. All the slats are mitered on their top ends for a decorative effect that moves upward from front to back.

TIP

Prevent copper from oxidizing and turning green by coating it with spray-on polyurethane.

1. Cut the front slats (B), short slats (C), middle slats (D) and back slats (E) to length.

2. On one slat, mark a point on one long edge, ½" in from an end. Draw a straight line from the point to the corner at the opposite edge. Cut along the line with a saw and miter box or with a power miter box. Using this slat as a guide, trace mitered cutting lines onto the tops of all the slats **(photo B).** Miter-cut the rest of the slats along the cutting lines.

ATTACH THE SLATS TO THE BASE.

1. Drill a pair of ⅛"-dia. pilot holes at the bottom of each slat. The pilot holes should be staggered to avoid splitting the base; drill one pilot hole ½" from the side and 1" up from the bottom; drill the other pilot hole ½" from the other side and 1½" up from the bottom. Countersink the pilot holes enough so the screw heads will be recessed.

2. Use four ¾"-wide spacers to align the slats to form a gradual upward slope (see *Diagram*, page 53). Set the base on a ½"-thick block to create a recess.

Use ¾"-wide spacers to maintain the gaps between the slats as you attach them to the base.

Then, arrange the four spacers in a stack so they form a hub over the center of the base (from above, the spacers should look like a pie cut into eight equal-sized pieces). Set the slats between the spacers so the bottoms are resting on the worksurface and they are flush against the base. Adjust the positions of the slats and spacers until each slat is opposite a slat across the base. Once you get the layout set, wrap a piece of masking tape around the slats, near the bottom, to hold them in place while you fasten them to the base.

3. Drive a 1⅝" deck screw through each pilot hole in each slat, and into the base **(photo C).** Do not overtighten the screws. Remove the spacers.

ATTACH THE STRAPS.

We wrapped two 1"-wide straps made of 22-gauge (fairly lightweight) copper around the luminary to brace the slats. Purchase copper strips that are 25" long or longer at your local craft store. If you cannot find strips that long, buy shorter ones and splice them together with a 1" overlapping seam.

1. Cut two 1"-wide copper strips to 25" in length to make the straps (F). Ordinary scissors will cut thin copper easily.

2. Test-fit the straps by taping them in place around the slats.

3. Mark drilling points for guide holes on the copper straps—one hole per slat, centered between the top and bottom of the strap.

4. Drill ¹⁄₁₆"-dia. pilot holes through the drilling points, then reposition the straps around the luminary. The bottom strap should conceal the screw heads at the bottoms of the slats. The second strap should be about 6¼" up from the bottom of the luminary.

5. Insert a 6"-long block of wood between two slats that are opposite one another, then drive a ¾" copper nail through the pilot holes in those straps to secure them to the slats **(photo D).** Move the block, and drive copper nails through the rest of the pilot holes.

APPLY FINISHING TOUCHES.

1. Cut a 1 × 3½"-long strip of copper to make a hanger (G).

2. Drill a ⅜"-dia. hole through the center, then nail the hanger to the outsides of the back slats, about 1" down from the tops.

3. Make a centering pin to hold a candle to the base by driving a 1¾" screw or a 6d nail up through the center of the base.

4. Apply a coat of exterior wood stain if you plan to keep the luminary outdoors.

Brace the slats with a spacer as you tack on the copper strips.

PROJECT
POWER TOOLS

Garden Bridge

Whether it's positioned over a small ravine or a swirl of stones, this handsome bridge will add romance and charm to your yard.

CONSTRUCTION MATERIALS

Quantity	Lumber
4	4 × 4" × 8' cedar
2	2 × 10" × 8' cedar
10	2 × 4" × 8' cedar
2	1 × 8" × 8' cedar
2	1 × 3" × 8' cedar
8	1 × 2" × 8' cedar
2	½" × 2" × 8' cedar lattice

A bridge can be more than simply a way to get from point A to point B without getting your feet wet. This striking cedar footbridge will be a design centerpiece in any backyard or garden. Even if the nearest trickle of water is miles from your home, this garden bridge will give the impression that your property is graced with a tranquil brook, and you'll spend many pleasurable hours absorbing the peaceful images it inspires. You can fortify the illusion of flowing water by laying a "stream" of landscaping stones beneath this garden bridge. If you happen to have a small ravine or waterway through your yard, this sturdy bridge will take you across it neatly and in high style.

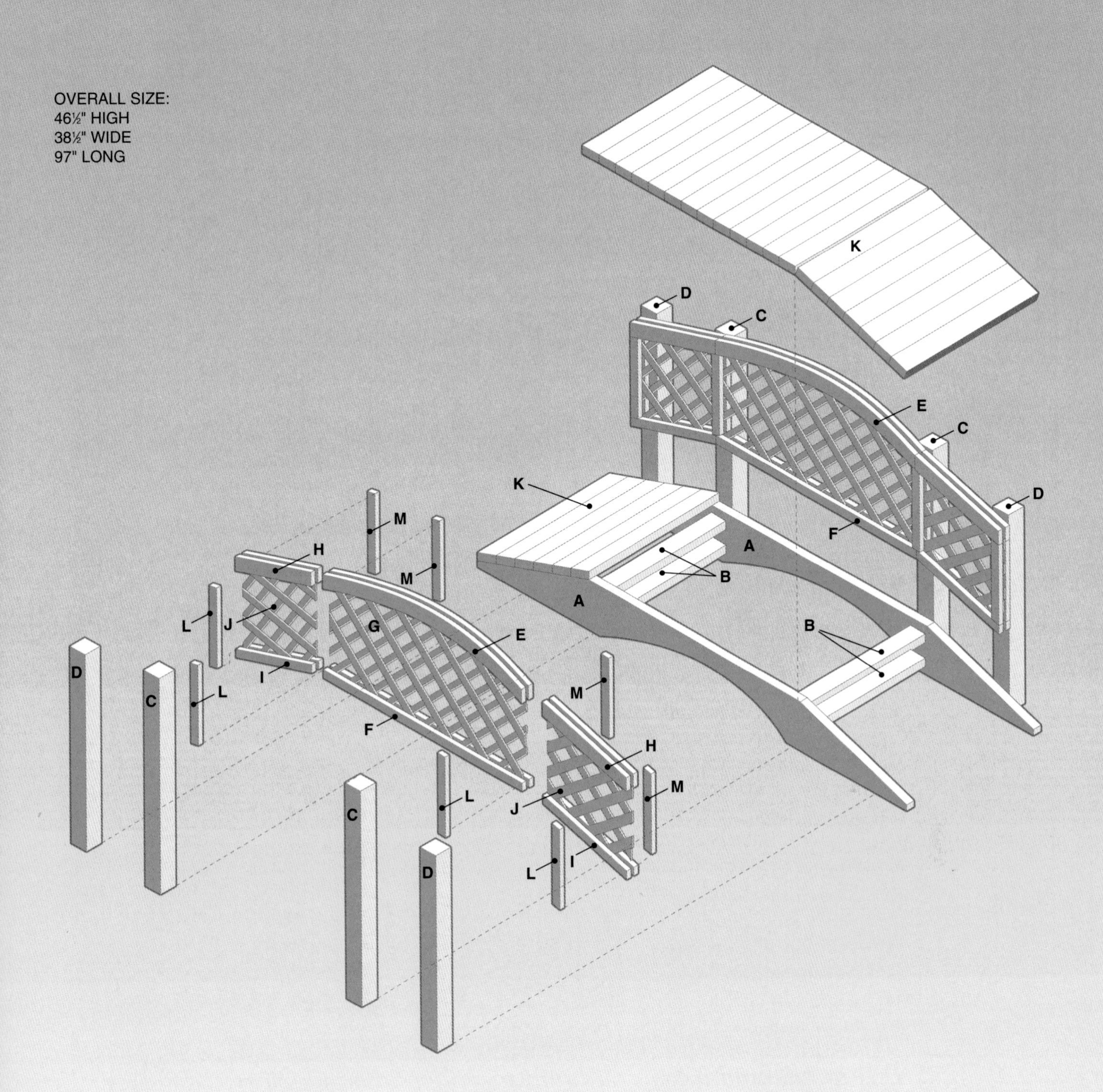

Cutting List

Key	Part	Dimension	Pcs.	Material
A	Stringer	1½ × 9¼ × 96"	2	Cedar
B	Stretcher	1½ × 3½ × 27"	4	Cedar
C	Middle post	3½ × 3½ × 42"	4	Cedar
D	End post	3½ × 3½ × 38"	4	Cedar
E	Center handrail	1½ × 7¼ × 44½"	4	Cedar
F	Center rail	⅞ × 1½ × 44½"	4	Cedar
G	Center panel	½ × 23½ × 44½"	2	Cedar lattice

Cutting List

Key	Part	Dimension	Pcs.	Material
H	End handrail	⅞ × 2¼ × 19½"	8	Cedar
I	End rail	⅞ × 1½×24"	8	Cedar
J	End panel	½ × 19 ×24"	4	Cedar lattice
K	Tread	1½ × 3½" ×30"	26	Cedar
L	Filler strip	⅞ × 1½" × 19"	8	Cedar
M	Trim strip	⅞ × 1½" ×21"	8	Cedar

Materials: ⅜ × 4" lag screws, 2" and 3" deck screws, finishing materials.

Note: Measurements reflect the actual size of dimension lumber.

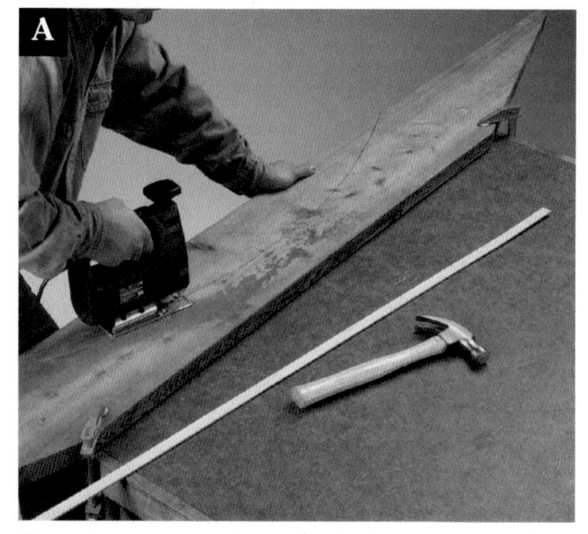

Use a jig saw to make the arched cutouts in the bottoms of the 2 × 10 stringers.

Attach pairs of stretchers between the stringers with 3" deck screws.

Directions: Garden Bridge

MAKE THE STRINGERS. The stringers are the main structural members of this bridge. Both stringers have arcs cut into their bottom edges, and the ends are cut at a slant to create the gradual tread incline of the garden bridge.

1. Draw several guidelines on the stringers (A) before cutting. First, draw a centerline across the width of each stringer; then mark two more lines across the width of each stringer, 24" to the left and right of the centerline; finally, mark the ends of each stringer, 1" up from one long edge, and draw diagonal lines from these points to the top of each line to the left and right of the center.

2. Use a circular saw to cut the ends of the stringers along the diagonal lines.

3. Tack a nail on the centerline, 5¼" up from the same long edge. Also tack nails along the bottom edge, 20½" to the left and right of the centerline.

4. Lay out the arc at the bottom of each stringer with a marking guide made from a thin, flexible strip of scrap wood or plastic.

5. Hook the middle of the marking guide over the center nail and slide the ends under the outside nails to form a smooth curve. Trace along the guide with a pencil to make the cutting line for the arc (you can mark both stringers this way, or mark and cut one, then use it as a template for marking the other).

6. Remove the nails and marking guide, and cut the arcs on the bottom edge of each stringer with a jig saw **(photo A).**

Cut the 4 × 4 posts to their finished height, then use lag screws to attach them to the outsides of the stringers.

ASSEMBLE THE BASE. Once the two stringers are cut

Attach the treads to the stringers with deck screws.

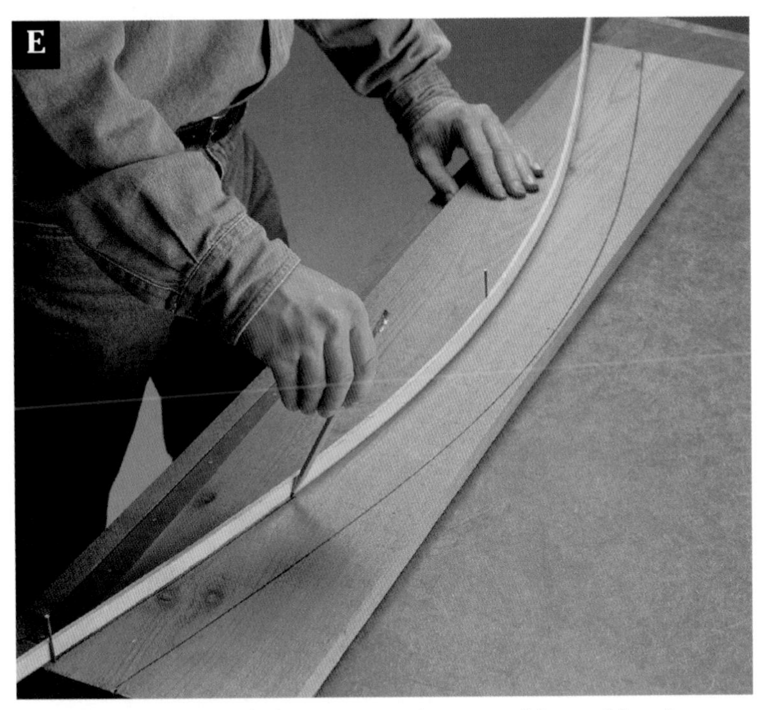

Use a flexible piece of plastic or wood as a marking guide when drawing the cutting lines for the center handrails.

to shape, they are connected with four straight boards, called stretchers (B), to form the base of the bridge.

1. Cut the stretchers (B), middle posts (C) and end posts (D) to size.

2. Mark the stretcher locations on the insides of the stringers, 1½" from the top and bottom of the stringers. The outside edges of the stretchers should be 24" from the centers of the stringers (see *Diagram*, page 57), leaving the inside edges flush with the bottoms of the arcs.

3. Stand the stringers upright and position the stretchers between them. Support the bottom stretchers with 1½"-thick spacer blocks for correct spacing. Fasten the stretchers between the stringers with countersunk 3" deck screws, driven through the stringers and into the ends of the stretchers.

4. Turn the stringer assembly upside down, and attach the top stretchers **(photo B).** The footbridge will get quite heavy at this stage: you may want to build the rest of the project on-site.

5. Clamp the middle posts to the outsides of the stringers so their outside edges are 24" from the center of the stringers. Make sure the middle posts are perpendicular to the stringers.

6. Drill ¼"-dia. pilot holes through the stringers and into the middle posts. Attach the middle posts with ⅜"-dia. × 4"-long lag screws, driven through the stringers and into the posts **(photo C).**

7. Clamp the end posts to the stringers, starting 7" from the stringer ends. Drill pilot holes and secure the end posts to the stringers with lag screws.

ATTACH THE TREADS.

1. Cut the treads (K) to size.

2. Position the treads on the stringers, making sure to space them evenly. The treads should be separated by gaps of about ¼".

3. Test-fit all the treads before you begin installing them. Then, secure the treads with 3"-long countersunk deck screws **(photo D).**

TIP

Lattice panels must be handled carefully, or they may fall apart. This is especially true when you are cutting the lattice panels. Before making any cuts, clamp two boards around the panel, close to the cutting line, to stabilize the lattice and protect it from the vibration of the saw. Always use a long, fine blade on your saw when cutting lattice.

Use a jig saw to cut the panels and center handrails to shape.

Fasten 1 × 2 filler strips to the posts to close the gaps at the sides of the lattice panels.

ATTACH THE CENTER HANDRAIL PANELS.

The center panels are made by sandwiching lattice sections between 1 × 2 cedar frames. Each center panel has an arc along its top edge. This arc can be laid out with a flexible marking guide, using the same procedure used for the stringers.

1. Cut the center handrails (E), center rails (F) and center panels (G) to size.

2. Using a flexible marking guide, trace an arc that begins 2½" up from one long edge of one center handrail. The top of this arc should touch the top edge of the workpiece.

3. Lower the flexible marking guide 2½" down on the center handrails. Trace this lower arc, starting at the corners, to mark the finished center handrail shape **(photo E).**

4. Cut along the bottom arc with a jig saw.

5. Trace the finished center handrail shapes onto the other workpieces, and cut along the bottom arc lines.

6. Cut the center panels (G) to size from ½"-thick cedar lattice.

7. Sandwich each center panel between a pair of center handrails so the top and side edges are flush. Clamp the parts, and gang-cut through the panel and center handrails along the top arc line with a jig saw **(photo F).**

8. Unfasten the boards, and sand the curves smooth.

9. Refasten the center panels between the arcs, ½" down from the tops of the arcs. Drive 2" deck screws through the inside center handrail and into the center panel and outside center handrail. Drive one screw every 4 to 6"—be sure to use pilot holes and make an effort to drive screws through areas where the lattice strips cross, so the screws won't be visible from above.

10. Fasten the center rails to the bottom of the center panels, flush with the bottom edges.

11. Center the panels between the middle posts, and fasten them to the posts so the tops of the handrails are flush with the inside corner of the middle posts at each end. The ends of the handrails are positioned at the center of the posts. Drive 3" deck screws through the center handrails and center rails to secure the panel to the center posts.

12. Cut the filler strips (L) to size. The filler strips fit between the center handrails and center rails, bracing the panel and providing solid support for the loose ends of the lattice.

13. Position the filler strips in the gaps between the center panels and the middle posts, and fasten them to the middle posts with 2" deck screws **(photo G).**

ATTACH THE END HANDRAIL PANELS.

Like the center panels, the end panels are made by sandwiching cedar lattice sections between board frames and fastening them to posts. The ends of the end panels and the joints between the end and center panels are covered by trim strips (M), which are

Clamp the rough end panels to the posts at the ends of the bridge, and draw alignment markers so you can trim them to fit exactly.

attached with deck screws.

1. Cut the end handrails (H), end rails (I), and end panels (J) to size.

2. Position an end handrail and an end rail on your worksurface, then place an end panel over the pieces. Adjust the end handrail and end rail so the top of the panel is ½" down from the top edge of the end handrail.

3. Sandwich the end panels between another set of end handrails and end rails, and attach the parts with 2" deck screws.

4. Repeat steps 2 and 3 for each end panel.

5. Clamp or hold the panels against the end posts and middle posts, and adjust the end panels so they are aligned with the center panel and the top inside corner of the end posts.

6. Draw alignment marks near the end of the panels along the outside of the end posts **(photo H)**, and cut the end panels to size.

7. Unclamp the panels, and draw cutting lines connecting the alignment marks. Cut along the lines with a jig saw.

8. Sand the end panels, and attach them to the posts with countersunk 3" deck screws, driven through the end handrails and end rails.

9. Slide filler strips between the end panels and the posts. Fasten the filler strips with 2" deck screws.

10. Cut the trim strips (M) to size.

11. Attach the trim strips over each joint between the end and center panels, and at the outside end of each end panel, with countersunk 3" deck screws **(photo I).**

APPLY FINISHING TOUCHES.

1. Sand all the surfaces to smooth out any rough spots.

2. Apply an exterior wood stain to protect the wood, if desired. You may want to consider leaving the cedar untreated, since that will cause the wood to turn gray—this aging effect may help the bridge blend better with other yard elements.

3. Get some help, and position the bridge in your yard.

4. For a dramatic effect, dig a narrow, meandering trench between two distinct points in your yard, line the trench with landscape fabric, then fill the trench with landscaping stones to simulate a brook.

Use deck screws to attach a trim strip over each joint between the end panels and center panels.

BLACK & DECKER
Cordless Jig Saw

Benches & Seats

Let's face it. Sometimes it's nice to just sit and admire the hard work you've put into your yard and garden. What better place to do that than on a custom-made bench or seat? No matter what style you prefer, you'll find plans here to create a beautiful seating option for your porch, patio, or garden. From ceramic-tiled benches to dreamy porch swings, there's bound to be a seat that beckons you.

PROJECT
POWER TOOLS

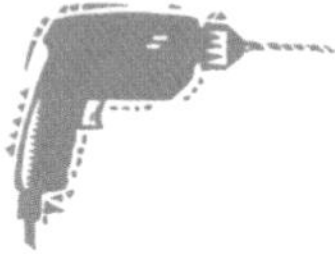

Tiled Garden Bench

Ornamental, weather-resistant tiles take your garden bench to a whole new level of beauty.

CONSTRUCTION MATERIALS

Quantity	Lumber
2	2 × 4" × 8' cedar
1	2 × 6" × 8' cedar
1	4 × 4" × 8' cedar
1	¾" × 4' × 4' exterior plywood
1	½" × 4' × 4' cementboard

Here's a splendid example of the term "return on investment." Four decorative tiles and a handful of accent tiles produce quite an impact. In fact, those accents and a few dozen 4 × 4" tiles transform a plain cedar bench into a special garden ornament. And you can accomplish the whole thing over one weekend. In addition to the standard *Outdoor Wood Projects* tools, you'll need some specialized, but inexpensive tile setting tools. A notched trowel, grout float and sponge are necessary for setting the tiles. A tile cutter is a more expensive tool, but typically can be rented at most tile retailers and home centers. These stores often have free tile setting classes, if you feel unsure of your skills or want to hone your technique.

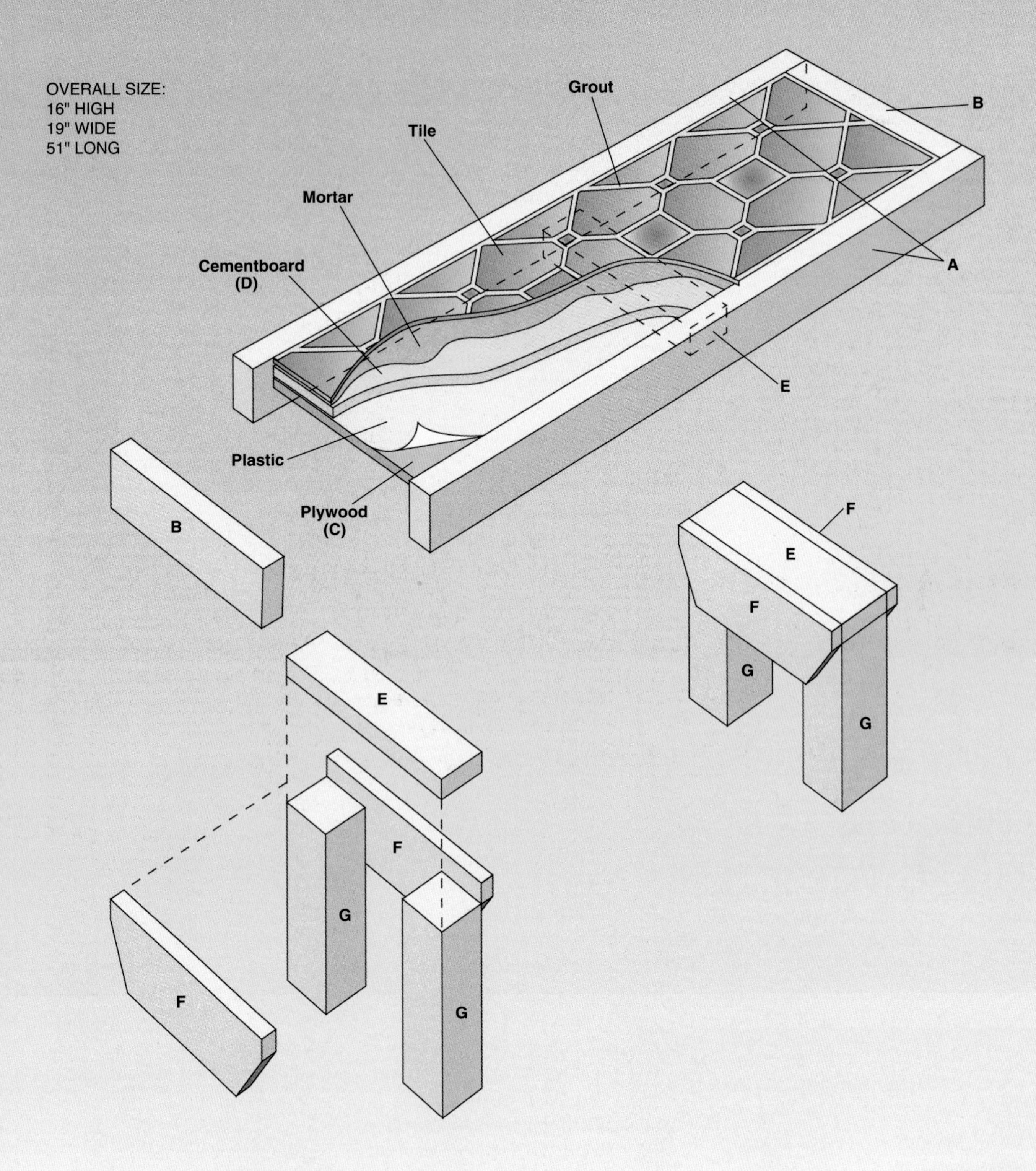

Cutting List

Key	Part	Dimension	Pcs.	Material
A	Sides	1½ × 3½ × 51"	2	Cedar
B	Ends	1½ × 3½ × 16"	2	Cedar
C	Core	15 × 48"	1	Ext. Plywood
D	Core	15 × 48"	1	Cementboard

Cutting List

Key	Part	Dimension	Pcs.	Material
E	Stretchers	1½ × 3½ × 16"	3	Cedar
F	Braces	1½ × 5½ × 16"	4	Cedar
G	Legs	3½ × 3½ × 13"	4	Cedar

Materials: plastic sheeting, 2" and 3" coated or galvanized deck screws, 1¼" cementboard screws, clear wood sealer, field and accent tiles, thin-set mortar, tile spacers, grout, grout sealer.

Note: Measurements reflect the actual size of dimension lumber.

Use 1½" blocks to support the stretchers. Drill pilot holes and fasten the stretchers to the sides with 3" screws.

Position the frame over the plywood/cementboard core. Drill pilot holes and then drive 2" galvanized deck screws through the stretchers and into the plywood.

Position each leg between a set of braces and against the sides of the bench frame. Drill pilot holes through each brace and attach the leg to the braces.

Directions: Tiled Garden Bench

BUILD THE FRAME.

1. Cut two sides and two ends, then position the ends between the sides so the edges are flush. Make sure the frame is square. Drill ⅛" pilot holes through the sides and into the ends. Drive 3" screws through the pilot holes.

2. Cut three stretchers. Mark the sides, 4½" from the inside of each end. Using 1½" blocks beneath them as spacers, position the stretchers and make sure they're level. Drill pilot holes and fasten the stretchers to the sides with 3" screws **(photo A).**

MAKE THE TILE BASE.

Cementboard is the typical substrate for tile setting. It has no structural strength, so it must be supported with a layer of exterior plywood.

1. Cut one core (C) from ¾" exterior-grade plywood. Cut the other (D) from cementboard. To cut the cementboard, use a utility knife or cementboard knife and a straightedge to score through the mesh on one side. Snap the panel back and cut through the mesh on the back side.

2. Staple plastic sheeting over the plywood, draping it over the edges. Lay the cementboard rough-side up on the plywood and attach it with 1¼" cementboard screws driven every 6". Make sure the screw heads are flush with the surface.

3. Turn the plywood/cementboard core so the cementboard is on the bottom. Position the bench frame upside down and over the plywood/cementboard core. Drill pilot holes and then drive 2" galvanized deck screws through the stretchers and into the plywood **(photo B).**

BUILD THE LEGS.

The legs are made of cedar 4 × 4s braced between two cedar 2 × 6s. The braces are angled to be more asthetically pleasing.

1. Cut four braces from a cedar 2 × 6. Mark the angle on each end of each brace by measuring down 1½" from the top edge and 1½" along the bottom edge. Draw a line between the two points and cut along that line, using a circular saw.

2. On each brace, measure down ¾" from the top edge and draw a reference line across the stretcher for the screw positions. Drill ⅛" pilot holes along the reference line. Position a brace on each side of the end stretchers and fasten them with 3" galvanized deck screws driven through the braces and into the stretchers.

3. Cut four 13" legs from a 4 × 4. Position each leg between a set of braces and against the sides of the bench frame. Drill pilot holes through each brace and attach the leg to the braces by driving 3" screws through the braces and into the leg **(photo C).** Repeat the process for each leg.

4. Sand all surfaces with 150-grit sandpaper, then seal all wood surfaces with clear wood sealer.

LAY OUT THE TILE.

Field tiles are the main tiles in a design. In this project, square field tiles are cut to fit around four decorative picture tiles and bright blue accent tiles.

1. Snap perpendicular reference lines to mark the center of the length and width of the bench. Beginning at the center of the bench, dry-fit the field tiles; using spacers. Set the ac-

Dry-fit the field tiles, using spacers. Set the accent tile in place and mark the field tile for cutting.

Apply thin-set mortar over the cementboard, using a notched trowel.

cent tile in place and mark the field tile for cutting **(photo D).**

2. Cut the field tile and continue dry-fitting the bench top, including the accent and border tiles.

SET THE TILE.

Tile is set into a cement product called thin-set mortar.

1. Mix the mortar, starting with the dry powder and gradually adding water. Stir the mixture to achieve a creamy consistency. The mortar should be wet enough to stick to the tiles and the cementboard, but stiff enough to hold the ridges made when applied with the notched trowel.

2. Remove the tiles from the bench and apply thin-set mortar over the cementboard, using a notched trowel **(photo E).** Apply only as much mortar as you can use in 10 minutes. (If the mortar begins to dry before you have set the tile, throw it away and spread new mortar.)

3. Set the tile into the thin-set mortar, using a slight twisting motion. Continue adding thin-set and setting the tile until the bench top is covered **(photo F).** Remove the spacers, using a needlenose pliers. Let the mortar dry according to manufacturer's directions.

GROUT THE TILE.

Grout fills the gaps between the tiles. A latex additive makes the grout more resistant to stains.

1. Apply masking tape around the wood frame to prevent the grout from staining the wood. Mix the grout and latex grout additive according to package instructions. Use a grout float to force the grout into the joints surrounding the tile, holding the float at an angle **(photo G).** Do not apply grout to the joint between the tile and the wood frame.

2. Wipe the excess grout from the tiles with a damp sponge. Rinse the sponge between each wipe. When the grout has dried slightly, polish the tiles with a clean, dry cloth to remove the slight haze of grout.

3. After the grout has dried (2-3 days) apply grout sealer to the grout lines, using a small foam brush or paintbrush. Take care to keep the sealer off the tiles. Caulk the gap between the tiles and the wood frame, using caulk that matches the grout color.

Set the tile into the thin-set mortar, using a slight twisting motion.

Mix grout and use a grout float to force it into the joints surrounding the tile.

Adirondack Chair

You will find dozens of patterns and plans for building popular Adirondack chairs in just about any bookstore, but few are simpler to make or more attractive than this clever project.

PROJECT POWER TOOLS

Adirondack furniture has become a standard on decks, porches and patios throughout the world. It's no mystery that this distinctive furniture style has become so popular. Attractive—but rugged—design and unmatched stability are just two of the reasons, and our Adirondack chair offers all of these benefits, and more.

But unlike most of the Adirondack chair designs available, this one is also very easy to build. There are no complex compound angles to cut, no intricate details on the back and seat slats, and no mortise-and-tenon joints. Like all of the projects in this book, our Adirondack chair can be built by any do-it-yourselfer, using basic tools and simple techniques. And because this design features all the elements of the classic Adirondack chair, your guests and neighbors may never guess that you built it yourself.

We made our Adirondack chair out of cedar and finished it with clear wood sealer. But you may prefer to build your version from pine (a traditional wood for Adirondack furniture), especially if you plan to paint the chair. White, battleship gray and forest green are popular color choices for Adirondack furniture. Be sure to use quality exterior paint with a glossy or enamel finish.

CONSTRUCTION MATERIALS

Quantity	Lumber
1	2 × 6" × 8' cedar
1	2 × 4" × 10' cedar
1	1 × 6" × 14' cedar
1	1 × 4" × 8' cedar
1	1 × 2" × 10' cedar

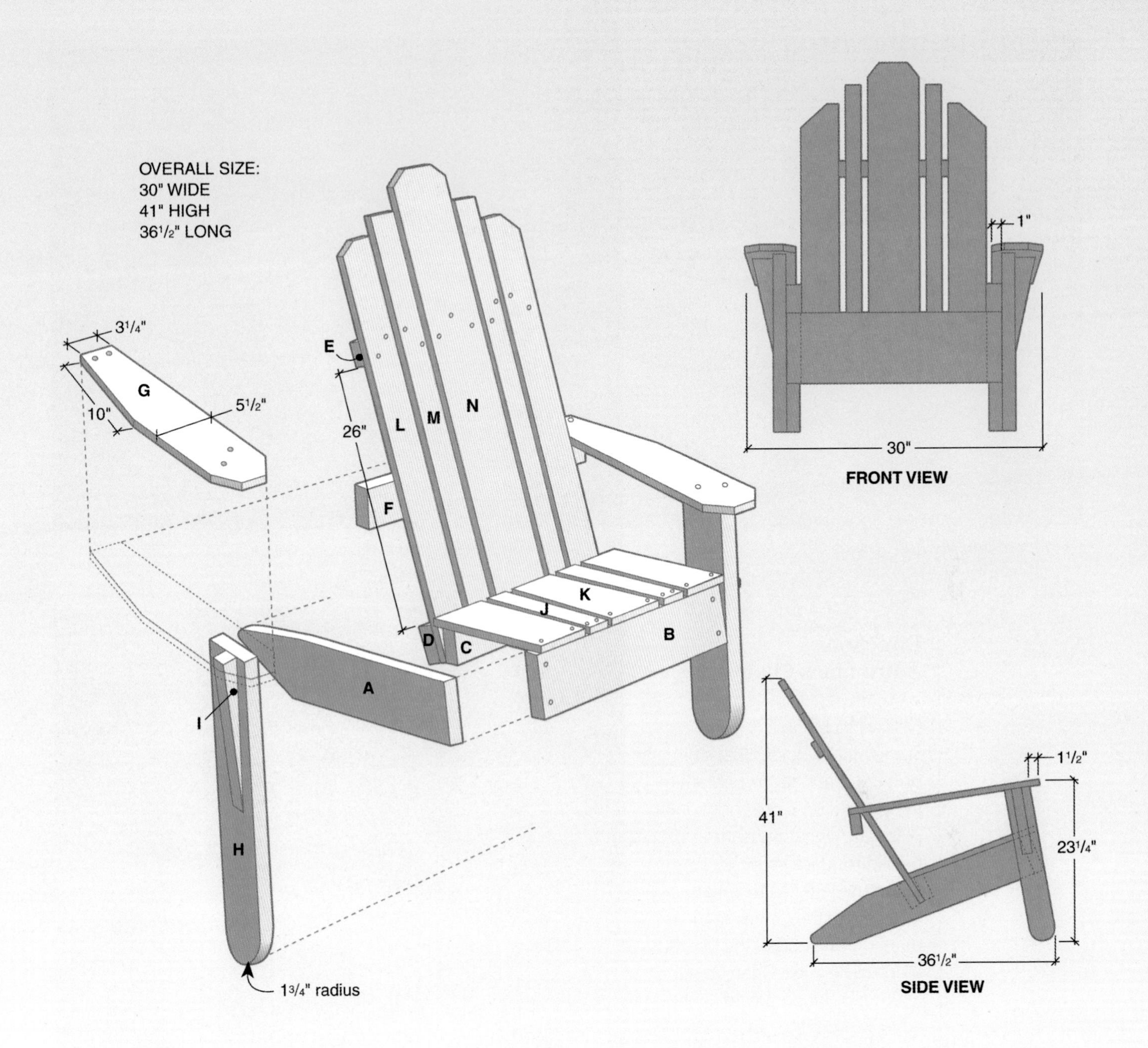

Cutting List

Key	Part	Dimension	Pcs.	Material
A	Leg	1½ × 5½ × 34½"	2	Cedar
B	Apron	1½ × 5½ × 21"	1	Cedar
C	Seat support	1½ × 3½ × 18"	1	Cedar
D	Low back brace	1½ × 3½ × 18"	1	Cedar
E	High back brace	¾ × 1½ × 18"	1	Cedar
F	Arm cleat	1½ × 3½ × 24"	1	Cedar
G	Arm	¾ × 5½ × 28"	2	Cedar
H	Post	1½ × 3½ × 22"	2	Cedar

Cutting List

Key	Part	Dimension	Pcs.	Material
I	Arm brace	1½ × 2¼ × 10"	2	Cedar
J	Narrow seat slat	¾ × 1½ × 20¼"	2	Cedar
K	Wide seat slat	¾ × 5½ × 20¼"	3	Cedar
L	End back slat	¾ × 3½ × 36"	2	CedaM
M	Narrow back slat	¾ × 1½ × 38"	2	Cedar
N	Center back slat	¾ × 5½ × 40"	1	Cedar

Materials: Moisture-resistant glue, 1¼", 1½", 2" and 3" deck screws, ⅜ × 2½" lag screws with washers, finishing materials.

Note: Measurements reflect the actual size of dimension lumber.

Cut tapers into the back edges of the legs.

Round the sharp slat edges with a router or a power sander.

Directions: Adirondack Chair

CUT THE LEGS.
Sprawling back legs that support the seat slats and stretch to the ground on a near-horizontal plane are signature features of the Adirondack style.
1. Cut the legs (A) to length.
2. To make the tapers, mark a point on one end of the board, 2" from the edge. Then, mark another point on the adjacent edge, 6" from the end. Connect the points with a straightedge.
3. Mark a point on the same end, 2¼" in from the other edge. Then, mark a point on that edge, 10" from the end. Connect these points to make a cutting line for the other taper.
4. Cut the two taper cuts with a circular saw.
5. Use the tapered leg as a template to mark and cut identical tapers on the other leg of the chair **(photo A).**

Make decorative cuts on the fronts of the arms (shown) and the tops of the back slats, using a jig saw.

BUILD THE SEAT.
The legs form the sides of the box frame that supports the seat slats. Where counterbores for deck screws are called for, drill holes ¼" deep with a counterbore bit.
1. Cut the apron (B) and seat support (C) to size.
2. Attach the apron to the front ends of the legs with glue and 3" deck screws, in the manner described above.
3. Position the seat support so the inside face is 16½" from the inside edge of the apron. Attach the seat support between the legs, making sure the tops of the parts are flush.
4. Cut the seat slats (J) and (K) to length, and sand the ends smooth. Arrange the slats on

Attach the square ends of the posts to the undersides of the arms, being careful to position the part correctly.

top of the seat box, and use wood scraps to set ⅜" spaces between the slats. The slats should overhang the front of the seat box by ¾".

5. Fasten the seat slats by drilling counterbored pilot holes and driving 2" deck screws through the holes and into the tops of the apron and seat support. Keep the counterbores aligned so the cedar plugs form straight lines across the front and back of the seat.

6. Once all the slats are installed, use a router with a ¼" roundover bit (or a power sander) to smooth the edges and ends of the slats **(photo B).**

MAKE THE BACK SLATS.

The back slats are made from three sizes of dimension lumber.

1. Cut the back slats (L), (M) and (N), to size.

2. Trim the corners on the wider slats. On the 1 × 6 slat (N), mark points 1" in from the outside, top corners. Then, mark points on the outside edges, 1" down from the corners. Connect the points and trim along the lines with a jig saw. Mark the 1 × 4 slats 2" from one top corner, in both directions. Draw cutting lines and trim.

ATTACH BACK SLATS.

1. Cut the low back brace (D) and high back brace (E) and set them on a flat surface.

2. Slip ¾"-thick spacers under the high brace so the tops of the braces are level. Then, arrange the back slats on top of the braces with ⅝" spacing between slats. The untrimmed ends of the slats should be flush with the bottom edge of the low back brace. The bottom of the high back brace should be 26" above the top of the low brace. The braces must be perpendicular to the slats.

3. Drill pilot holes in the low brace and counterbore the holes. Then, attach the slats to the low brace by driving 2" deck screws through the holes. Follow the same steps for the high brace and attach the slats with 1¼" deck screws.

CUT THE ARMS.

The broad arms of the chair are supported by posts in front, and a cleat attached to the backs of the chair slats.

1. Cut the arms (G) to size.

2. To create decorative angles at the outer end of each arm, mark points 1" from each corner along both edges. Use the points to draw a pair of 1½" cutting lines on each arm. Cut along the lines using a jig saw or circular saw **(photo C).**

3. As an option, mark points for cutting a tapered cut on the inside, back edge of each arm (see *Diagram*). First, mark points on the back of each arm, 3¼" in from each inside edge. Next, mark the outside edges 10" from the back. Then, connect the points and cut the tapers with a circular saw or jig saw. Sand the edges smooth.

ORIGINS OF THE ADIRONDACK STYLE

The Adirondack style originated in Westport, New York, around the turn of the century, when the "Westport chair" made its first appearance. It was very similar to the modern Adirondack chair, except that the back and seat were usually made from single pieces of very wide and clear hemlock or basswood. These chairs became very popular along the East Coast, with many different models cropping up. Because of deforestation, however, the wide boards in the Westport-style furniture became hard to obtain. So designers came up with versions of the chair that featured narrower back and seat slats, and the prototype for the Adirondack chair, as we know it today, was born.

Drive screws through each post and into an arm brace to stabilize the arm/post joint.

Clamp wood braces to the parts of the chair to hold them in position while you fasten the parts together.

ASSEMBLE THE ARMS, CLEATS AND POSTS.

1. Cut the arm cleat (F) and make a mark 2½" in from each end of the cleat.

2. Set the cleat on edge on your work surface. Position the arms on the top edge of the cleat so the back ends of the arms are flush with the back of the cleat and the untapered edge of each arm is aligned with the 2½" mark. Fasten the arms to the cleats with glue.

3. Drill pilot holes in the arms and counterbore the holes. Drive 3" deck screws through the holes and into the cleat.

4. Cut the posts (H) to size. Then, use a compass to mark a 1¾"-radius roundover cut on each bottom post corner (the roundovers improve stability).

5. Position the arms on top of the square ends of the posts. The posts should be set back 1½" from the front ends of the arm, and 1" from the inside edge of the arm. Fasten the arms to the posts with glue.

6. Drill pilot holes in the arms and counterbore the holes. Then, drive 3" deck screws through the arms and into the posts **(photo D).**

7. Cut tapered arm braces (I) from wood scraps, making sure the grain of the wood runs lengthwise (see page 69). Position an arm brace at the outside of each arm/post joint, centered side to side on the post. Attach each brace with glue.

8. Drill pilot holes in the inside face of the post near the top and counterbore the holes. Then, drive deck screws through the holes and into the brace **(photo E).** Drive a 2" deck screw down through each arm and into the top of the brace.

ASSEMBLE THE CHAIR.

All that remains is to join the back, seat/leg assembly and arm/post assembly to complete construction. Before you start, gather scrap wood to brace the parts while you fasten them.

1. Set the seat/leg assembly on your work surface, clamping a piece of scrap wood to the front apron to raise the front of the assembly until the bottoms of the legs are flush on the surface (about 10").

2. Use a similar technique to

TIP

Making tapered cuts with a circular saw is not difficult if the alignment marks on your saw base are accurate. Before attempting to make a tapered cut where you enter the wood at an angle, always make test cuts on scrap wood to be sure the blade starts cutting in alignment with the alignment marks on your saw. If not, either re-set your alignment marks, or compensate for the difference when you cut the tapers.

brace the arm/post assembly so the bottom of the back cleat is 20" above the work surface. Arrange the assembly so the posts fit around the front of the seat/leg assembly, and the bottom edge of the apron is flush with the front edges of the posts.
3. Drill a ¼"-dia. pilot hole through the inside of each leg and partway into the post. Drive a ⅜ × 2½" lag screw and washer through each hole, but do not tighten completely **(photo F).** Remove the braces.
4. Position the back so the low back brace is between the legs, and the slats are resting against the front of the arm cleat. Clamp the back to the seat support with a C-clamp, making sure the top edge of the low brace is flush with the tops of the legs.
5. Tighten the lag screws at the post/leg joints. Then, add a second lag screw at each joint.
6. Drill three evenly spaced pilot holes near the top edge of the arm cleat and drive 1½" deck screws through the holes and into the back slats **(photo G).** Drive 3" deck screws through the legs and into the ends of the low back brace.

APPLY FINISHING TOUCHES.
1. Glue ¼"-thick, ⅜"-dia. cedar wood plugs into visible counterbores **(photo H).**
2. After the glue dries, sand the plugs even with the surrounding surface. Finish-sand all exposed surfaces with 120-grit sandpaper.
3. Finish the chair as desired—we simply applied a coat of clear wood sealer.

Drive screws through the arm cleat, near the top, and into the slats.

Glue cedar plugs into the counterbores to conceal the screw holes.

PROJECT
POWER TOOLS

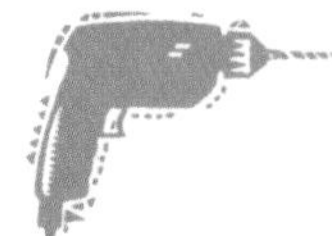

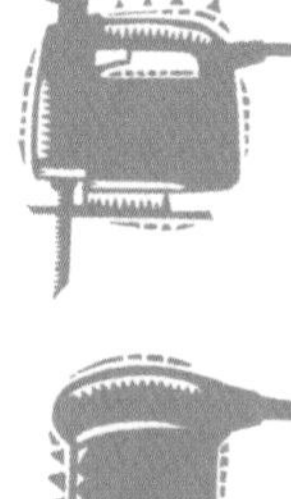

Porch Swing

You'll cherish the pleasant memories created by this porch swing built for two.

CONSTRUCTION MATERIALS

Quantity	Lumber
9	1 × 2" × 8' pine
1	1 × 4" × 4' pine
2	2 × 4" × 10' pine
1	2 × 6" × 10' pine

Nothing conjures up pleasant images of a cool summer evening like a porch swing. When the porch swing is one that you've built yourself, those evenings will be all the more pleasant. This porch swing is made from sturdy pine to provide years and years of memory making. The gentle curve of the slatted seat and the relaxed angle of the swing back are designed for your comfort. When you build your porch swing, pay close attention to the spacing of the rope holes drilled in the back, arms and seat of the swing. They are arranged to create perfect balance when you hang your swing from your porch ceiling.

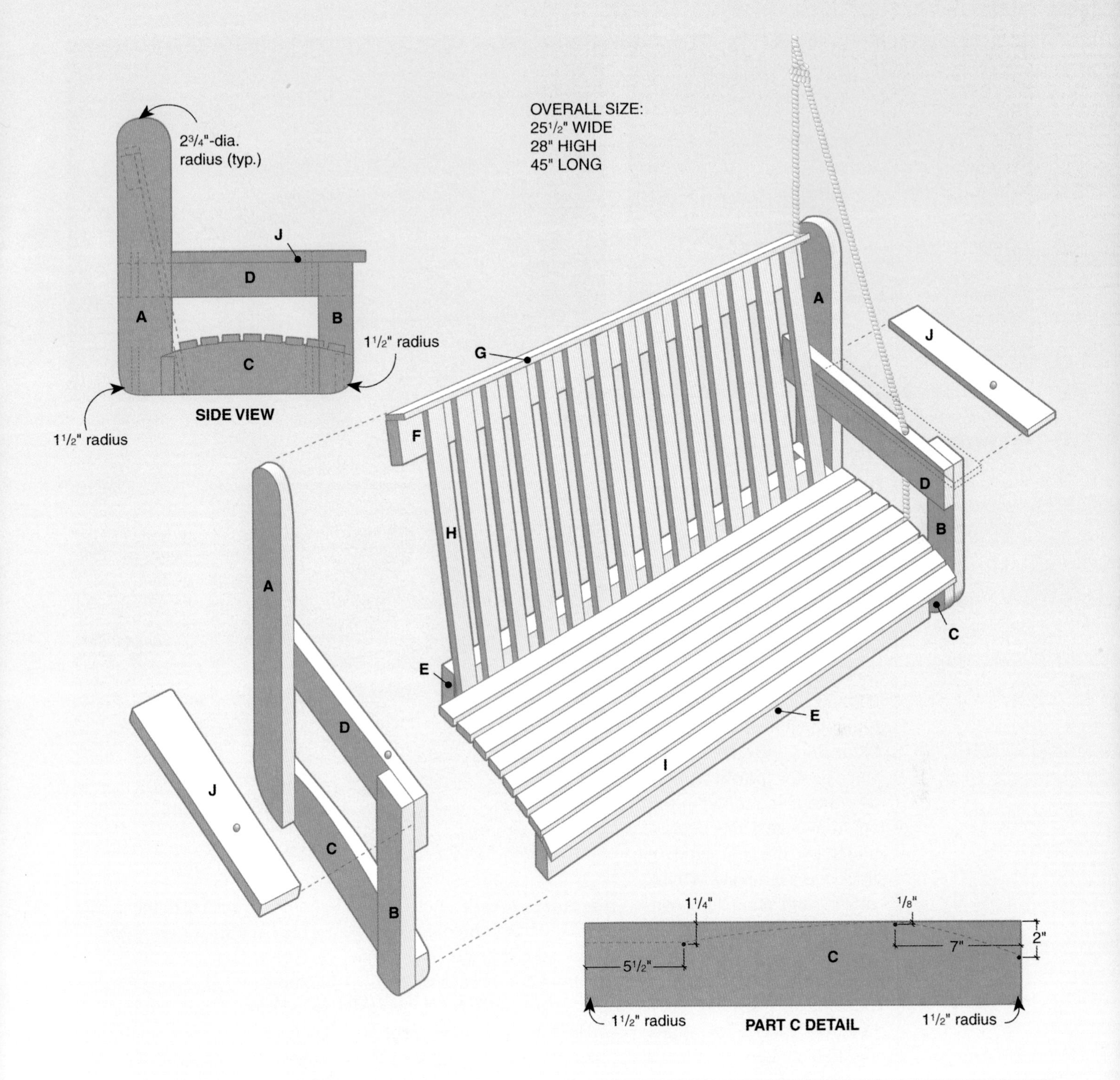

Cutting List

Key	Part	Dimension	Pcs.	Material
A	Back upright	1½ × 5½ × 28"	2	Pine
B	Front upright	1½ × 3½ × 13½"	2	Pine
C	Seat support	1½ × 5½ × 24"	2	Pine
D	Arm rail	1½ × 3½ × 24"	2	Pine
E	Stretcher	1½ × 3½ × 39"	2	Pine

Cutting List

Key	Part	Dimension	Pcs.	Material
F	Back cleat	1½ × 3½ × 42"	1	Pine
G	Top rail	¾ × 1½ × 42"	1	Pine
H	Back slat	¾ × 1½ × 25"	14	Pine
I	Seat slat	¾ × 1½ × 42"	8	Pine
J	Arm rest	¾ × 3½ × 20"	2	Pine

Materials: Moisture-resistant glue, ½"-dia. nylon rope (20'), #8 x 2", #10 x 2½" and #10 x 3" wood screws.

Note: Measurements reflect the actual size of dimension lumber.

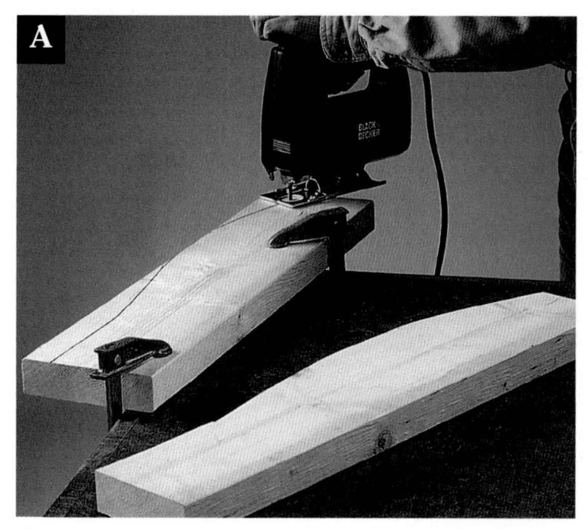

Use a jig saw to cut the contours into the tops of the seat supports.

Use a ⅝" spade bit and a right-angle drilling guide when drilling rope holes through the seat supports.

Directions: Porch Swing

MAKE THE SEAT SUPPORTS.
1. Cut the seat supports (C) to length. Using the pattern on page 75 as a guide, lay out the contour on one of the seat supports. Use a flexible ruler, bent to follow the contour, to ensure a smooth cutting line.
2. Cut along the cutting line with a jig saw **(photo A).** Sand the contour and round the bottom front edge with a belt sander. Use the contoured seat support as a template to mark, cut and sand a matching contour on the other seat support.

BUILD THE SEAT FRAME.
1. Cut the arm rails (D) and stretchers (E) to length. Attach one stretcher between the seat supports, ¾" from the front edges and ½" from the bottom edges, using glue and 2½" wood screws. Fasten the other stretcher between the supports so the front face of the stretcher is 6" from the backs of the supports, and all bottom edges are flush.
2. Use a ⅝" spade bit and drill guide holes for the ropes through the seat supports and the arm rails. Drill a hole 1½" from the back end of each piece. Also drill a hole 4½" from the front end of each piece. Use a right-angle drill guide to make sure holes stay centered all the way through **(photo B).**

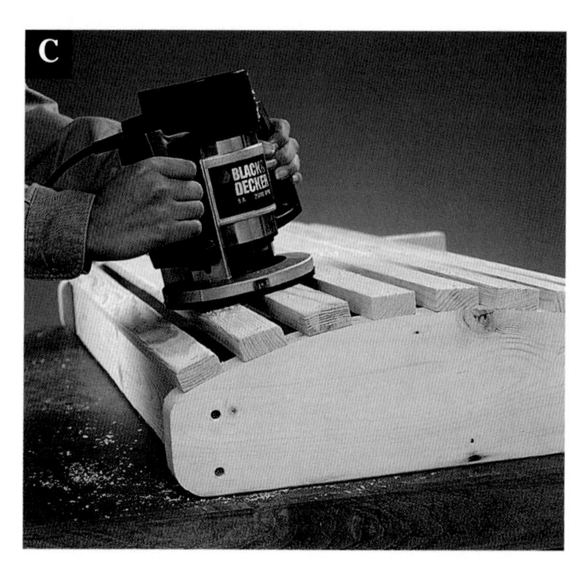

Smooth out the top exposed edges of the seat slats with a router and ¼" roundover bit (or use a power sander).

INSTALL THE SEAT SLATS.
1. Cut the seat slats (I) to length. (Make sure you buy full-sized 1 × 2s, not 1 × 2 furring strips.) Arrange the slats across the seat supports, using ½"-thick spacers to make sure the gaps are even. The front slat should overhang the front stretcher by about ¼", and the back slat should be flush with the front of the back stretcher.
2. Fasten the slats to the seat support with glue and #8 × 2" wood screws (one screw at each slat end). Smooth the top edges of the slats with a router and ¼" roundover bit, or a power sander **(photo C).**

BUILD THE BACK.
1. Cut the back cleat (F) and the back slats (H) to length.
2. Fasten the slats to the back cleat, leaving a 1½" gap at each end, and spacing the slats at 1⅜" intervals **(photo D).** The tops of the slats should be flush

Use 1 × 2 spacers to align the back slats, then fasten the slats to the back cleat.

Fasten the top rail to the back cleat, so the front edge of the rail is flush with the fronts of the slats.

Slide the back assembly against the seat assembly and attach.

with the top of the cleat.
3. Cut the top rail (G) to length. Fasten it to the cleat so the front edge of the rail is flush with the fronts of the slats **(photo E).** Drill a ⅝"-dia. rope hole at each end of the top cleat, directly over the back holes in the arm rails.

ATTACH THE UPRIGHTS AND ARM REST.
1. Cut the back uprights (A) and front uprights (B). Make a round profile cut at the tops of the back uprights (see pattern, page 75), using a jig saw. Attach the uprights to the outside faces of the seat supports, flush with the ends of the supports, using glue and 3" wood screws.
2. Round the bottom front edges of the front uprights with a sander so they are flush with the seat supports.
3. Use 2½" screws to attach arm rails between the uprights, flush with the tops and with rope holes aligned.
4. Slide the back slat assembly behind the seat assembly **(photo F).** Attach the back cleat to the back uprights with 3" screws, so the upper rear corners of the cleat are flush with the back edges of the uprights.
5. Cut and sand the arm rests (J) and set them on the arm rail, centered side to side and flush with the back uprights.
6. Mark the locations of the rope holes in the arm rails onto the arm rests. Drill matching holes into the arm rests. Attach the arm rests to the rails with glue and 2" screws.

APPLY FINISHING TOUCHES.
Sand and paint swing. Thread ½"-dia. nylon rope through all four sets of rope holes. Tie them to hang the swing.

TIP

Use heavy screw eyes driven into ceiling joists to hang porch swings. If the ropes don't line up with the ceiling joists, lag-screw a 2 × 4 cleat to the ceiling joists and attach screw eyes to the cleat.

PROJECT
POWER TOOLS

Garden Bench

Graceful lines and trestle construction make this bench a charming complement to porches, patios and decks—as well as gardens.

Construction Materials

Quantity	Lumber
1	2 × 8" × 6' cedar
4	2 × 2" × 10' cedar
1	2 × 4" × 6' cedar
1	2 × 6" × 10' cedar
1	2 × 2" × 6' cedar
1	1 × 4" × 12' cedar

Casual seating is a welcome addition to any outdoor setting. This lovely garden bench sits neatly at the borders of any porch, patio or deck. It creates a pleasant resting spot for up to three adults without taking up a lot of space. Station it near your home's rear entry and you'll have convenient seating for removing shoes or setting down grocery bags while you unlock the door.

The straightforward design of this bench lends itself to accessorizing. Station a rustic cedar planter next to the bench for a lovely effect. Or, add a framed lattice trellis to one side of the bench to cut down on wind and direct sun.

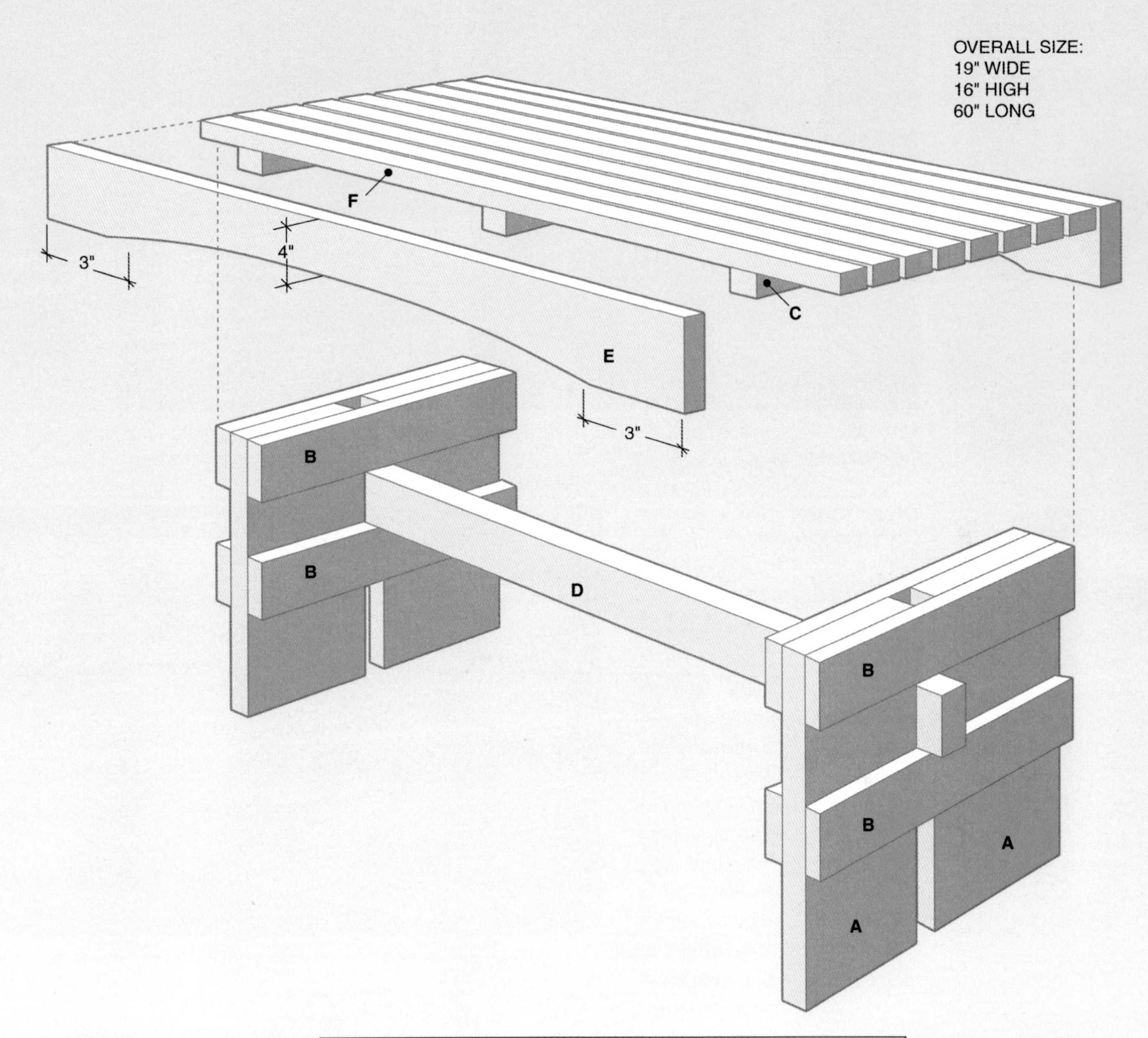

Cutting List

Key	Part	Dimension	Pcs.	Material
A	Leg half	1½ × 7¼ × 14½"	4	Cedar
B	Cleat	¾ × 3½ × 16"	8	Cedar
C	Brace	1½ × 1½ × 16"	3	Cedar
D	Trestle	1½ × 3½ × 60"	1	Cedar
E	Apron	1½ × 5½ × 60"	2	Cedar
F	Slat	1½ × 1½ × 60"	8	Cedar

Materials: Moisture-resistant glue, wood sealer or stain, 1½" and 2½" deck screws.

Note: Measurements reflect the actual size of dimension lumber.

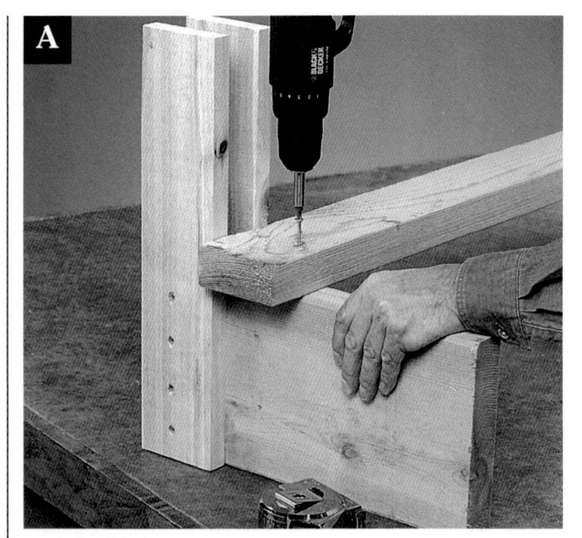

Make sure the trestle is positioned correctly against the cleats, and attach it to the leg.

Attach the remaining leg half to the cleats on both ends to complete the leg assembly.

Directions: Garden Bench

BUILD THE BASE.

1. Cut the leg halves (A), cleats (B) and trestle (D) to length. Sandwich one leg half between two cleats so the cleats are flush with the top and the outside edge of the leg half. Then, join the parts by driving four 1½" deck screws through each cleat and into the leg half. Assemble two more cleats with a leg half in the same fashion.

2. Stand the two assemblies on their sides, with the open ends of the cleats pointing upward. Arrange the assemblies so they are roughly 4' apart. Set the trestle onto the inner edges of the leg halves, pressed flush against the bottoms of the cleats. Adjust the position of the assemblies so the trestle overhangs the leg half by 1½" at each end. Fasten the trestle to each leg half with glue and 2½" deck screws **(photo A).**

Attach the outer brace for the seat slats directly to the inside faces of the cleats.

3. Attach another pair of cleats to each leg half directly below the first pair, positioned so each cleat is snug against the bottom of the trestle.

4. Slide the other leg half between the cleats, keeping the top edge flush with the upper cleats. Join the leg halves with the cleats using glue and 2½" deck screws **(photo B).**

5. Cut the braces (C) to length. Fasten one brace to the inner top cleat on each leg assembly, so the tops are flush **(photo C).**

MAKE THE APRONS.

1. Cut the aprons (E) to length.

2. Lay out the arch onto one apron, starting 3" from each end. The peak of the arch, located over the midpoint of the apron, should be 1½" up from the bottom edge.

3. Draw a smooth, even arch by driving a casing nail at the peak of the arch and one at each of the starting points. Slip a flexible ruler behind the nails at the starting points and in front of the nail at the peak to create a smooth arch. Then,

TIP

Take extra care to countersink screw heads completely whenever you are building furnishings that will be used as seating. When sinking galvanized deck screws, use a counterbore bit or a standard ⅜"-dia. bit to drill ¼"-deep counterbores, centered around ⅛"-dia. pilot holes.

trace along the inside of the ruler to make a cutting line **(photo D).**

4. Cut along the line with a jig saw and sand the cut smooth.

5. Trace the profile of the arch onto the other apron and make and sand the cut.

6. Cut the slats (F) to length. Attach a slat to the top, inside edge of each apron with glue and deck screws **(photo E).**

INSTALL THE APRONS AND SLATS.

1. Apply glue at each end on the bottom sides of the attached slats. Flip the leg and trestle assembly and position it flush with the aprons so that it rests on the glue on the bottoms of the two slats. The aprons should extend 1½" beyond the legs at each end of the bench. Drive 2½" deck screws through the braces and into both slats.

2. Position the middle brace (C) between the aprons, centered end to end on the project. Fasten it to the two side slats with deck screws.

3. Position the six remaining slats on the braces, using ½"-thick spacers to create equal gaps between them. Attach the slats with glue and drive 2½" deck screws up through the braces and into each slat **(photo F).**

APPLY FINISHING TOUCHES.

Sand the slats smooth with progressively finer sandpaper. Wipe away the sanding residue with a rag dipped in mineral spirits. Let the bench dry. Apply a finish of your choice—a clear wood sealer protects the cedar without altering the color.

TIP

Sometimes our best efforts produce furniture that wobbles because it is not quite level. One old trick for leveling furniture is to set a plastic wading pool on a flat plywood surface that is set to an exact level position with shims. Fill the pool with about ¼" of water. Set the furniture in the pool, then remove it quickly. Mark the tops of the waterlines on the legs, and use them as cutting lines for trimming the legs to level.

Use a flexible ruler pinned between casing nails to trace a smooth arch onto the aprons.

Attach a 2 × 2 slat to the top, inside edge of each apron, using 2½" deck screws and glue.

Attach the seat slats with glue and 2½" deck screws. Insert ½"-thick spacers to set gaps between the slats.

Patio Chair

You won't believe how comfortable plastic tubes can be until you sit in this unique patio chair. It's attractive, reliable and very inexpensive to build.

PROJECT POWER TOOLS

For solid support, you can't go wrong with this patio chair. Crashing painfully to the ground just when you're trying to sit and relax outdoors is nobody's idea of fun. This patio chair is designed for durability and comfort. It uses rigid plastic tubing for cool, comfortable support that's sure to last through many fun-filled seasons. Say good-bye to expensive or highly-specialized patio furniture with this outdoor workhorse.

This inventive seating project features CPVC plastic tubing that functions like slats for the back and seat assemblies. The ½"-dia. tubes have just the right amount of flex and support, and can be purchased at any local hardware store. Even though the tubing is light, there is no danger of this chair blowing away in the wind. It has a heavy, solid frame that will withstand strong gusts of wind and fearsome summer showers. For even greater comfort, you can throw a favorite pillow, pad or blanket over the tubing and arms and relax in the sun.

The materials for this project are inexpensive. All the parts except the seat support are made from 2 × 4 cedar. The seat support is made from 1 × 3 cedar. For a companion project to this patio chair, see *Gate-Leg Picnic Tray,* pages 130-133.

CONSTRUCTION MATERIALS

Quantity	Lumber
3	2 × 4" × 10' cedar
1	1 × 3" × 2' cedar
7	½" × 10' CPVC tubing

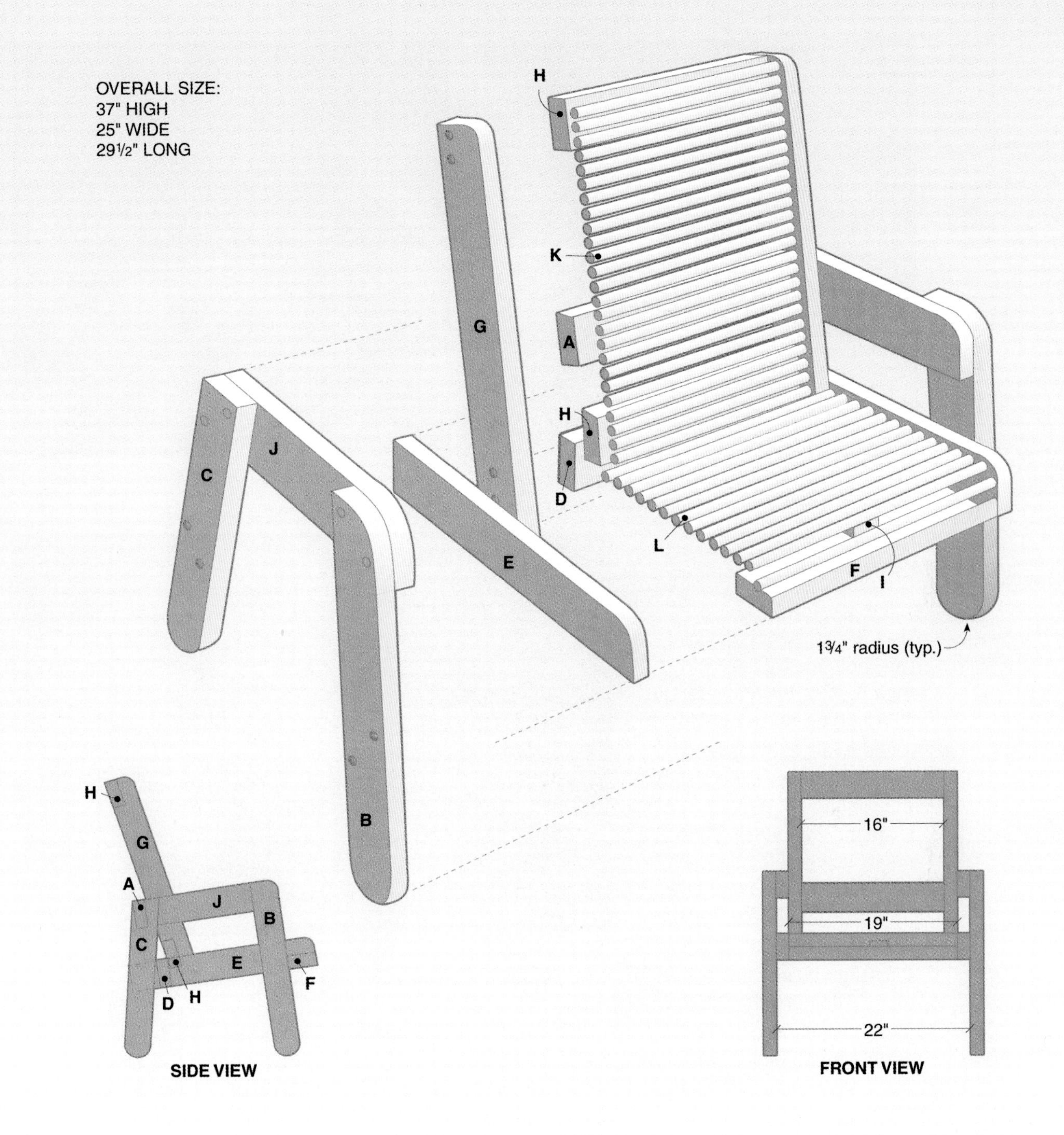

Cutting List

Key	Part	Dimension	Pcs.	Material
A	Back support	1½ × 3½ × 19"	1	Cedar
B	Front leg	1½ × 3½ × 22½"	2	Cedar
C	Rear leg	1½ × 3½ × 21"	2	Cedar
D	Seat stop	1½ × 3½ × 19"	1	Cedar
E	Seat side	1½" × 3½ × 24½"	2	Cedar
F	Seat front	1½ × 3½ × 19"	1	Cedar

Cutting List

Key	Part	Dimension	Pcs.	Material
G	Back side	1½ × 3½ × 28"	2	Cedar
H	Back rail	1½ × 3½ × 16"	2	Cedar
I	Seat support	¾ × 2½ × 17¾"	1	Cedar
J	Arm rail	1½ × 3½ × 19½"	2	Cedar
K	Back tube	½-dia. × 17½"	25	CPVC
L	Seat tube	½-dia. × 20½"	14	CPVC

Materials: Moisture-resistant glue, 1¼", 2½" and 3" deck screws, ⅜"-dia. cedar plugs, finishing materials.

Note: Measurements reflect the actual size of dimension lumber.

Use a portable drilling guide when drilling the holes for the tubes in the seat sides.

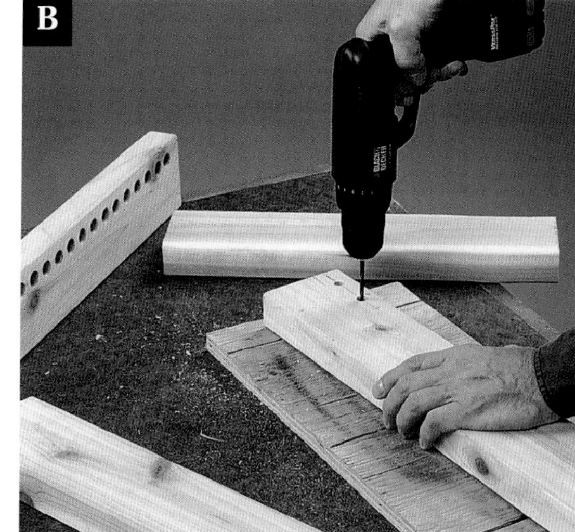

Drill pilot holes before attaching the back rails and sides.

Directions: Patio Chair

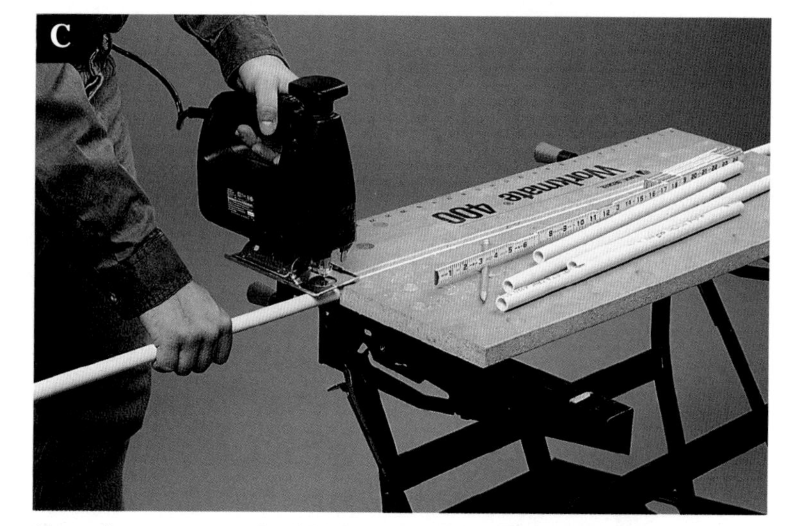

Use a jig saw to cut the CPVC tubing slats. For stability, arrange the tubing so the saw blade is very close to the work surface.

MAKE THE BACK SIDES.
The back sides of the patio chair provide the frame for the CPVC tubing. Make sure all your cuts are accurate and smooth to achieve good, snug-fitting joints.

1. Cut the back sides (G) to length, using a circular saw.

2. Drill the stopped holes for the plastic tubes on the inside faces of the back sides. These holes must be accurately positioned and drilled. Use a pencil with either a combination square or a straightedge to draw a centering line to mark the locations for the holes. Make the centering line ⅝" from the front edge of each back side.

3. Drill ⅝-dia. × ¾"-deep holes and center them exactly 1" apart along the centerline. Start the first hole 3" from the bottom end of each back side. Use a portable drilling guide and a square to make sure the holes are straight and perfectly aligned **(photo A).** A portable drilling guide fits easily onto your power drill to ensure quick and accurate drilling. Some portable drilling guides are equipped with depth stops, making them the next best thing to a standard drill press.

4. Cut 1"-radius roundovers on the top front corner of each back side.

BUILD THE BACK FRAME.

1. Use a circular saw to cut the back rails (H) to length. These pieces will be attached to the inside faces of the back sides.

2. To eliminate the sharp edges, clamp the pieces to a stable work surface and use a sander or a router to soften the edges on the top and bottom of the back rails, and the top edges of the back sides.

3. Dry-fit the back rails and back sides and mark their positions with a pencil.

TIP

Plastic tubing is in many ways an ideal outdoor furniture material. It's a cool, strong material that is extremely lightweight and long lasting. It won't bleed onto the wood or stain as it ages like metal frames and fasteners, and when it gets dirty, all you need to do is wash it off with some soap and water.

Attach the remaining side to complete the back assembly.

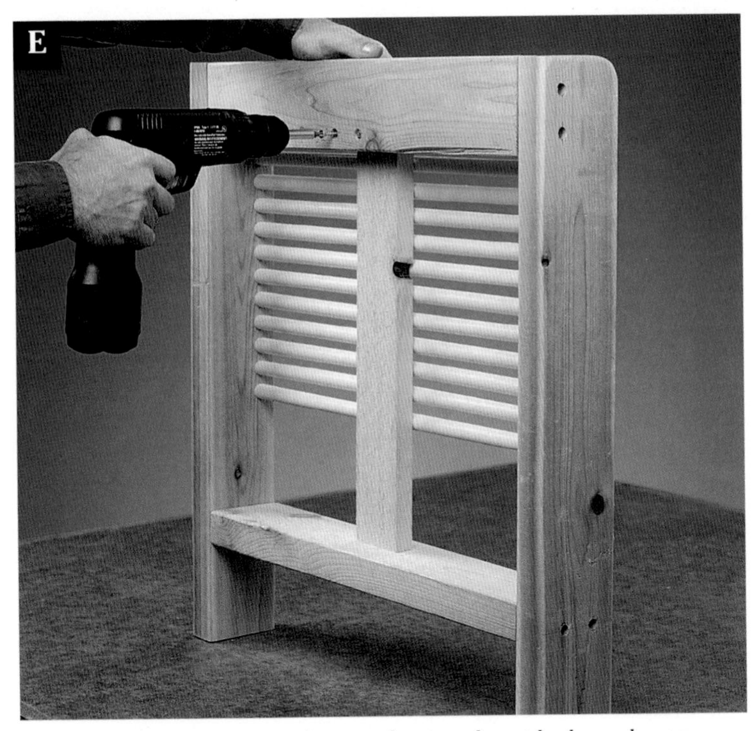

Attach the seat support to the seat front and seat lock as shown.

4. Drill ⅛" pilot holes in the back side and counterbore the holes to a ¼" depth, using a counterbore bit. **(photo B).**

5. Apply moisture-resistant glue to one end of each rail and fasten the rails to a single back side with 3" deck screws

COMPLETE THE BACK ASSEMBLY.

Before assembling the back, you need to prepare the CPVC tubing for the frame holes. Make sure the tubing is ½"-dia. CPVC, which is rated for hot water. This plastic tubing is usually available in 10' lengths. (Standard PVC tubing is not usually sold in small diameters that will fit the ⅝"-dia. holes you have drilled.)

1. Use a jig saw to cut 25 pieces of the ½"-dia. CPVC tubing. Remember, these pieces will be used for the back seat assembly only. The seat assembly requires additional pieces. Cut the back tubes to 17½" lengths **(photo C).**

2. Wash the grade stamps off the tubing with lacquer thinner. (Wear gloves and work in a well-ventilated area when using lacquer thinner.) Rinse the tubing with clean water.

3. Once the pipes are clean and dry, insert them into the holes on one of the back sides. Slide the remaining back side into place, positioning the plastic tubes into the holes.

4. Attach the rails to the back side by driving 3" deck screws through the pilot holes **(photo D).**

BUILD THE SEAT FRAME.

One important difference between the seat frame and the back frame is the positioning of the CPVC tubing. On the seat frame, one tube is inserted into the sides slightly out of line at the front to make the chair more comfortable for your legs.

1. Cut the seat sides (E), seat front (F), seat stop (D) and seat support (I) to length. Use the

TIP

The easiest way to cut CPVC tubing is with a power miter box, but no matter what kind of saw you are using, remember to work in a well-ventilated room. Although plastic tubing generally cuts easily without melting or burning, it can release some toxic fumes as it is cut. When you're finished, you might consider treating the tubes with some automotive plastic polish to help preserve them.

same methods as with the back frame to draw the centering line for the plastic tubing on the seat sides. Drill the tube holes into each seat side. Start the holes 2" from the front end of the seat sides.

2. Position a single tube hole on the seat frame ⅞" below the top edge and 1" from the front end of each seat side. This front tube provides a gradual downward seat profile for increased leg comfort.

3. To eliminate the sharp edges on the seat assembly, round the seat sides, seat support edges and seat front edges with a sander or router. Cut 1"-radius roundovers on the top front corners of the seat sides.

4. Use a combination square to mark a line across the width of the inside of the seat sides, 3½" from the back edges. This is where the back face of the seat stop is positioned. Test-fit the pieces to make sure their positions are correct. Lay out and mark the position of the seat stop and seat front on each seat side.

5. Drill pilot holes to fasten one of the seat sides to the seat stop and seat front, as you did with the back assembly. Counterbore the holes. Connect the parts with moisture-resistant glue and deck screws.

Make identical radius cuts on the bottoms of the legs.

Use a square to make sure the seat is perpendicular to the leg.

COMPLETE THE SEAT FRAME.

1. Cut 14 pieces of ½"-dia. CPVC pipe. Each piece should be 20½" long. Once again, clean the grade stamps off the tubes with lacquer thinner and rinse them with clean water. Let them dry and insert them into the holes on one seat side.

2. Carefully slide the remaining seat side into place and fasten the pieces with moisture-resistant glue and deck screws.

3. Position the seat support (I) under the tubing in the center of the seat. Attach the seat support to the middle of the seat front and seat stop with moisture-resistant glue and 1¼" deck screws **(photo E).**

BUILD THE ARMS AND LEGS.

The arms and legs are all that remain for the patio chair assembly. When you make the radius cuts on the bottom edges of the front and back legs, make sure the cuts are exactly the same on each leg (see *Diagram,* page 83). Otherwise, the legs may be uneven and rock

Slide the back frame into the seat frame so the back sides rest against the seat stop and seat support.

back and forth when you sit.
1. Cut the back support (A), front legs (B), rear legs (C) and arm rails (J) to length.
2. Use a jig saw to cut a full-radius roundover on the bottoms of the legs **(photo F).** Cut a 1"-radius roundover on the top front corners of the arms and the front legs.
3. To attach the front legs to the outsides of the arm rails, drill pilot holes in the front legs and counterbore the holes. Then, attach the parts at a 90° angle, using 2½" deck screws. The legs should be flush with the front ends of the rails.
4. Attach the leg/arm rail assembly to the seat frame so that the top edge of the seat frame is 15" from the bottom of the leg. The front of the seat should extend exactly 3½" past the leg. Use a square to make sure the seat is perpendicular to the legs **(photo G).**
5. To attach the rear legs, drill pilot holes in the rear legs and counterbore the holes. Attach the rear legs to the arm rails and seat sides with glue. Then, drive 2½" deck screws through the rear legs and into the arm rails and seat sides. The back edge of the legs should be flush with the ends of the arm rails and seat sides. Trim the excess material from the tops of the legs so they are flush with the tops of the arm rails.
6. To attach the back support, drill pilot holes in the arm rails. Counterbore the holes. Then, attach the back support between the rails with glue and drive 2½" deck screws through the arm rails and into the back support. The back support should be flush with the ends of the arm rails.
7. Round and sand all rough edges smooth.

ATTACH THE BACK FRAME.

1. Slide the back frame into the seat frame **(photo H)** so that the back sides rest against the seat stop and the back rail rests on the seat support.
2. Drill pilot holes in the seat stop and counterbore the holes. Apply glue, and attach the back frame by driving deck screws through the seat stop and into the back rail.

APPLY FINISHING TOUCHES.

1. For a refined look, apply glue to the bottoms of ⅜"-dia. cedar wood plugs, and insert the plugs into the screw counterbores. Sand the tops of the plugs until they are flush with the surrounding surface.
2. Wipe the chair with mineral spirits and finish the chair with a clear wood sealer.

TIP

When using a jig saw, it is tempting to speed up a cut by pushing the tool with too much force. When cutting curves or roundovers, this is likely to cause the saw blade to bend. This often causes irregular cuts and burns, especially when working with cedar. You can achieve smoother curves and roundovers with multiple gentle passes with the saw, until the proper curve is achieved. Finish the job by sanding the curves smooth.

PROJECT
POWER TOOLS

Trellis Seat

Spice up your patio or deck with this sheltered seating structure. Set it in a secluded corner to create a warm, inviting outdoor living space.

CONSTRUCTION MATERIALS

Quantity	Lumber
1	4 × 4" × 6' cedar
2	2 × 8" × 8' cedar
5	2 × 4" × 12' cedar
1	1 × 6" × 10' cedar
11	1 × 2" × 8' cedar
2	½" × 4 × 4' cedar lattice

Made of lattice and cedar boards, our trellis seat is ideal for conversation or quiet moments of reading. The lattice creates just the right amount of privacy for a small garden or patio. It's an unobtrusive structure that is sure to add some warmth to your patio or deck. Position some outdoor plants along the top cap or around the frame sides to dress up the project and bring nature a little closer to home. For a cleaner appearance, conceal visible screw heads on the seat by counterboring the pilot holes for the screws and inserting cedar plugs (available at most woodworking stores) into the counterbores.

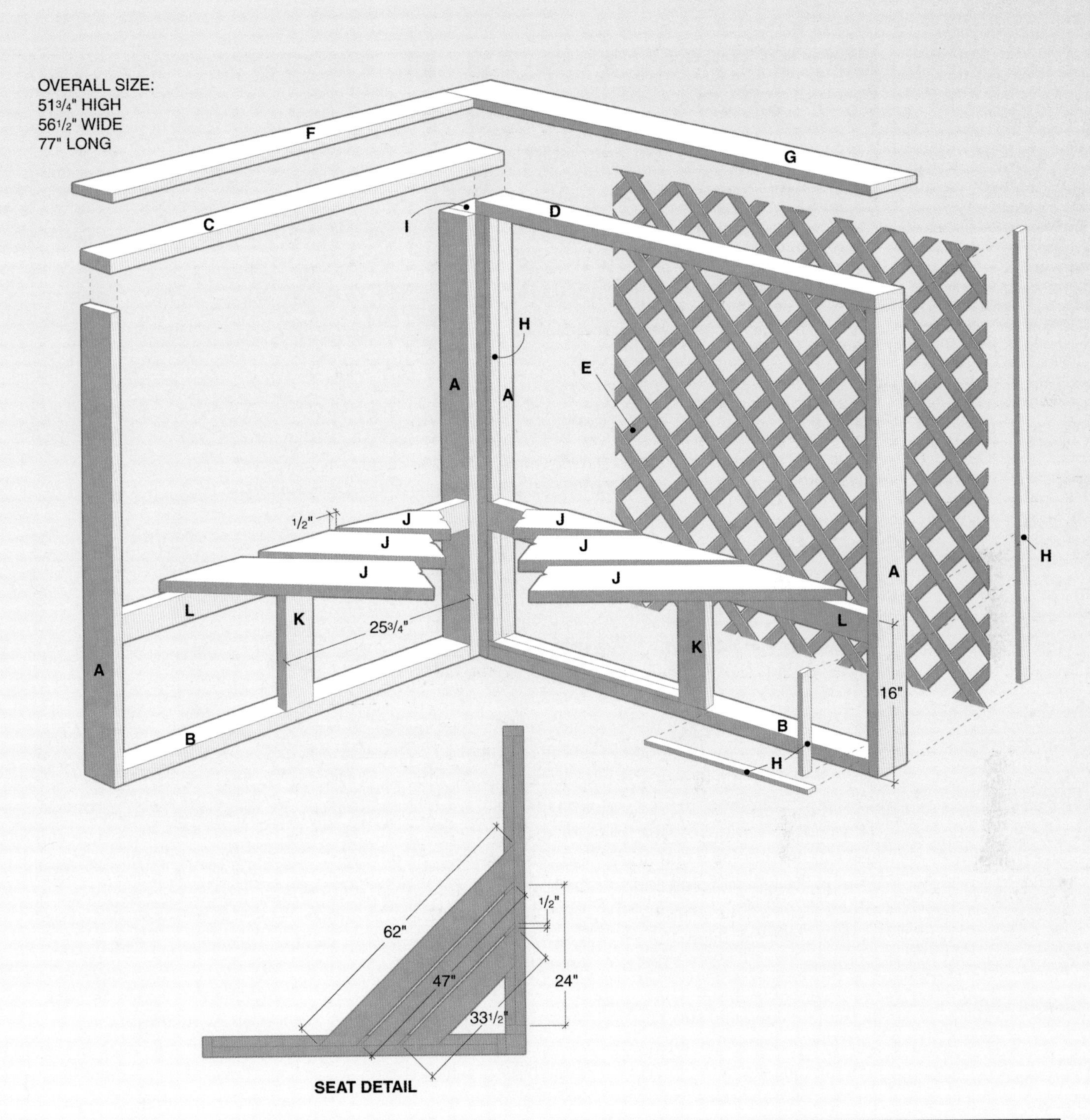

Key	Part	Dimension	Pcs.	Material
A	Frame side	1½ × 3½ × 49½"	4	Cedar
B	Frame bottom	1½ × 3½ × 48"	2	Cedar
C	Long rail	1½ × 3½ × 56½"	1	Cedar
D	Short rail	1½ × 3½ × 51"	1	Cedar
E	Lattice	½" × 4 × 4'	2	Cedar
F	Short cap	¾ × 5½ × 51"	1	Cedar

Cutting List

Key	Part	Dimension	Pcs.	Material
G	Long cap	¾ × 5½ × 56½"	1	Cedar
H	Retaining strip	¾ × 1½" cut to fit	22	Cedar
I	Post	3½ × 3½ × 49½"	1	Cedar
J	Seat board	1½ × 7¼ × *	3	Cedar
K	Brace	1½ × 3½ × 11"	2	Cedar
L	Seat support	1½ × 3½ × 48"	2	Cedar

Materials: Moisture-resistant glue, 1¼", 2", 2½" and 3" deck screws, 4d galvanized casing nails, finishing materials.

Note: Measurements reflect the actual size of dimension lumber.

*Cut one each: 32", 49", 63"

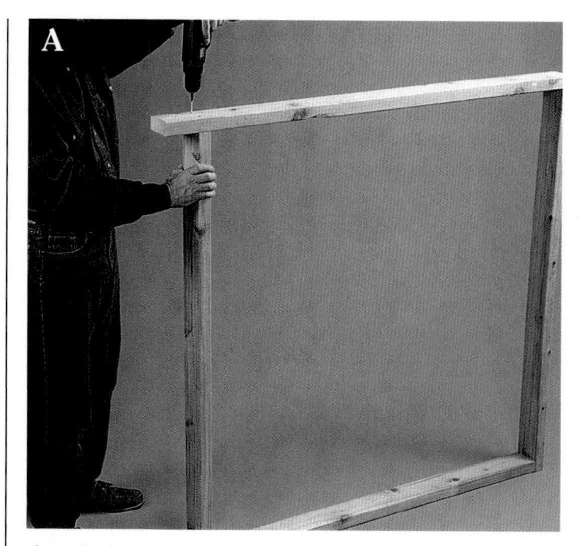

Attach the long rail at the top of one trellis frame with a 3½" overhang at one end to cover the post.

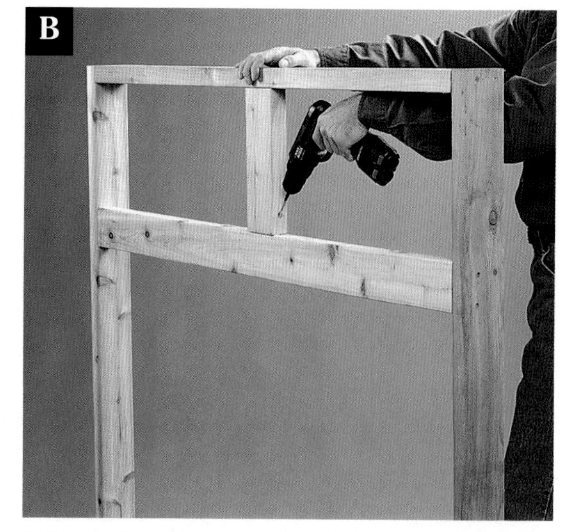

Drive deck screws toenail-style through the braces and into the seat supports.

Directions: Trellis Seat

MAKE THE TRELLIS FRAME.

1. Cut the frame sides (A), frame bottoms (B), long rail (C), short rail (D), braces (K) and seat supports (L) to length. To attach the frame sides and frame bottoms, drill two evenly spaced 3⁄16" pilot holes in the frame sides. Counterbore the holes ¼" deep, using a counterbore bit. Fasten with glue and drive 2½" deck screws through the frame sides and into the bottoms.

2. Drill pilot holes in the top faces of the long and short rails. Counterbore the holes. Attach the long and short rails to the tops of the frame sides with glue. Drive deck screws through the rails and into the ends of the frame sides. The long rail should extend 3½" past one end of the frame **(photo A).**

3. Mark points 22¼" from each end on the frame bottoms to indicate position for the braces. Turn the frame upside-down. Drill pilot holes in the frame bottoms where the braces will be attached. Counterbore the holes. Position the braces flush with the inside frame bottom edges. Attach the pieces by driving 3" deck screws through the frame bottoms and into the ends of the braces.

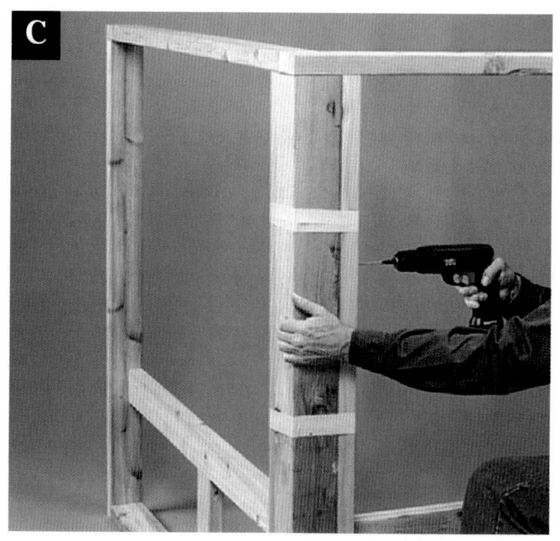

Fasten the trellis frames to the post at right angles.

4. Position the seat supports 16" up from the bottoms of the frame bottoms, resting on the braces. Make sure the supports are flush with the inside edges of the braces. Attach with glue and 3" deck screws driven through the frame sides and into the ends of the seat supports.

5. Attach the braces to the seat supports by drilling angled 3⁄16" pilot holes through each brace edge. Drive 3" deck screws toenail style through the braces and into the top edges of the seat supports **(photo B).**

JOIN THE TRELLIS FRAMES TO THE POST.

1. Cut the post (I) to length.

2. Attach the two frame sections to the post. First, drill pilot holes in the frame sides. Counterbore the holes. Drive evenly spaced 3" deck screws through the frame sides and into the

TIP

Fabricated lattice panels are sold at any building center in standard ¾" thickness. For our trellis seat project, we found and used ½"-thick lattice panels. If you cannot locate any of the thinner panels, use ¾"-thick lattice, and substitute ½"-thick retaining strips at the backs of the trellis frames.

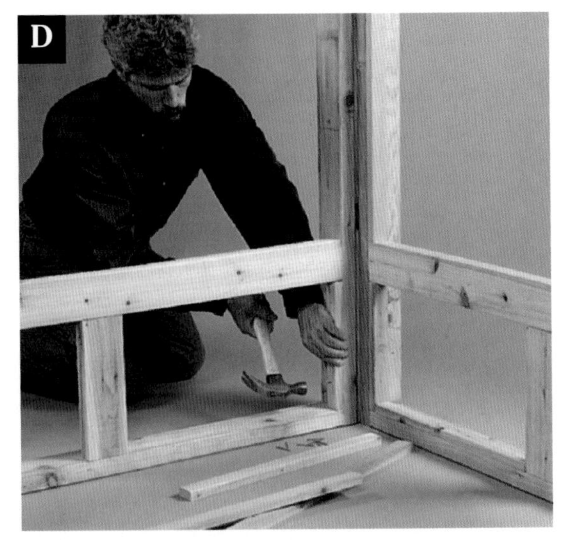

Nail 1 × 2 retaining strips for the lattice panels to the inside faces of the trellis frames.

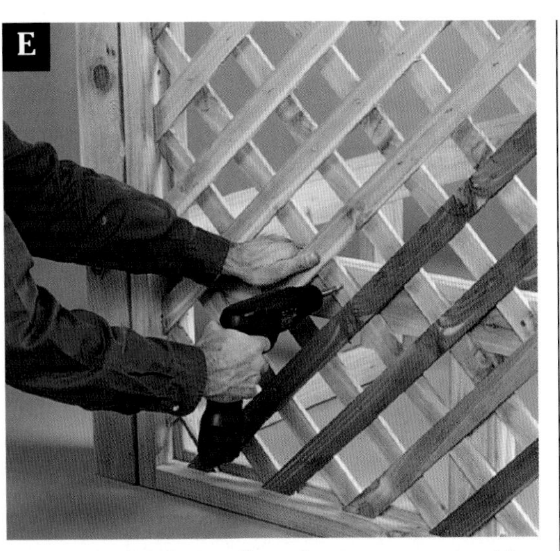

Fasten the lattice panels to the seat supports with 1¼" deck screws, then attach outer retaining strips.

post **(photo C).** Make sure the overhang of the long rail fits snugly over the top of the post.

ATTACH THE LATTICE RETAINING STRIPS.

1. Cut the lattice retaining strips (H) to fit along the inside faces of the trellis frames (but not the seat supports or braces).

2. Nail the strips to the frames, flush with the inside frame edges, using 4d galvanized casing nails **(photo D).**

CUT AND INSTALL THE LATTICE PANELS.

1. Since you will probably be cutting through some metal fasteners in the lattice, fit your circular saw with a remodeler's blade. Sandwich the lattice panel between two boards near the cutting line to prevent the lattice from separating. Clamp the boards and the panel together, and cut the lattice panels to size. Always wear protective eyewear when operating power tools.

2. Position the panels into the frames against the retaining strips, and attach them to the seat supports with 1¼" deck screws **(photo E).** Secure the panels by cutting retaining strips to fit along the outer edges of the inside faces of the trellis frame. Nail strips in place.

BUILD THE SEAT.

1. Cut the seat boards (J) to length. On a flat work surface, lay the seat boards together, edge to edge. Insert ½"-wide spacers between the boards.

2. Draw cutting lines to lay out the seat shape onto the boards as if they were one board (see *Seat Detail,* page 89, for seat board dimensions). Gang-cut the seat boards to their finished size and shape with a circular saw.

3. Attach the seat boards to the seat supports with evenly spaced deck screws, maintaining the ½"-wide gap. Smooth the seat board edges with a sander or router.

INSTALL THE TOP CAPS.

1. Cut the short cap (F) and long cap (G).

2. Attach the caps to the tops of the long and short rails with deck screws **(photo F).**

APPLY FINISHING TOUCHES.

Brush on a coat of clear wood sealer to help preserve the trellis seat.

Attach the long and short caps to the tops of the trellis frames. The long cap overlaps the long rail and the post.

PROJECT
POWER TOOLS

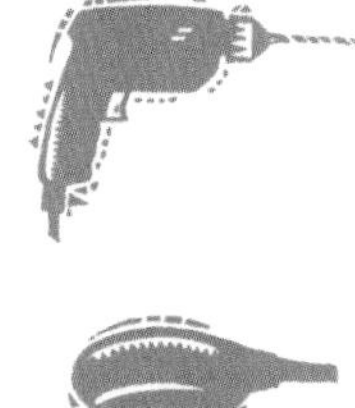

Fire Pit Bench

With seating for three and storage room below, this versatile bench will be at home anywhere in your yard.

Construction Materials

Quantity	Lumber
2	2 × 2" × 8' cedar
4	1 × 4" × 8' cedar
4	2 × 4" × 8' cedar
1	1 × 2" × 8' cedar

Summer cookouts, moonlit bonfires or even a midwinter warm-up are all perfect occasions to use this cedar fire pit bench. If you are extremely ambitious, you can build four benches to surround your fire pit on all sides. If you don't need that much seating, build only two and arrange them to form a cozy conversation area around the fire. Even without a fire pit, you can build a single bench as a stand-alone furnishing for your favorite spot in the yard or garden.

This solid cedar bench will seat up to three adults comfortably. The slats below give the bench strength, while providing a convenient spot for storing and drying firewood.

OVERALL SIZE:
18" HIGH
18½" WIDE
48" LONG

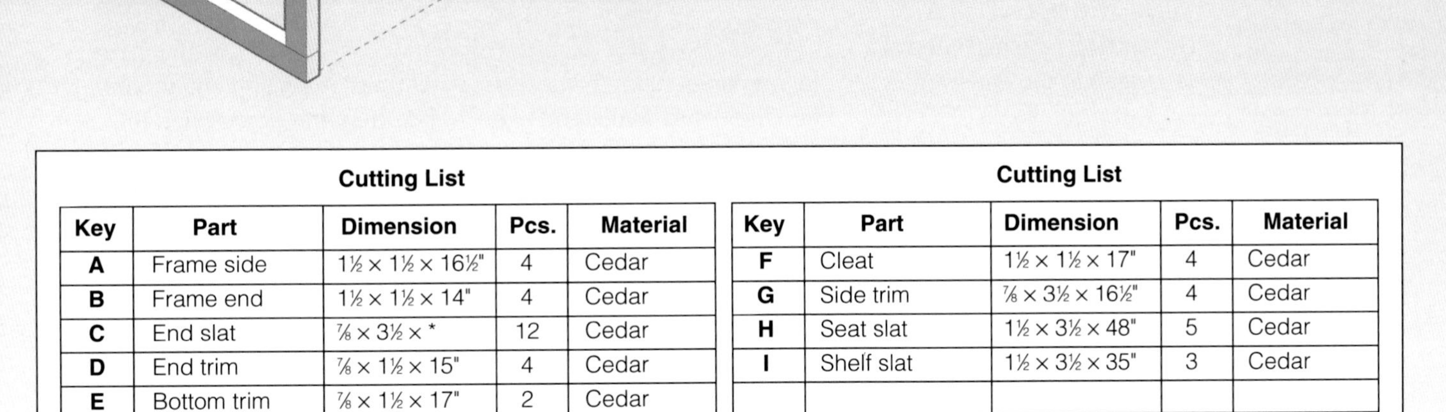

Cutting List

Key	Part	Dimension	Pcs.	Material
A	Frame side	1½ × 1½ × 16½"	4	Cedar
B	Frame end	1½ × 1½ × 14"	4	Cedar
C	End slat	⅞ × 3½ × *	12	Cedar
D	End trim	⅞ × 1½ × 15"	4	Cedar
E	Bottom trim	⅞ × 1½ × 17"	2	Cedar
F	Cleat	1½ × 1½ × 17"	4	Cedar
G	Side trim	⅞ × 3½ × 16½"	4	Cedar
H	Seat slat	1½ × 3½ × 48"	5	Cedar
I	Shelf slat	1½ × 3½ × 35"	3	Cedar

Materials: 1½" and 2½" deck screws, finishing materials.

Note: Measurements reflect the actual size of dimension lumber.

* Cut to fit

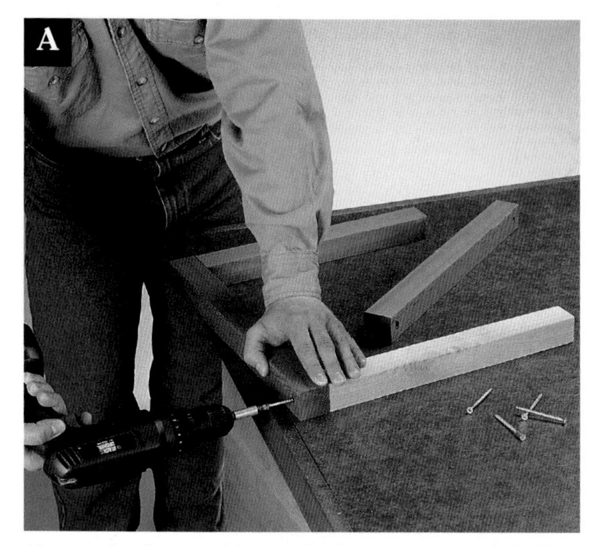

Fasten the frame sides to the frame ends with 2½" galvanized deck screws.

Trim off the ends of the slats so the ends are flush with the outside edges of the end frames.

Directions: Fire Pit Bench

BUILD THE END FRAMES.

1. Cut the frame sides (A) and frame ends (B) to length. Place a frame end between two frame sides. Drill ⅛" pilot holes in the frame sides. Counterbore the holes ¼" deep, using a counterbore bit. Drive 2½" deck screws through the frame sides and into the ends of the frame end **(photo A).** Attach another frame end between the free ends of the frame sides.

2. Follow the same procedure to build the second end frame.

ATTACH THE END SLATS.

The end slats are mounted at 45° angles to the end frames.

1. Lay the frames on a flat surface. Use a combination square as a guide for drawing a reference line at a 45° angle to one corner on each frame, starting 3½" in from the corner.

2. To measure and cut the end slats (C), lay the end of a full-length 1 × 4 cedar board across one frame so one edge meets the corner and the other edge follows the reference line. Position the board so the end overhangs the frame by an inch or two. Mark a point with an equal overhang on the other side of the frame.

3. Cut the 1 × 4, then fasten the cut-off piece to the frame by driving pairs of 1½" deck screws into the end frame. Lay a 1 × 4 back across the frame, butted up against the attached slat, and mark and cut another slat the same way. Attach the slat. Continue cutting and attaching the rest of the slats to cover the frame. Attach slats to the other end frame.

4. Draw straight cutting lines on the tops of the slats, aligned with the outside edges of the end frames. Using a straightedge and circular saw, trim off the ends of the slats along the cutting lines **(photo B).**

Use shelf slats to set the correct distance between the end-frame assemblies, then attach the end frames to the bottoms of the seat slats.

Fasten the bottom cleats to the shelf slats, keeping the ends of the slats flush with the outside edges of the cleats.

Attach the bottom cleats to the end frames with deck screws. Use a spacer to keep the cleat 1½" up from the bottom of the bench.

COMPLETE THE END FRAMES.

1. Cut the end trim (D) and bottom trim (E) to length. Fasten them to the outside faces of the end slats to create a frame the same length and width as the end frame. Cut the side trim (G) pieces to length and fasten to the frame assembly with 1½" deck screws, making sure the edges of the side trim are flush with the outside edges of the end frames and trim frames.

2. Cut the cleats (F). Fasten a top cleat to the inside of each frame with 2½" deck screws. The top cleats should be flush with the tops of the end frames. (The bottom cleats will be attached later.)

ATTACH THE SEAT SLATS.

1. Cut the seat slats (H) to length. Lay them on a flat surface with the ends flush and ⅛" spaces between slats. Cut the shelf slats (I). Set the end frame assemblies on top of the seat slats. Slip two of the shelf slats between the ends to set the correct distance.

2. Fasten the end-frame assemblies to the seat slats by driving 1½" deck screws through the cleats on the end frames **(photo C).**

ATTACH THE SHELF SLATS.

1. Arrange the shelf slats on your work surface so the ends are flush, with 1½" gaps between the slats. Lay the remaining two cleats across the ends of the slats. Fasten the cleats to the slats with 2½" deck screws **(photo D).**

2. Set the shelf assembly between the ends of the bench, resting on a 1½" spacer. Attach the shelf by driving 2½" screws through the cleats and into the end frames **(photo E).**

APPLY FINISHING TOUCHES.

1. With a compass, draw a 1½"-radius roundover at the corners of the seat. Cut the roundovers with a jig saw. Sand the entire fire pit bench—especially the edges of the seat slats—to eliminate any possibility of slivers. Or, use a router with a roundover bit to trim off the sharp edges.

2. Apply exterior wood stain to all exposed surfaces.

TIP

When storing firewood, it is tempting to cover the wood with plastic tarps to keep it dry. But more often than not, tarps will only trap moisture and keep the firewood permanently damp. With good ventilation wood dries out quickly, so your best bet is to store it uncovered or in an open shelter.

PROJECT
POWER TOOLS

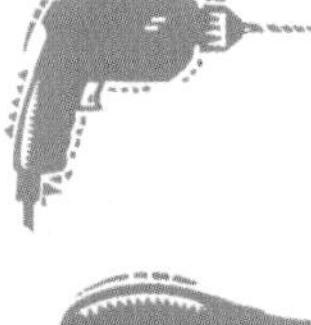

Fold-up Lawn Seat

With this fold-up seat built for two, you won't have to sacrifice comfort and style for portability.

CONSTRUCTION MATERIALS

Quantity	Lumber
1	2 × 8" × 6' cedar
4	2 × 4" × 8' cedar
2	1 × 6" × 8' cedar

Even though this cedar lawn seat folds up for easy transport and storage, it is sturdier and more attractive than just about any outdoor seating you are likely to make or buy. The backrest and legs lock into place when the seat is in use. To move or store this two-person seat, simply fold the backrest down and tuck the legs into the seat frame to convert the seat into a compact package.

Because it is portable and stores in a small space, you can keep the fold-up lawn seat tucked away in a garage or basement and set it up for extra seating when you are entertaining. Or, if security around your home is an issue, you can bring it inside easily during times when you're not home.

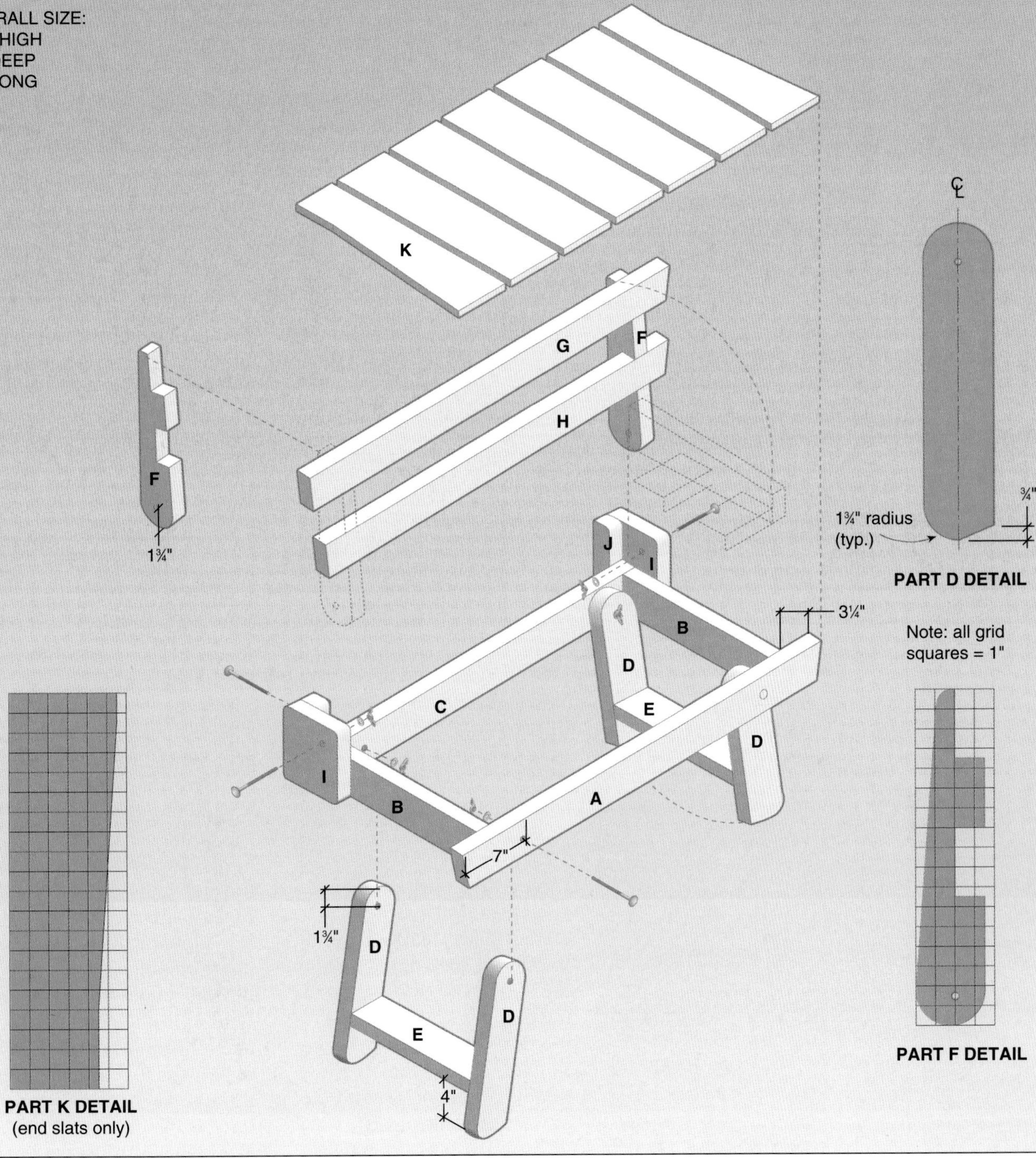

Cutting List

Key	Part	Dimension	Pcs.	Material
A	Front seat rail	1½ × 3½ × 42"	1	Cedar
B	Side seat rail	1½ × 3½ × 17"	2	Cedar
C	Back seat rail	1½ × 3½ × 35½"	1	Cedar
D	Leg	1½ × 3½ × 16¼"	4	Cedar
E	Stretcher	1½ × 3½ × 13⅞"	2	Cedar
F	Backrest post	1½ × 3½ × 17"	2	Cedar

Cutting List

Key	Part	Dimension	Pcs.	Material
G	Top rest	1½ × 3½ × 42"	1	Cedar
H	Bottom rest	1½ × 3½ × 40"	1	Cedar
I	Cleat	1½ × 7¼ × 6"	2	Cedar
J	Stop	1½ × 7¼ × 2"	2	Cedar
K	Slat	⅞ × 5½ × 20"	7	Cedar

Materials: Moisture-resistant glue, ⅜ × 4" carriage bolts (6) with washers and wing nuts, 1¼", 2" and 2½" deck screws.

Note: Measurements reflect the actual size of dimension lumber.

Directions: Fold-up Lawn Seat

MAKE THE LEGS.
The lawn seat is supported by two H-shaped legs that fold up inside the seat.
1. Cut the legs (D) to length. Mark a point 1¾" in from one end of each leg. Make sure the point is centered on the leg.
2. Set the point of a compass on each point and draw a 1¾"-radius semicircle to make a cutting line for a roundover at the top of each leg. Cut the roundovers with a jig saw. Then, drill a ⅜"-dia. pilot hole through each point.
3. At the other end of each leg, mark a centerpoint measured from side to side. Measure in ¾" from the end along one edge, and mark another point. Connect the points with a straight line, and cut along the line with a jig saw to create the flat leg bottoms.
4. Mark 1¾"-radius roundovers at the opposite edges of the leg bottoms, using the approach from Steps 1 and 2. Cut the roundovers with a jig saw.
5. Measure 5½" from the flat end of each leg and drill two pilot holes, 1" from the edge. Counterbore the holes to ¼" depth, using a counterbore bit.
6. Cut the stretchers (E) to length. Attach one stretcher between each pair of legs so the bottoms of the stretchers are 4" from the bottoms of the legs **(photo A).** Attach the legs with glue and drive 2½" deck screws through the pilot holes. Check that the flat ends of the legs are at the same end.

A

Fasten the stretchers between the legs with glue and deck screws.

B

Smooth out the post notches with a wood file.

MAKE THE BACKREST POSTS.
Two posts are notched to hold the two boards that form the backrest.
1. Cut the backrest posts (F) to size. On one edge of each post, mark points 6½", 10" and 13½" from the end of the post. Draw a line lengthwise on each post, 1½" in from the edge with the marks. Extend lines from each point across the lengthwise line. These are the outlines for the notches in the posts (see *Diagram,* page 97). Use a jig saw to cut the notches, then file or sand the cuts smooth **(photo B).**
2. Use a compass to draw a semicircle with a 1¾" radius at the bottom of each post. Measure 1¾" from the bottoms, and mark drilling points centered side to side. Drill a ⅜"-dia. pilot hole at each point.
3. Mark 1" tapers on the back edges of the posts (see *Part F Detail,* page 97). Cut the tapers with a jig saw. Then, sand the posts smooth. Make sure to sand away any sharp edges.

ASSEMBLE THE BACKREST.
1. Cut the top rest (G) and bottom rest (H). Mark trim lines at the ends, starting ½" in from the ends on one edge and tapering to the opposite corner. Check that the trim lines on each end are symmetrical. Trim the ends with a jig saw.
2. Position the posts on their back (tapered) edges, and insert the top and bottom rests into their notches. Position the posts 32½" apart. Center the rests on the posts. The overhang should be equal on each rest. Then, attach the rests to the posts with glue and 2" deck screws **(photo C).**

BUILD THE SEAT FRAME.
The seat frame is made by attaching two side rails between a front rail and back rail. The front rail is tapered to match the backrest.
1. Cut the front seat rail (A), side seat rails (B) and back seat rail (C) to length. Sand the parts smooth.
2. Drill three evenly spaced ⅛" pilot holes, 4" in from each end of the front rail to attach the side rails. Counterbore the holes, using a ¼" drill bit. Make a ½" taper cut at each end of the front rail.
3. Drill centered, ⅜"-dia. holes, 7" in from each end for the leg assemblies. Also drill ⅜"-dia. pilot holes for carriage bolts through the back rail, centered 3¾" in from each end.
4. Apply moisture-resistant glue to one end of each side rail, and position the side rails against the front rails. Fasten the side rails to the back of the front rail by driving deck screws through the pilot holes in the front rail and into the ends of the side rails.
5. Fasten the back rail to the free ends of the side rails with glue and screws. Check that the ends are flush. Then, sand

Center the top and bottom rests in the post notches, and fasten them with glue and deck screws.

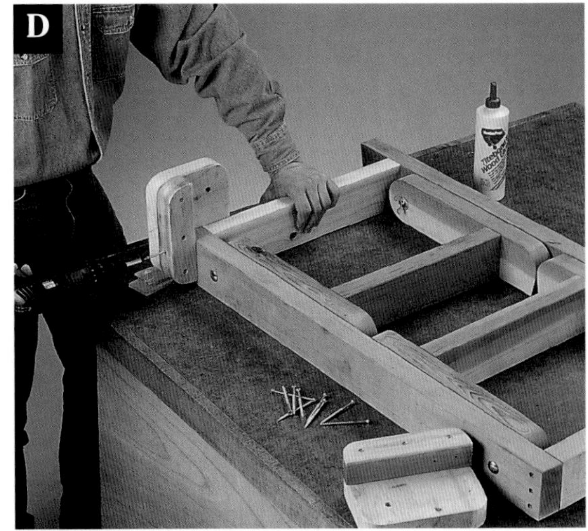

Attach the cleats and stops to the rear edges of the seat frame.

the frame to round the bottom outside edges.

JOIN THE LEGS AND SEAT FRAME.

1. Position the leg assemblies inside the seat frame. Make certain the rounded corners face the ends of the frame.
2. Apply paste wax to four carriage bolts. Align the pilot holes in the legs and seat frame, and attach the parts with the carriage bolts (see *Diagram,* page 97).

ATTACH THE CLEATS AND STOP.

The cleats (I) and stops (J) are attached to each other on the back corner of the seat frame to provide an anchor for the backrest. Once the cleats and stops are attached, carriage bolts are driven through the cleats and into the posts on the backrest. The stops fit flush with the back edges of the cleats to prevent the backrest from folding all the way over.
1. Cut the cleats and stops to size. Then, position a stop against a cleat face, flush with one long edge. The top and bottom edges must be flush. Attach the stop to the cleat with glue and 2½" deck screws.
2. Drill a ⅜"-dia. hole through each cleat, centered 1¾" in from the front and top edges.
3. Smooth the edges of the cleats and stops with a sander. Attach them to the rear corners of the seat frame with glue and 2½" deck screws **(photo D).** Make certain the bottom edges of the cleats and stops are ½" above the bottom of the frame.

ATTACH THE SEAT SLATS.

The seat slats are all the same length, but the end slats are tapered from front to back.
1. Cut the slats (K). Then, plot a 1"-grid on two of the slats (see *Part K Detail,* page 97).
2. Draw cutting lines at the edges of the two slats (see *Part K Detail*). NOTE: The taper straightens 4" from the back of the slat. Cut the tapers with a circular saw or jig saw. Smooth the edges with a router and roundover bit, or a sander.
3. Attach the slats to the seat frame, using glue and 1¼" screws. Make sure the wide ends of the end slats are flush with the ends of the frame, and the back ends of all slats are flush with the back edge of the frame. The gaps between slats should be equal.

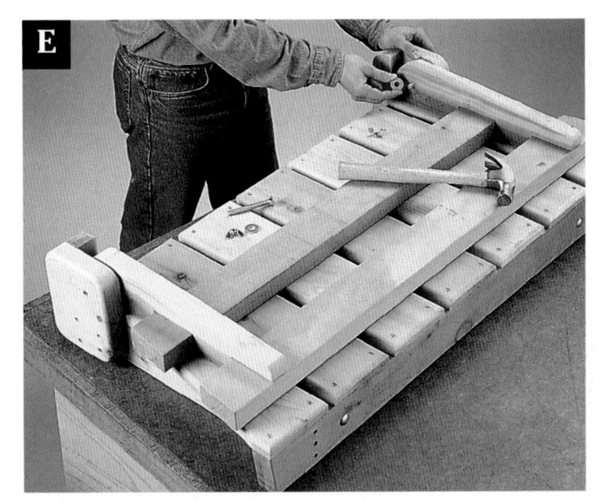

Attach the backrest to the seat frame with carriage bolts and wing nuts.

ASSEMBLE THE LAWN SEAT.

1. Finish all of the parts with an exterior wood stain.
2. Fit the backrest assembly between the cleats. Align the holes in the posts and cleats, and insert the bolts.
3. Place washers and wing nuts on the ends of the bolts to secure the backrest to the seat frame **(photo E).** Hand-tighten the wing nuts to lock the backrest and legs in position. Loosen the wing nuts when you want to fold the lawn seat for transport or storage.

PROJECT POWER TOOLS

Sun Lounger

Designed for the dedicated sun worshipper, this sun lounger has a backrest that can be set in either a flat or an upright position.

Construction Materials

Quantity	Lumber
3	2 × 2" × 8' pine
1	2 × 4" × 8' pine
5	2 × 4" × 10' pine
2	2 × 6" × 10' pine

Leave your thin beach towel and flimsy plastic chaise lounge behind, as you relax and soak up the sun in this solid wood sun lounger. Set the adjustable backrest in an upright position while you make your way through your summer reading list. Then, for a change of pace, set the backrest in the flat position and drift off in a pleasant reverie. If you're an ambitious suntanner, take comfort in the fact that this sun lounger is lightweight enough that it can be moved easily to follow the path of direct sunlight. Made almost entirely from inexpensive pine or cedar, this sun lounger can be built for only a few dollars—plus a little sweat equity.

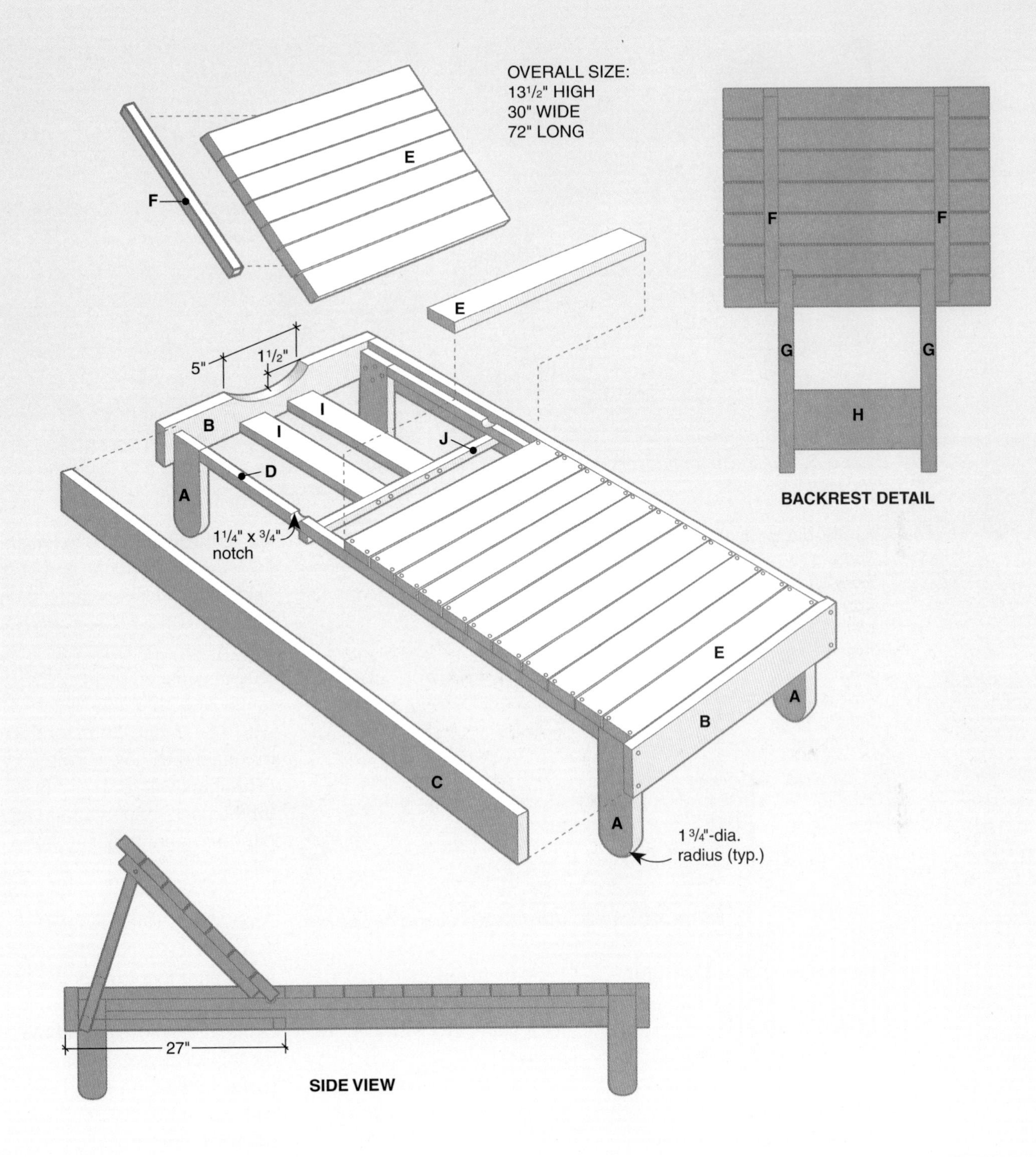

Cutting List

Key	Part	Dimension	Pcs.	Material
A	Leg	1½ × 3½ × 12"	4	Pine
B	Frame end	1½ × 5½ × 30"	2	Pine
C	Frame side	1½ × 5½ × 69"	2	Pine
D	Ledger	1½ × 1½ × 62"	2	Pine
E	Slat	1½ × 3½ × 27"	19	Pine

Cutting List

Key	Part	Dimension	Pcs.	Material
F	Back brace	1½ × 1½ × 22"	2	Pine
G	Back support	1½ × 1½ × 20"	2	Pine
H	Cross brace	1½ × 5½ × 13"	1	Pine
I	Slide support	1½ × 3½ × 24"	2	Pine
J	Slide brace	1½ × 1½ × 27"	1	Pine

Materials: Moisture-resistant wood glue, 2½" deck screws, ¼"-dia. × 3½" carriage bolts (2) with washers and nuts.

Note: Measurements reflect the actual size of dimension lumber.

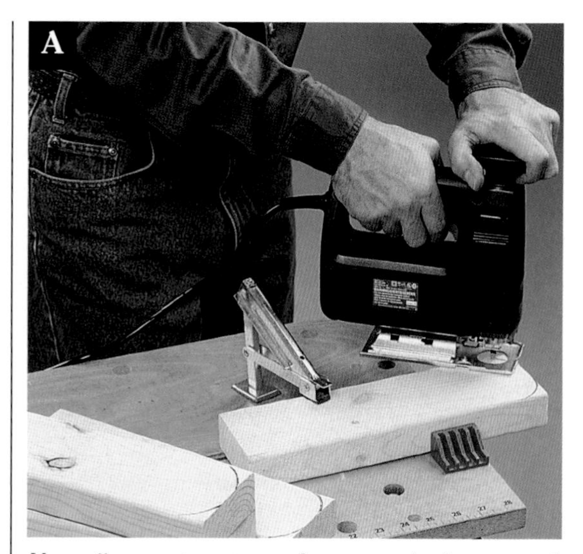

Use a jig saw to cut roundovers on the bottoms of the legs.

Assemble the frame pieces and legs, then add the support boards for the slats and backrest.

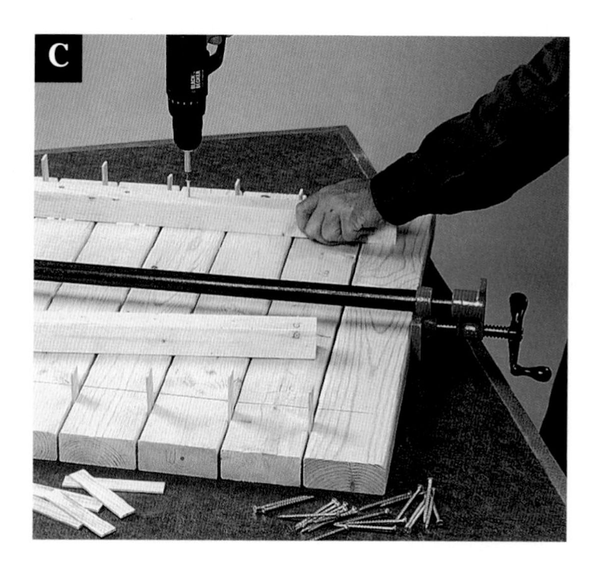

Use ⅛"-thick spacers to keep an even gap between slats as you fasten them to the back braces and the ledgers in the bed frame.

Directions: Sun Lounger

MAKE THE LEGS.

The rounded leg bottoms help the sun lounger rest firmly on uneven surfaces.

1. Cut the legs (A) to length. To ensure uniform length, cut four 2 × 4s to about 13" in length. Clamp them together edge to edge and gang-cut them to final length (12") with a circular saw.

2. Use a compass to scribe a 3½"-radius roundoff cut at the bottom corners of each leg. Make the roundoff cuts with a jig saw **(photo A).** Sand smooth.

CUT THE FRAME PIECES AND LEDGERS.

1. Cut the frame ends (B) and frame sides (C) to length. Use a jig saw to cut a 5"-wide, 1½"-deep notch into the top edge of one frame end, centered end to end, to create a handgrip.

2. Cut the ledgers (D) to length. Measure 24" from one end of each ledger. Place a mark, then cut a 1¼"-wide, ¾"-deep notch into the top edge of each ledger, centered on the 24" mark. Smooth out the notch with a 1½"-radius drum sander mounted on a power drill. (This mark will serve as a pivot for the back support.) Sand all parts and smooth out all sharp edges.

ASSEMBLE THE FRAME.

1. Attach the frame sides and frame ends to form a box around the legs, with the tops of the frame pieces 1½" above the tops of the legs to leave space for the 2 × 4 slats. Fasten with glue and drive 2½" deck screws through the legs and into the frame sides. Also drive screws through the frame ends and into the legs.

2. Attach the ledgers to the frame sides, fitted between the legs, using glue and 2½" deck screws. Make sure the ledger tops are flush with the tops of the legs and the notches are at the same end as the notch in the frame.

TIP

For a better appearance, always keep the screws aligned. In some cases, you may want to add some screws for purely decorative purposes: in this project, we drove 1" deck screws into the backrest slats to continue the lines created by the screw heads in the lower lounge slats.

CUT AND INSTALL THE BACKREST SUPPORTS.

1. Cut the slide brace (J) to length. Position the slide brace between the frame sides, 24" from the notched frame end, fitted against the bottom edges of the ledgers. Glue and screw the slide brace to the bottom edges of the ledgers.

2. Cut the slide supports (I) to length. Position the supports so they are about 3" apart, centered below the notch in the frame end. The ends of the supports should fit neatly against the frame end and the slide brace. Attach with glue and drive screws through the frame end and the slide brace, and into the ends of the slide supports **(photo B).**

FASTEN THE SLATS.

1. Cut all the slats (E) to length. Use a straightedge guide to ensure straight cuts (the ends will be highly visible). Or, simply hold a speed square against the edges of the boards and run your circular saw along the edge of the speed square.

2. Cut the back braces (F) to length. Lay seven of the slats on a flat work surface, and slip ⅛"-wide spacers between the slats. With the ends of the slats flush, set the back braces onto the faces of the slats, 4" in from the ends. Drive a 2½" deck screw through the brace and into each slat **(photo C).**

3. Install the remaining slats in the lounge frame, spaced ⅛" apart, by driving two screws through each slat end and into the tops of the ledgers. One end slat should be ⅛" from the inside of the uncut frame end, and the other 27" from the outside of the notched frame end.

ASSEMBLE THE BACKREST SUPPORT FRAMEWORK.

The adjustable backrest is held in place by a small framework attached to the back braces. The framework can either be laid flat so the backrest lies flat, or raised up and fitted against the inside of the notched frame end to support the upright backrest.

1. Cut the back supports (G) to length. Clamp the pieces together face to face, with the ends flush. Clamp a belt sander to your work surface, and use it as a grinder to round off the supports on one end.

2. Cut the cross brace (H) to length. Position the cross brace between the back supports, 2" from the non-rounded ends. Attach with glue and drive 2½" deck screws through the supports and into the cross brace.

3. Position the rounded ends of the supports so they fit between the ends of the back braces, overlapping by 2½" when laid flat. Drill a ¼" guide hole through the braces and the supports at each overlap joint.

4. Thread ¼"-dia. × 3½"-long carriage bolts through the guide holes, with a flat washer between each support and brace. Hand-tighten a washer and nylon locking nut onto each bolt end (see **photo D).**

INSTALL THE BACKREST.

1. Set the backrest onto the ledger boards near the notched end of the frame.

2. With the backrest raised, tighten the locking nut on the backrest support framework until it holds the framework together securely while still allowing the joint to pivot **(photo D).**

APPLY FINISHING TOUCHES.

Sand all surfaces and edges **(photo E)** to eliminate slivers. Apply two coats of water-based, exterior polyurethane for a smooth, protective finish. Or, use a primer and a light-colored exterior paint.

Use a washer and nylon locking nut to fasten the back braces to the back supports.

Sand all surfaces carefully to eliminate splinters, and check to make sure all screw heads are set below the wood surface.

PROJECT
POWER TOOLS

Tree Surround

Turn wasted space beneath a mature tree into a shady seating area.

Construction Materials

Quantity	Lumber
11	2 × 4" × 8' cedar
2	1 × 6" × 8' cedar
24	1 × 4" × 8' cedar

This tree surround with built-in benches provides ample seating in your yard, while protecting the base of the tree trunk. Situated in a naturally shady area, the surround/bench creates an ideal spot to relax with a good book or spend a quiet moment alone.

The tree surround can be built in four pieces in your garage or basement, then assembled on-site to wrap around the tree. As shown, the tree surround will fit a tree trunk up to 25" in diameter. But with some basic math, it's easy to adjust the sizes of the pieces so the surround fits just about any tree in your yard.

Unlike most tree bench designs, this project is essentially freestanding and does not require you to set posts (digging holes at the base of a tree can be next to impossible in some cases). And because it is cedar, it will blend right into most landscapes.

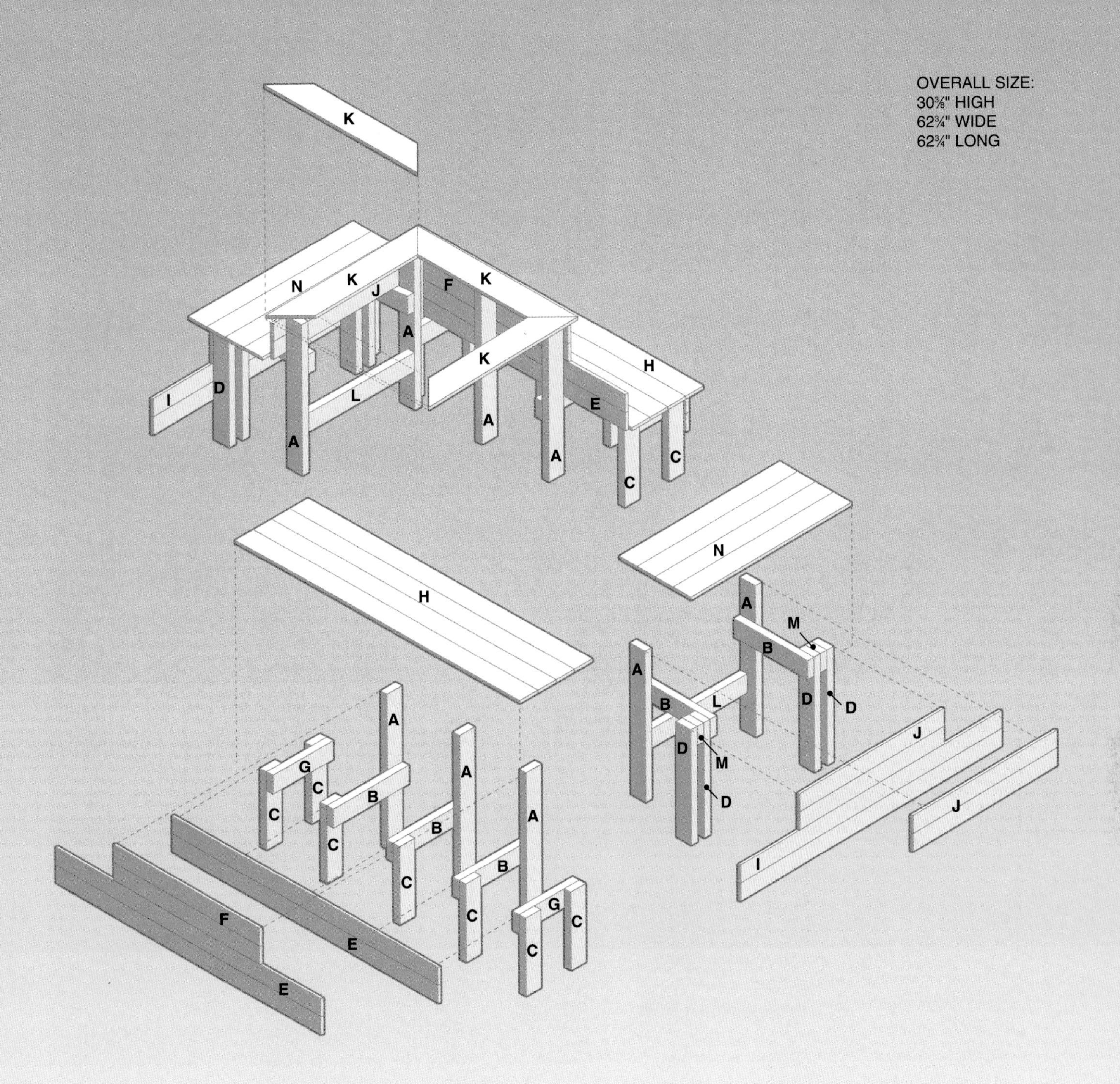

Cutting List

Key	Part	Dimension	Pcs.	Material
A	Inside post	1½ × 3½ × 29½"	10	Cedar
B	Seat rail	1½ × 3½ × 16¾"	10	Cedar
C	Short post	1½ × 3½ × 15"	14	Cedar
D	Long post	1½ × 3½ × 22¼"	8	Cedar
E	Face board	⅞ × 3½ × 60½"	8	Cedar
F	Face board	⅞ × 3½ × 34"	4	Cedar
G	Side seat rail	1½ × 3½ × 13¼"	4	Cedar

Cutting List

Key	Part	Dimension	Pcs.	Material
H	Bench slat	⅞ × 3½ × 62¾"	8	Cedar
I	Face board	⅞ × 3½ × 58¾"	4	Cedar
J	Face board	⅞ × 3½ × 32¼"	8	Cedar
K	End cap	⅞ × 5½ × 36"	4	Cedar
L	Stringer	1½ × 3½ × 22¼"	2	Cedar
M	Nailer	1½ × 3½ × 3½"	4	Cedar
N	Bench slat	⅞ × 3½ × 36¼"	8	Cedar

Materials: Moisture-resistant glue, 1½" and 2½" deck screws, finishing materials.

Note: Measurements reflect the actual size of dimension lumber.

Directions: Tree Surround

BUILD THE SHORT BENCH FRAMES.

The tree surround is built as two short benches on the sides, and two taller benches on the ends. The benches are joined together to wrap around the tree. Drill a ⅛" pilot hole for every screw used in this project. Counterbore the holes to a ¼" depth, using a counterbore bit.

1. To build the support frames for the short benches, cut the inside posts (A), seat rails (B) and short posts (C) to length. Lay a short post on top of an inside post, with the bottom ends flush. Trace a reference line onto the face of the inside post, following the top of the short post.

2. Separate the posts. Lay a seat rail across the faces of the two posts so it is flush with the outside edge and top of the short post, and just below the reference line on the inside post. Use a square to make sure the seat rails are perpendicular to the posts and their ends are flush with the post edges. Join the pieces with moisture-resistant glue. Drive 2½" deck screws through the seat rails and into the short posts and inside posts. Make six of these assemblies **(photo A).**

3. Cut the four side seat rails (G) to length. Attach them to pairs of short posts so the tops and ends are flush.

TIP

Leave room for the tree to grow in trunk diameter when you build and install a tree surround. Allow at least 3" between the tree and the surround on all sides. Adjust the dimensions of your tree surround, if needed, to create the additional space.

Seat rails are attached to the short posts and inside posts to make the bench frames.

The face boards attached at the fronts of the short posts on the short benches should extend ⅞" past the edges of the posts.

ATTACH THE SHORT BENCH FACE BOARDS.

1. Cut the face boards (E) to length for the fronts of the short benches. Draw lines on the inside faces of these face boards, ⅞" and 14⅞" from each end, and at their centers. These reference lines will serve as guides when you attach the face boards to the short bench frames.

2. Lay the two frames made with two short posts on your work surface, with the back edges of the back posts down. Attach a face board to the front edges of the front posts, with 1½" deck screws, so the ends of the face board extend ⅞" be-

Attach face boards to the inside posts to create the backrest. The lowest board should be ⅛" above the seat rails.

yond the outside edges of the frames (the seat rail should be on the inside of the frame). Attach another face board ⅛" below the top face board, making sure the reference lines are aligned **(photo B).**

ASSEMBLE THE SHORT BENCHES.

1. Stand the frame and face board assemblies on their feet. Fit the short bench frames made with the inside posts against the inside faces of the face boards. Center the short posts of the frames on the reference lines drawn on the face boards. Attach these frames to the face boards with 1½" deck screws.

2. Set another face board at the backs of the seat rails, against the inside posts. Slip a 10d finish nail under the face board where it crosses each seat rail to create a ⅛" gap. Make sure the ends of the face board extend ⅞" beyond the edges of the end frames. Attach the face board to the inside posts with 1½" deck screws **(photo C).** Attach another face board ⅛" up on the inside posts.

3. Cut the face boards (F) to length. Fasten two of these shorter face boards to each bench assembly so the ends overhang the inside posts by ⅞". Maintain a ⅛" gap between the face boards. The top edge of the highest face board on each bench assembly should be flush with the tops of the inside posts.

4. Cut the bench slats (H) to length. Position the front bench slat so it overhangs the front of the face board below it by 1⅛" and both ends of the face board by 1⅛" **(photo D).** Attach the front slat by driving two 1½" deck screws through the slat and into each seat rail. Fasten the back seat slat so it butts against the inside posts. Attach the remaining bench slats so the spaces between the slats are even.

TIP

Cover the ground at the base of a tree with a layer of landscaping stone or wood bark before you install a tree surround. To prevent weeds from growing up through the ground cover, lay landscaping fabric in the area first. Add a border of landscape edging to keep everything contained. If the ground at the base of the tree is not level, you can make installation of the tree surround easier by laying a base of landscaping rock, then raking it and tamping it until it is level.

Measure to make sure the front bench slat overhangs the face board below it by 1⅛".

MAKE THE TALL BENCHES. The two tall benches are built much like the short benches, but with doubled posts at the front, for extra strength, and a stringer to support the frames.

1. Cut the long posts (D) and four nailers (M) to length. Arrange the long posts in pairs, with nailers in between at the tops. Fasten the doubled posts and nailers together with glue and 2½" deck screws, making sure the nailers are aligned with the fronts and tops of the posts.

2. Attach a seat rail to the doubled posts **(photo E).** Then, attach the free end of each seat rail to an inside post, as you did for the short benches.

3. Cut the stringers (L) to length. Position a stringer between each pair of inside posts, flush with the back edges and 8" up from the bottoms of the posts. Attach them with glue and 2½" deck screws driven through the inside posts and into the ends of the stringers.

4. Cut the face boards for the tall benches (I, J) to length. Use 10d finish nails to leave ⅛" gaps between the face boards, as before, including the gap above the back ends of the seat rails. Use 1½" deck screws to attach two of the shorter boards (J) to the long posts so the top board is flush with the tops of the posts and seat rails, and the ends overhang the outside edges of the doubled posts by ½" **(photo F).**

5. Attach the longer face boards (I) below the shorter face boards, so they overhang the doubled posts by the same amount on each end **(photo G).** The overhang portions will cover the sides of the short bench frames after assembly. Attach two of the shorter face boards (J) to the front edges of the inside posts so their ends overhang the outside faces of the posts by 3½".

6. Cut the bench slats for the tall benches (N) to length. Position the slats on the seat rails. Fasten the front slat so it overhangs the front of the face board below it by 1⅛" and the ends of the face board by 2". Fasten a slat flush with the

After making doubled posts for the tall benches, attach the seat rails.

The shorter face boards for the tall benches are attached so the ends are flush with the outsides of the doubled posts.

The longer face boards attached to the tall benches overhang the doubled posts so they cover the sides of the short bench frames when the tree surround is assembled.

back of the bench. Attach the remaining slats on each tall bench so the spaces between the slats are even.

APPLY THE FINISH.
Now is a good time to apply a finish to the benches. Sand all the surfaces smooth and wipe the wood clean. Apply at least two coats of exterior wood stain to protect the wood.

ASSEMBLE THE TREE SURROUND.
1. If necessary, prepare the ground around the tree where the tree surround will stand (see *Tip Box*, page 107). When the ground is roughly level, you can assemble the tree surround and shim beneath the posts to level it.
2. Set all four benches around the tree so the overhang on the tall bench face boards covers the end frames of the short benches. The ends of the face boards should butt against the backs of the face boards on the short benches. Clamp or tack the benches together. Don't fasten the pieces together until you've made adjustments to level the tree surround.
3. Use a carpenter's level to check the tree surround. Set the level on each of the benches to determine whether adjustments are needed. For shims, use flat stones, such as flagstone, or prefabricated concrete pavers. If you don't want to use shims, mark the spots on the ground that need raising or lowering, and separate the benches to make the required adjustments.
4. When the tree surround is level and the benches fit together squarely, attach the tall benches to the short benches by driving 2½" deck screws through the face boards on the tall benches and into the posts on the short benches.

ATTACH THE CAP.
1. Cut the end caps (K) to length. Draw 45° miter lines at each end of one cap, with both miter lines pointing inward. Make the miter cuts with a circular saw **(photo H)** or—even better—a power miter box.
2. Tack or clamp the end cap in place. Mark and cut the three remaining end caps one at a time to ensure even joints. Attach the caps with 1½" deck screws driven through the caps and into the ends of the inside posts. Sand the parts smooth and apply the same finish to the caps that you applied to the tree surround benches.

The 1 × 6 caps are mitered to make a square frame around the top of the tree surround after it is assembled around your tree.

PROJECT
POWER TOOLS

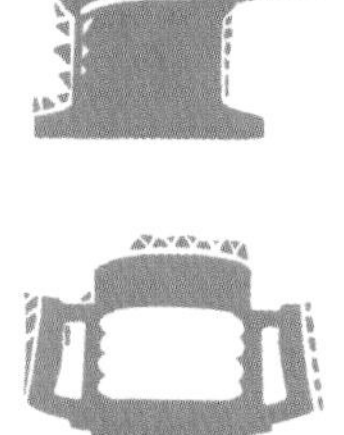

Park Bench

This attention-grabbing park bench is a real showpiece that can transform even a plain yard into a formal garden.

CONSTRUCTION MATERIALS

Quantity	Lumber
5	2 × 4" × 8' pine
1	2 × 2" × 4' pine
4	1 × 6" × 8' pine

Add color and style to your backyard or garden with this bright, elegant park bench. Some careful jig saw work is all it takes to make the numerous curves and contours that give this bench a sophisticated look. But don't worry if your cuts aren't all perfect—the shapes are decorative, so the bench will still work just fine. In fact, if you prefer a simpler appearance, you can build the park bench with all straight parts, except the roundovers at the bottoms of the legs. But if you are willing to do the extra work, you're sure to be pleased with the final result. You may even want to finish it with bright red paint so no one will miss it.

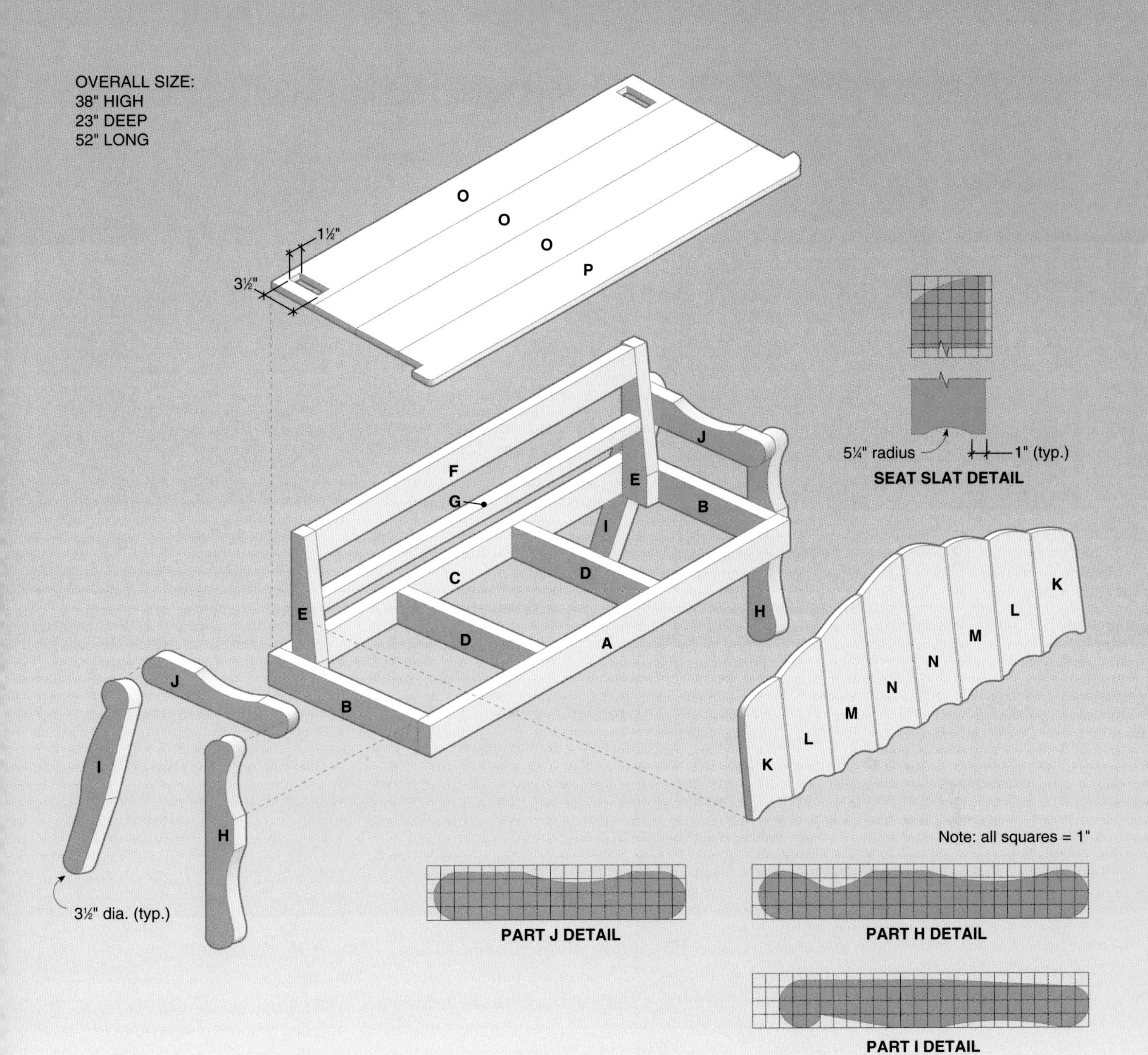

Cutting List

Key	Part	Dimension	Pcs.	Material
A	Front rail	1½ × 3½ × 49"	1	Pine
B	Side rail	1½ × 3½ × 20¼"	2	Pine
C	Back rail	1½ × 3½ × 46"	1	Pine
D	Cross rail	1½ × 3½ × 18¾"	2	Pine
E	Post	1½ × 3½ × 18"	2	Pine
F	Top rail	1½ × 3½ × 43"	1	Pine
G	Bottom rail	1½ × 1½ × 43"	1	Pine
H	Front leg	1½ × 3½ × 24½"	2	Pine

Cutting List

Key	Part	Dimension	Pcs.	Material
I	Rear leg	1½ × 3½ × 23"	2	Pine
J	Armrest	1½ × 3½ × 18½"	2	Pine
K	End slat	¾ × 5½ × 14"	2	Pine
L	Outside slat	¾ × 5½ × 16"	2	Pine
M	Inside slat	¾ × 5½ × 18"	2	Pine
N	Center slat	¾ × 5½ × 20"	2	Pine
O	Seat slat	¾ × 5½ × 49"	3	Pine
P	Seat nose slat	¾ × 5½ × 52"	1	Pine

Materials: Moisture-resistant glue, 1¼", 2½" deck screws, finishing materials.

Note: All measurements reflect the actual size of dimension lumber.

Use a router or sander to round over the sharp bottom edges and corners of the completed seat frame.

Attach the seat slats and nose slat to the top of the seat frame with glue and deck screws.

Directions:
Park Bench

BUILD THE SEAT FRAME.
The seat frame is made by assembling rails and cross rails to form a rectangular unit.
1. Cut the front rail (A), side rails (B), back rail (C) and cross rails (D) to length. Sand rough spots with medium-grit sandpaper.
2. Fasten the side rails to the front rail with moisture-resistant glue. Drill ⅛" pilot holes in the front rail. Counterbore the holes to accept ⅜"-dia. wood plugs. Drive 2½" deck screws through the front rail and into the side rail ends. Make sure the top and bottom edges of the side rails are flush with the top and bottom edges of the front rail.
3. Attach the back rail between the side rails. Drill pilot holes in the side rails. Counterbore the holes. Fasten with glue and drive screws through the side rails and into the ends of the back rail. Keep the back rail flush with the ends of the side rails.
4. Use glue and deck screws to fasten the two cross rails between the front and back rails, 14½" in from the inside face of each side rail.
5. Complete the seat frame by rounding the bottom edges and corners with a router and a ⅜"-dia. roundover bit **(photo A)** or a hand sander.

MAKE THE SEAT SLATS.
The seat nose slat has side cutouts to accept the front legs. The back seat slat has cutouts, called mortises, to accept the posts that support the backrest.
1. Cut the seat nose slat (P) and one seat slat (O) to length. To mark the 2 × 4 cutouts, use the end of a 2 × 4 as a template.
2. Position the 2 × 4 on the seat slat at each end, 1½" in from the back edge and 1½" in from the end. The long sides of the 2 × 4 should be parallel to the ends of the back seat slat. Trace the outline of the 2 × 4 onto the slat. Drill a starter hole within the outline on the back seat slat. Make the cutout with a jig saw.
3. Use a jig saw to cut a 3"-long × 1½"-wide notch at each end of the nose slat, starting at the back edge (see *Diagram,* page 111). Sand the notches and mortises with a file or a thin sanding block. Use a router with roundover bit to shape the front edge of the nose slat.

ATTACH THE SEAT SLATS.
1. Cut the rest of the seat slats (O) to length. Lay the slats on the seat frame so the ends of the slats are flush with the frame, and the nose slat overhangs equally at the sides of the frame.
2. Draw reference lines

TIP

Making smooth contour cuts with a jig saw can be a little tricky. To make it easier, install fairly thick saw blades, because they are less likely to "wander" with the grain of the wood. Using a scrolling jig saw will also help, since they are easier to turn than standard jig saws.

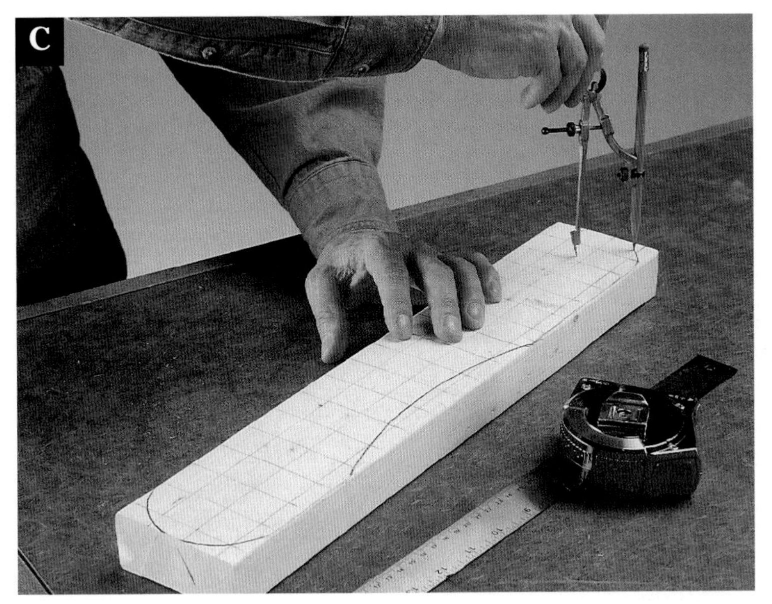
After drawing a 1" grid on the legs and armrests, draw the finished shape of the parts, following the Grid Patterns on page 111.

onto the tops of the seat slats and nose slat, directly over the top of each rail in the frame. Mark two drilling points on each slat on each line—on all but the front of the nose slat. Points should be ¾" in from the front and back of the slats. On the nose slat, mark drilling points 1½" in from the front of the slat. Drill pilot holes and counterbore the holes.

3. Sand the seat slats and nose slat. Attach them to the seat frame with glue. Drive 1¼" deck screws through the slats and into the frame and cross rails **(photo B).** Start with the front and back slats, and space the inner slats evenly.

MAKE THE LEGS AND ARMRESTS.

The front legs (H), rear legs (I) and armrests (J) are shaped using the grid patterns on page 111.

1. Cut workpieces for the parts to the full sizes shown in the *Cutting List.* Use a pencil to draw a 1"-square grid pattern on each workpiece.

2. Using the grid patterns as a reference, draw the shapes onto the workpieces **(photo C).** (It will help if you enlarge the patterns on a photocopier or draw them to a larger scale on a piece of graph paper first.)

3. Cut out the shapes with a jig saw. Sand the contour cuts smooth. Use a drum sander mounted in your electric drill for best results.

ATTACH THE LEGS.

The front and rear legs are attached to the armrests, flush with the front and rear ends.

1. Fasten the front legs to the outside faces of the armrests. Drill pilot holes in the legs. Counterbore the holes. Drive deck screws through the holes and into the armrests. Use a framing square to make sure the legs are perpendicular to the armrests.

2. Temporarily fasten the rear legs to the outside faces of the armrests by drilling centered pilot holes and driving a screw through each rear leg and into the armrest. The rear leg must, for now, remain adjustable.

3. Clamp the seat to the legs. The front of the edge of the seat should be 16¾" up from the bottoms of the front legs. The back of the seat should be 14¼" up from the bottoms of the rear legs. Position square wood spacers between the seat and each armrest to keep the armrest parallel to the frame.

4. Adjust the rear legs so their back edges are flush with the top corners of the side rails. The rear legs extend slightly beyond the back of the seat frame.

5. Drill pilot holes in the front and rear legs. Counterbore the holes. Drive deck screws through the front and rear legs and into the side. Drive an additional screw through each

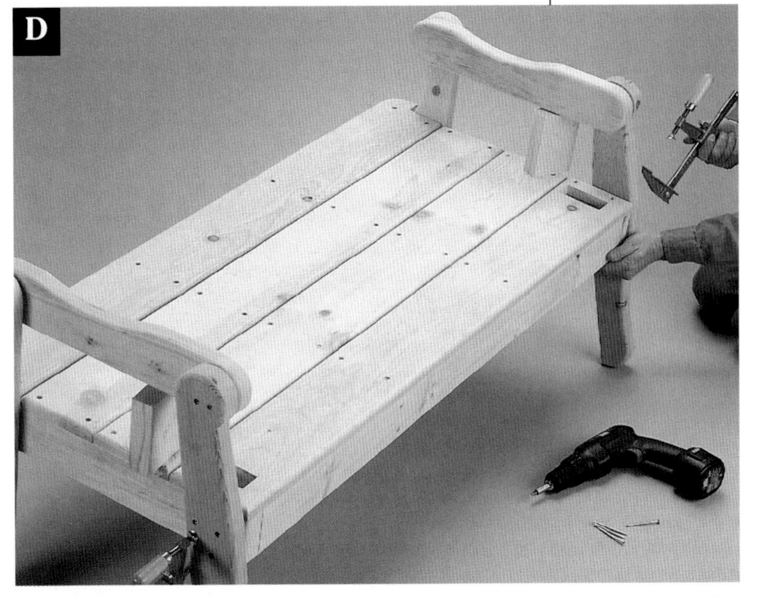
Carefully clamp the leg frames to the seat, and attach them with glue and screws, driven through centered, counterbored pilot holes.

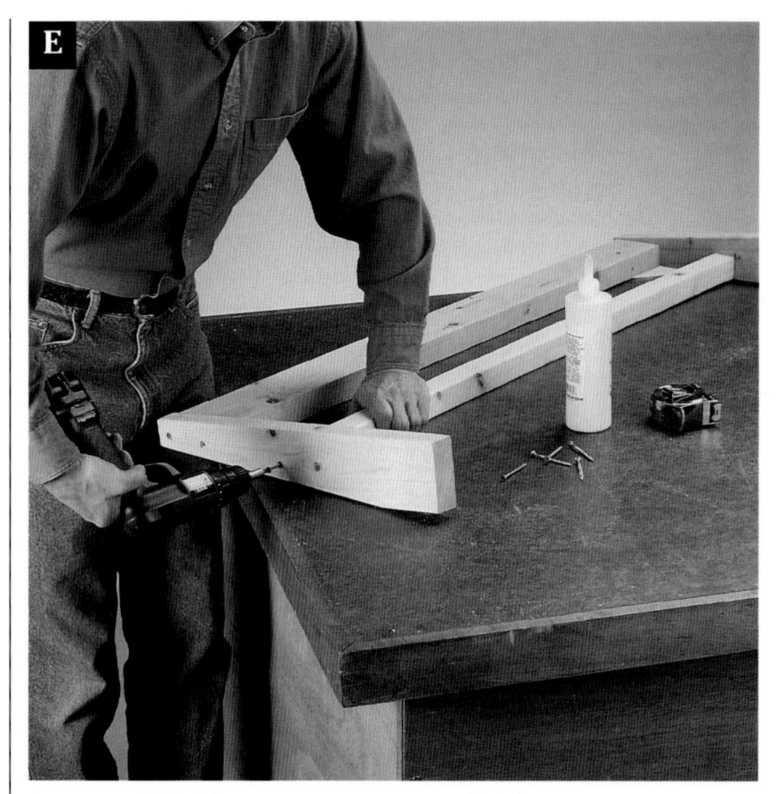

Glue the ends of the top and bottom rails, then drive deck screws through the posts to attach them to the rails.

rear leg into the armrests to secure the rear legs in position.
6. Unclamp the leg assemblies, and remove them from the frame **(photo D).** Apply glue to the leg assemblies where they join the frame and reattach the legs and armrests using the same screw holes.

BUILD THE BACK FRAME.
The back frame is made by attaching a top rail (F) and bottom rail (G) between two posts (E). Once the back frame has been built, it is inserted into the mortises in the rear seat slats. Tapers cut on the front edges of the posts will create a backward slope so the back slats make a more comfortable backrest. When you attach the rails between the posts, make sure they are flush with the front edges of the posts.
1. Cut the posts, top rail and bottom rail to length. Mark a tapered cutting line on each post, starting 1½" in from the back edge, at the top. Extend the line so it meets the front edge 3½" up from the bottom. Cut the taper in each post, using a circular saw or jig saw.
2. Use glue and deck screws to fasten the top rail between the posts so the front face of the top rail is flush with the front (tapered) edge of each post. The top front corner of the top rail should be flush with the top of the posts.
3. Position the bottom rail between the posts so its bottom edge is 9" up from the bottoms of the posts. Make sure the front face of the bottom rail is flush with the front edges of the posts. Drill pilot holes in the posts. Counterbore the holes. Attach the parts with glue and 2½" deck screws **(photo E).** Use a router with a ⅜"-dia. roundover bit or a hand sander to round over the back edges of the back frame.

MAKE THE BACK SLATS.
The back slats are shaped on their tops and bottoms to create a scalloped effect. If you'd rather not spend the time cutting these contours, you can simply cut the slats to length and round over the top edges.
1. Cut the end slats (K), outside slats (L), inside slats (M) and center slats (N) to length. Draw a 1"-square grid pattern on one slat. Then, draw the shape shown in the back slat detail on page 111 onto the slat. Use a compass to mark a 5¼"-radius scalloped cutout at the bottom of the slat.
2. Cut the slat to shape with a jig saw and sand smooth.
3. Use the completed slat as a template to trace the same profile on the tops and bottoms of the remaining slats **(photo F).** Cut them to shape with a jig saw and sand the cuts smooth.

ATTACH THE BACK SLATS.
1. Before attaching the back slats to the back frame, clamp a straight board across the fronts of the posts with its top edge 8½" up from the bottoms of the posts **(photo G).** Use this board as a guide to keep the slats aligned as you attach them.
2. Fasten the end slats to the

TIP

Depending on the location of your yard and garden furnishings, you may encounter some uneven terrain. If the ground is hard, you can install adjustable leveler glides to the bottoms of the legs to keep your projects level. If the project is resting on softer ground, the easiest solution is to use flat stones to shim under the lower sides or legs. Or, you can throw down a loose-stone base under the project, and tamp it until it is level.

Use the first completed back slat as a template for tracing cutting lines on the rest of the back slats.

Clamp a straight board to the back frame to help keep the back slats aligned along their bottom edges as you install them.

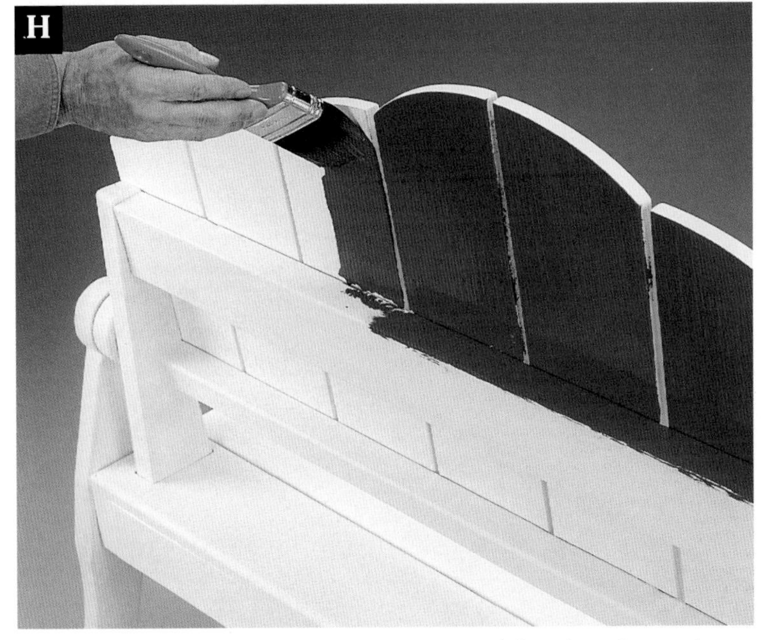

Apply two thin coats of exterior primer to seal the pine, then paint the park bench with two coats of enamel house trim paint.

back frame with glue and deck screws, making sure the bottoms are resting flat against the clamped guide board. (For more information on back slat positioning, see the *Diagram*, page 111.) Make sure the outside edges of the end slats are flush with the outside edges of the posts.

3. Attach the remaining slats between the end slats, spaced so the gaps are even.

ASSEMBLE THE BENCH.

1. Attach the rear frame by sliding the back into place inside the notches in the rear seat slat. The posts should rest against the back rail and side rails. Keep the bottoms of the posts flush with the bottom edges of the side rails.

2. Drill pilot holes in the posts and the back rail. Counterbore the holes. Drive 2½" screws through the posts and into the side rails, and through the back rail and into the posts.

APPLY FINISHING TOUCHES.

1. Apply moisture-resistant glue to ⅜"-dia. wood plugs, and insert them into each counterbored screw hole. Sand the plugs flush with the wood.

2. Sand all surfaces smooth with medium (100- or 120-grit) sandpaper. Finish-sand with fine (150- or 180-grit) sandpaper.

3. Finish as desired—try two thin coats of primer and two coats of exterior house trim paint **(photo H).** Whenever you use untreated pine for an outdoor project, be sure to use an exterior-rated finish to protect the wood.

16 ft
1
in
ft
2
3
4
5

Tables

Tables are the backbone of the outdoor entertainment scene. They hold food and beverages, display collections, and provide places for family and friends to gather. A structure that plays such a varied role should be given some thoughtful design consideration. Create a table that fills your needs—whether it be entertaining, intimate dining, or design—and you'll find your outdoor home is instantly transformed into a place of comfort and convenience.

PROJECT
POWER TOOLS

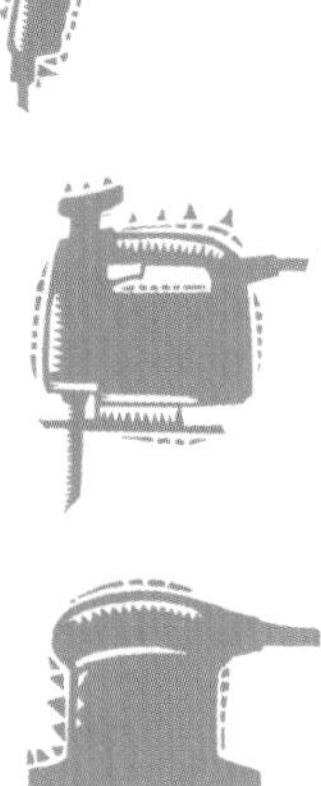

Outdoor Occasional Table

The traditional design of this deck table provides a stylishly simple addition to any porch, deck or patio.

CONSTRUCTION MATERIALS

Quantity	Lumber
2	1 × 3" × 8' cedar
6	1 × 4" × 8' cedar

Create a functional yet stylish accent for your porch, deck or patio with this cedar deck table. This table makes an ideal surface for serving cold lemonade on hot summer days, a handy place to set your plate during a family cookout, or simply a comfortable place to rest your feet after a long day. Don't be fooled by its lightweight design and streamlined features—the little table is extremely sturdy. Structural features such as middle and end stringers tie the aprons and legs together and transfer weight from the table slats to the legs. This attractive little table is easy to build and will provide many years of durable service.

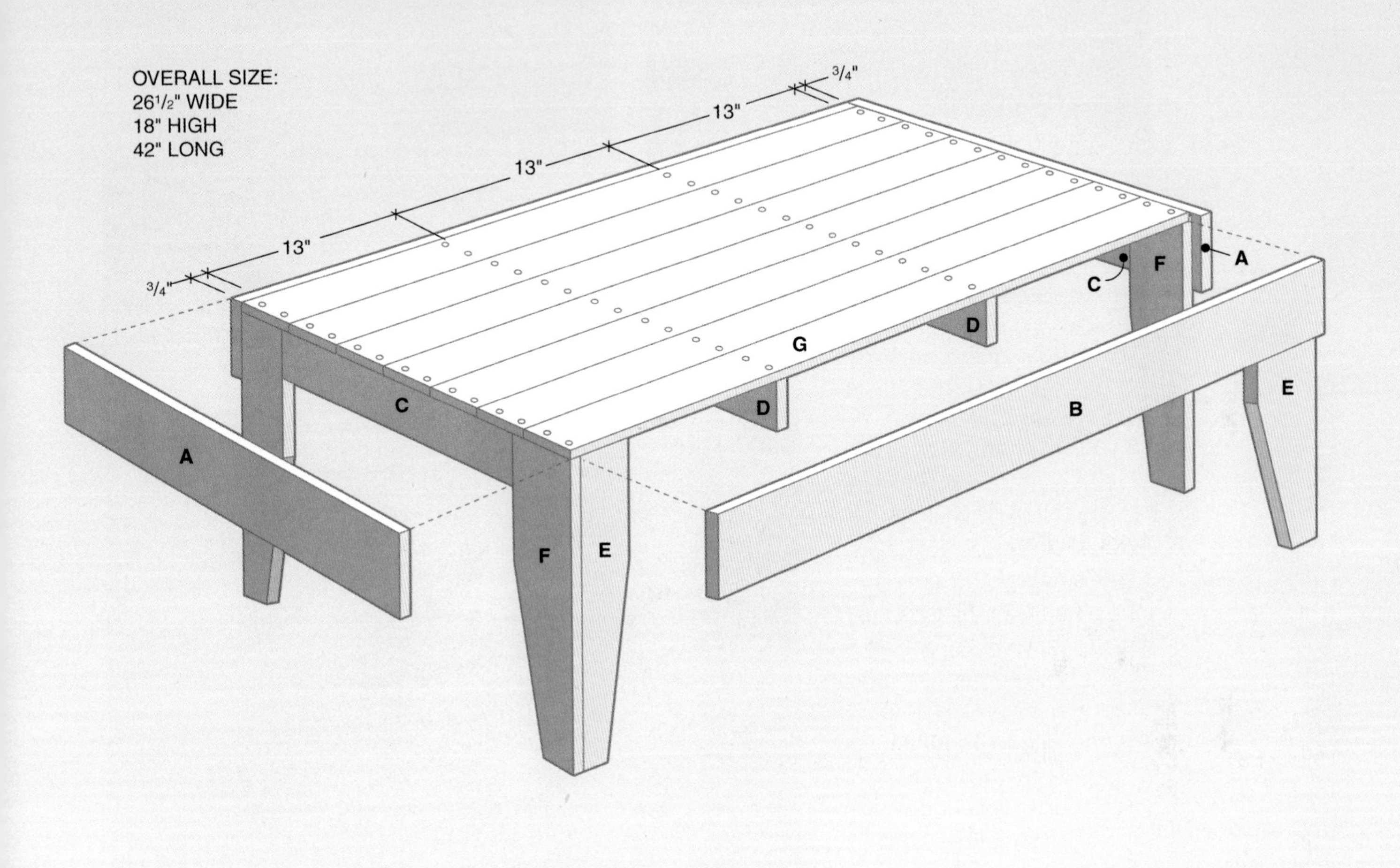

Cutting List

Key	Part	Dimension	Pcs.	Material
A	End apron	¾ × 3½ × 26½"	2	Cedar
B	Side apron	¾ × 3½ × 40½"	2	Cedar
C	End stringer	¾ × 2½ × 18"	2	Cedar
D	Middle stringer	¾ × 2½ × 25"	2	Cedar

Cutting List

Key	Part	Dimension	Pcs.	Material
E	Narrow leg side	¾ × 2½ × 17¼"	4	Cedar
F	Wide leg side	¾ × 3½ × 17¼"	4	Cedar
G	Slat	¾ × 3½ × 40½"	7	Cedar

Materials: Moisture-resistant glue, 1¼" deck screws.

Note: Measurements reflect the actual size of dimension lumber.

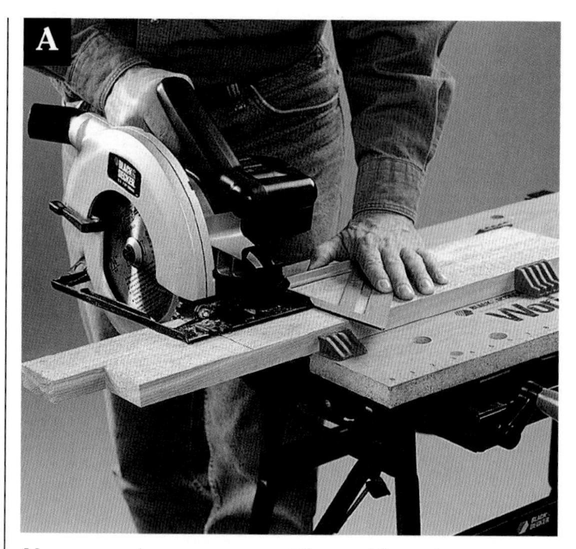

Use a speed square as a cutting guide and gang-cut the table parts when possible for uniform results.

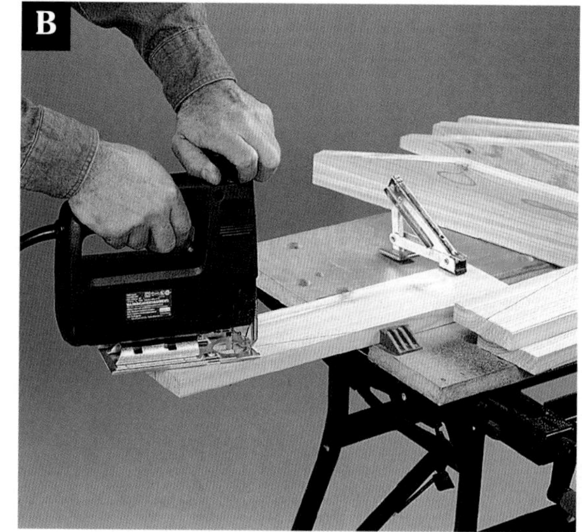

Mark the ends of the tapers on the leg sides, then connect the marks to make taper cutting lines.

Directions: Outdoor Occasional Table

MAKE THE STRINGERS AND APRONS.

1. The stringers and aprons form a frame for the tabletop slats. To make them, cut the end aprons (A) and side aprons (B) to length **(photo A).** For fast, straight cutting, use a speed square as a saw guide—the flange on the speed square hooks over the edge of the boards to hold it securely in place while you cut.

2. Cut the end stringers (C) and middle stringers (D) to length.

TIP

Rip-cut cedar 1 × 4s to 2½" in width if you are unable to find good clear cedar 1 × 3s (nominal). When rip-cutting, always use a straightedge guide for your circular saw. A straight piece of lumber clamped to your workpiece makes an adequate guide, or buy a metal straightedge guide with built-in clamps.

MAKE THE LEG PARTS.

1. Cut the narrow leg sides (E) and wide leg sides (F) to length.

2. On one wide leg side piece, measure 8¾" along one edge of the leg side and place a mark. Measure across the bottom end of the leg side 1½" and place a mark. Connect the two marks to create a cutting line for the leg taper. Mark cutting lines for the tapers on all four wide leg sides **(photo B).**

3. On the thin leg sides, measure 8¾" along an edge and ¾" across the bottom end to make endpoints for the taper cutting lines.

4. Clamp each leg side to your work surface. Cut along the taper cutoff line, using a jig saw or circular saw, to create the tapered leg sides **(photo C).** Sand all leg parts smooth.

Use a jig saw or circular saw to cut the leg tapers.

ASSEMBLE THE LEG PAIRS.

1. Apply a ½"-wide layer of moisture-resistant glue on the face of a wide leg side, next to the untapered edge. Then apply a thin layer of glue to the untapered edge of a narrow leg side. Join the leg sides together at a right angle to form a leg pair. Reinforce the joint with 1¼" deck screws.

2. Glue and screw the rest of the leg pairs in the same manner **(photo D).** Be careful not to use too much glue. Excess glue can get messy and could

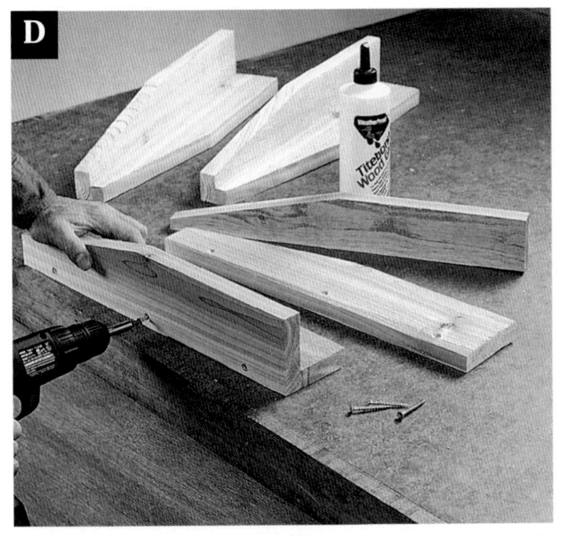

Fasten the leg pairs by driving deck screws through the face of the wide side and into the narrow edge.

Test the layout of the slats before you fasten them, adjusting as necessary to make sure gaps are even.

cause problems later if you plan to stain or clear-coat the finish.

MAKE THE TABLETOP FRAME.

1. Fasten the side aprons (B) to the leg pairs with glue and screws. Be sure to screw from the back side of the leg pair and into the side aprons so the screw heads will be concealed. The narrow leg side of each pair should be facing in toward the center of the side apron, with the outside faces of the wide leg sides flush with the ends of the side apron. The tops of the leg pairs should be ¾" down from the tops of the side aprons to create recesses for the tabletop slats.

2. Attach the end aprons (A) to the leg assemblies with glue. Drive screws from the back side of the leg pairs. Make sure the end aprons are positioned so the ends are flush with the outside faces of the side aprons.

3. Attach the end stringers (C) to the end aprons between the leg pairs with glue. Drive the screws from the back sides of the end stringers and into the end aprons.

4. Cut the middle stringers (D) to length. Measure 13" in from the inside face of each end stringer and mark reference lines on the side aprons for positioning the middle stringers.

5. Use glue to attach the middle stringers to the side aprons—centered on the reference lines. Drive deck screws through the side aprons and into the ends of the middle stringers. Make sure the middle stringers are positioned ¾" down from the tops of the side aprons.

CUT AND INSTALL THE SLATS.

1. Before you cut the slats (G), measure the inside dimension between the end aprons to be sure that the slat length is correct. Then cut the slats to length, using a circular saw and a speed square to keep the cuts square. It is extremely important to make square cuts on the ends of the slats since they're going to be the most visible cuts on the entire table.

2. Run a bead of glue along the top faces of the middle and end stringers. Screw the slats to the stringers leaving a gap of approximately 1/16" between each of the individual slats **(photo E).**

APPLY FINISHING TOUCHES.

1. Smooth all sharp edges by using a router with a roundover bit or a power sander with medium-grit (#100 to 120) sandpaper. Finish-sand the entire table and clean off the sanding residue.

2. Apply a finish, such as clear wood sealer. If you want, fill any screw counterbores with tinted wood putty.

TIP

Clamp all workpiece parts whenever possible during the assembly process. Clamping will hold glued-up and squared-up parts securely in place until you permanently fasten them with screws. Large, awkward assemblies will be more manageable with the help of a few clamps.

PROJECT POWER TOOLS

Picnic Table for Two

Turn a quiet corner of your yard into an intimate setting for dining alfresco with this compact picnic table.

CONSTRUCTION MATERIALS

Quantity	Lumber
1	2 × 8" × 6' cedar
1	2 × 6" × 8' cedar
4	2 × 4" × 8' cedar
3	1 × 6" × 8' cedar
4	1 × 2" × 8' cedar

A picnic table doesn't have to be a clumsy, uncomfortable family feeding trough. In this project, you'll create a unique picnic table that's just the right size for two people to enjoy. Portable and lightweight, it can be set in a corner of your garden, beneath a shade tree or on your deck or patio to enhance your outdoor dining experiences.

The generously proportioned tabletop can be set with full table settings for a formal meal in the garden. But it's intimate enough for sharing a cool beverage with a special person as you watch the sun set. Made with plain dimensional cedar, this picnic table for two is both sturdy and long-lasting.

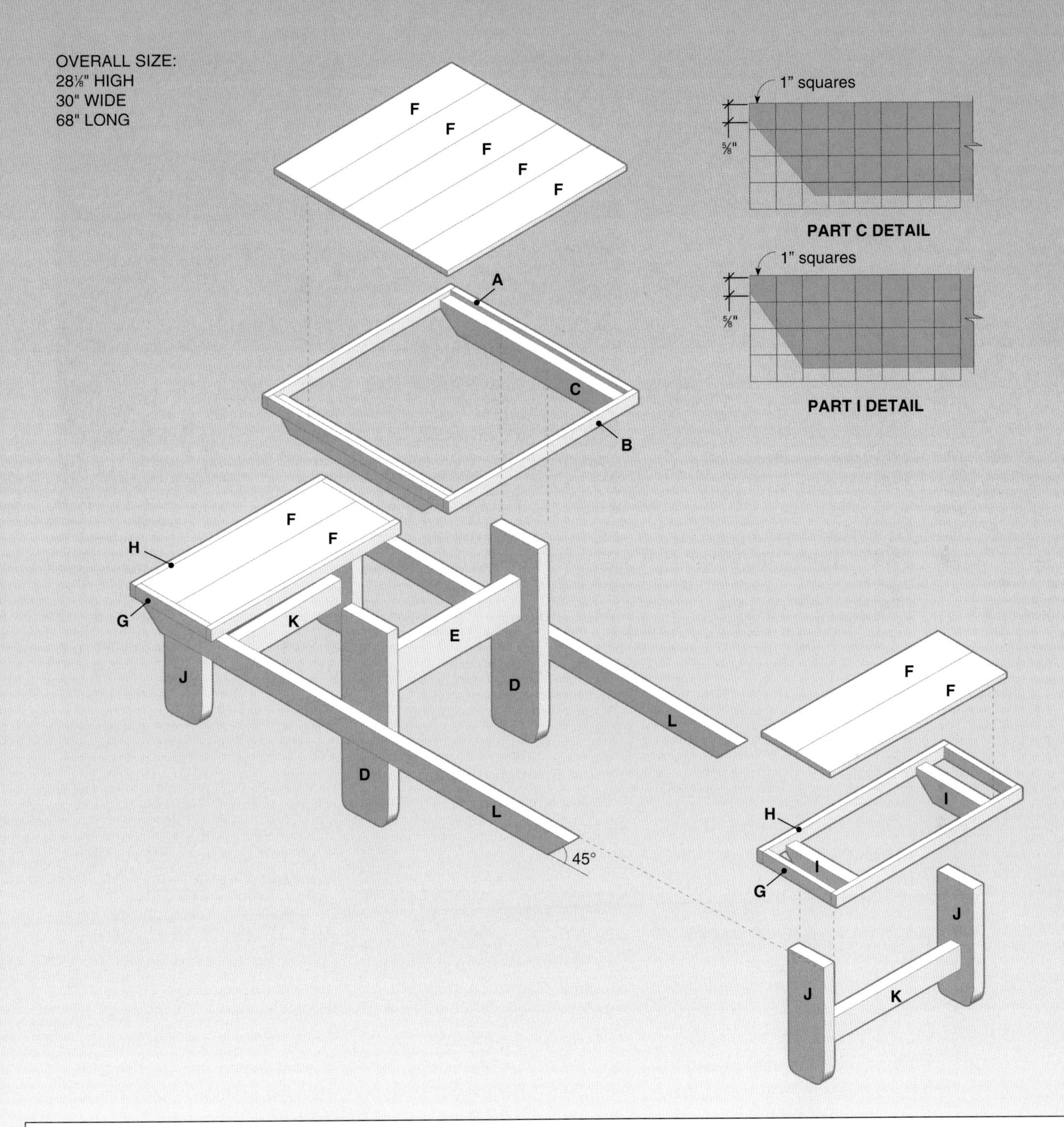

Cutting List

Key	Part	Dimension	Pcs.	Material
A	Tabletop frame	⅞ × 1½ × 27¾"	2	Cedar
B	Tabletop frame	⅞ × 1½ × 30"	2	Cedar
C	Table stringer	1½ × 3½ × 27¾"	2	Cedar
D	Table leg	1½ × 7¼ × 27¼"	2	Cedar
E	Table stretcher	1½ × 5½ × 22¼"	1	Cedar
F	Slat	⅞ × 5½ × 28¼"	9	Cedar

Cutting List

Key	Part	Dimension	Pcs.	Material
G	Bench frame	⅞ × 1½ × 11¼"	4	Cedar
H	Bench frame	⅞ × 1½ × 30"	4	Cedar
I	Bench stringer	1½ × 3½ × 11¼"	4	Cedar
J	Bench leg	1½ × 5½ × 15¼"	4	Cedar
K	Bench stretcher	1½ × 3½ × 22¼"	2	Cedar
L	Cross rail	1½ × 3½ × 68"	2	Cedar

Materials: Moisture-resistant glue, 1⅝" and 2½" brass or galvanized deck screws, finishing materials.

Note: Measurements reflect the actual size of dimension lumber.

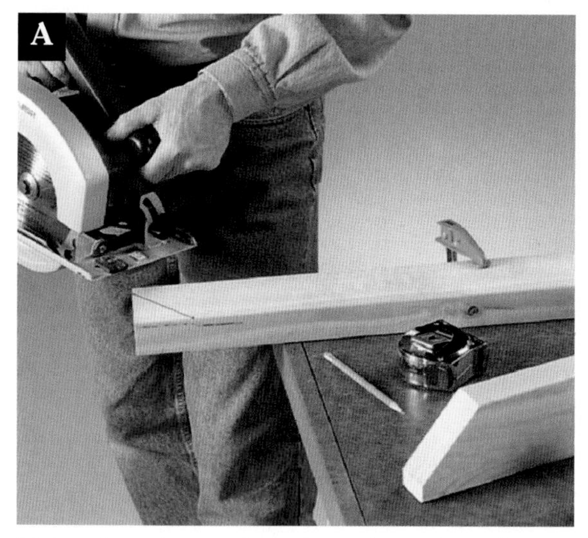

Make triangular cutoffs at the ends on the table stringers, using a circular saw.

Install the tabletop slats by driving screws through the tabletop frame and into the ends of the slats.

Directions: Picnic Table for Two

BUILD THE TABLETOP.

1. Cut the tabletop frame pieces (A, B), the table stringers (C) and the table slats (F) to length. Sand the parts.

2. Draw cutting lines that start 2½" from one end of each stringer and connect with a point at the same end, ⅝" in from the opposite edge of the board (see *Diagram*, page 123). Cut along the lines with a circular saw to make the cutoffs **(photo A).**

3. Fasten the shorter tabletop frame pieces (A) to the sides of the stringers. The tops of the frame pieces should extend ⅞" above the tops of the stringers, and the ends should be flush. First, drill ⅛" pilot holes in the frame pieces. Counterbore the holes ¼" deep, using a counterbore bit. Attach the pieces with glue and drive 1⅝" deck screws through the frame pieces and into the stringers.

4. Position the longer tabletop frame pieces (B) so they overlay the ends of the shorter frame pieces. Fasten them together with glue and 1⅝" deck screws to complete the frames.

5. Set the slats inside the frame so the ends of the slats rest on the stringers. Space the slats evenly. Drill two pilot holes through the tabletop frame by the ends of each slat. Counterbore the holes. Drive 1⅝" deck screws through the frame and into the end of each slat, starting with the two end slats **(photo B).**

MAKE AND ATTACH THE TABLE-LEG ASSEMBLY.

1. Cut the table legs (D) and table stretcher (E) to length. Use a compass to draw a 1½"-radius roundover curve on the corners of one end of each leg. Cut the curves with a jig saw.

2. Hold an end of the stretcher against the inside face of one of the table legs, 16" up from the bottom of the leg and centered side to side. Trace the outline of the stretcher onto the leg. Repeat the procedure on the other leg.

3. Drill two evenly spaced pilot holes through the stretcher outlines on the legs. Counterbore the holes on the outside faces of the legs. Attach the stretcher with glue and drive 2½" deck screws through the legs and into the ends of the stretcher.

4. Turn the tabletop upside down. Apply glue to the table stringers where they will contact the legs. Position the legs in place within the tabletop frame. Attach them by driving 2½" deck screws through the legs and into the table stringers **(photo C).**

BUILD THE BENCH TOPS.

1. Cut the bench slats (F), bench frame pieces (G, H) and bench stringers (I). Cut the ends of the bench stringers in the same way you cut the table stringers, starting ⅝" from the top edge and 2" from the ends on the bottom edges.

2. Assemble the frame pieces into two rectangular frames by driving 1⅝" deck screws through the longer frame pieces and into the ends of the shorter pieces.

3. Turn the bench frames upside down. Center the bench slats inside them so the outer edges of the slats are flush

Position the table legs inside the tabletop frame, and attach them to the table stringers.

Set the bench legs against the outer faces of the stringers. Attach the legs to the stringers, then attach the stretcher between the legs.

against the frame. Attach the slats by driving 1⅝" deck screws through the frames and into the ends of the slats.

4. Fasten the stringers inside the frame so the tops of the stringers are flat against the undersides of the slats, 3" from the inside of each frame end. Attach with glue and drive 1⅝" deck screws through the angled ends of the stringers and into the undersides of the slats. Locate the screws far enough away from the ends of the stringers so they don't stick out through the tops of the slats. The stringers are not attached directly to the bench frames.

BUILD THE BENCH LEGS.

1. Cut the bench legs (J) and bench stretchers (K) to length. With a compass, draw a roundover curve with a 1½" radius on the corners of one end of each leg. Cut the roundovers with a jig saw.

2. Center the tops of the bench legs against the outside faces of the bench stringers. Drill pilot holes in the stringers. Counterbore the holes. Attach the legs to the stringers with glue, and drive 2½" deck screws through the stringers and into the legs.

3. Drill pilot holes in the bench legs and counterbore the holes in similar fashion to the approach described in "Make and Attach the Table-Leg Assembly," above. Glue the bench stretchers and attach them between the legs with 2½" deck screws **(photo D).**

JOIN THE TABLE AND BENCHES.

1. Cut the cross rails (L) to length, miter-cutting the ends at a 45° angle (see *Diagram*). Position the benches so the ends of the cross rails are flush with the outside ends of the bench frames. Drill pilot holes in the cross rails. Counterbore the holes. Apply glue and attach the cross rails to the bench legs with 2½" deck screws.

2. Stand the benches up and center the table legs between the cross rails. Apply glue to the joints between the cross rails and legs. Clamp the table legs to the cross rails, making sure the parts are perpendicular **(photo E).** Secure the parts by driving several 2½" deck screws through the cross rails and into the outside face of each leg.

APPLY FINISHING TOUCHES.

Sand all the sharp edges and flat surfaces of the table. Apply a nontoxic wood sealant.

Center the table within the cross rails, and clamp it in place.

PROJECT
POWER TOOLS

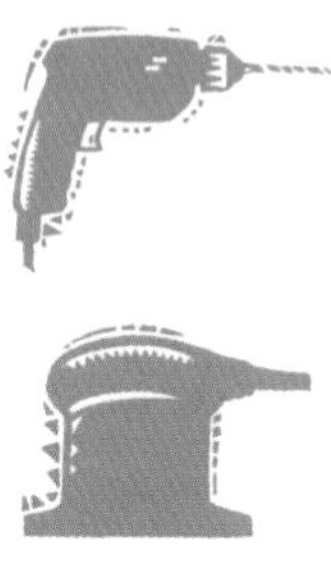

Patio Table

This patio table blends sturdy construction with rugged style to offer many years of steady service.

CONSTRUCTION MATERIALS

Quantity	Lumber
2	4 × 4" × 10' cedar
3	2 × 2" × 8' cedar
2	1 × 4" × 8' cedar
4	1 × 6" × 8' cedar

Everyone knows that a shaky, unstable patio table is a real headache. But you won't be concerned about wobbly legs with this patio table. It's designed for sturdiness and style. As a result, it's a welcome addition to any backyard patio or deck.

This all-cedar patio table is roomy enough to seat six, and strong enough to support a large patio umbrella—even in high wind. The legs and cross braces are cut from solid 4 × 4 cedar posts, then lag-bolted together. If you can find it at your local building center, buy heartwood cedar posts. Heartwood, cut from the center of the tree, is valued for its density, straightness and resistance to decay. Because it's used for an eating surface, you'll want to apply a natural, clear linseed-oil finish.

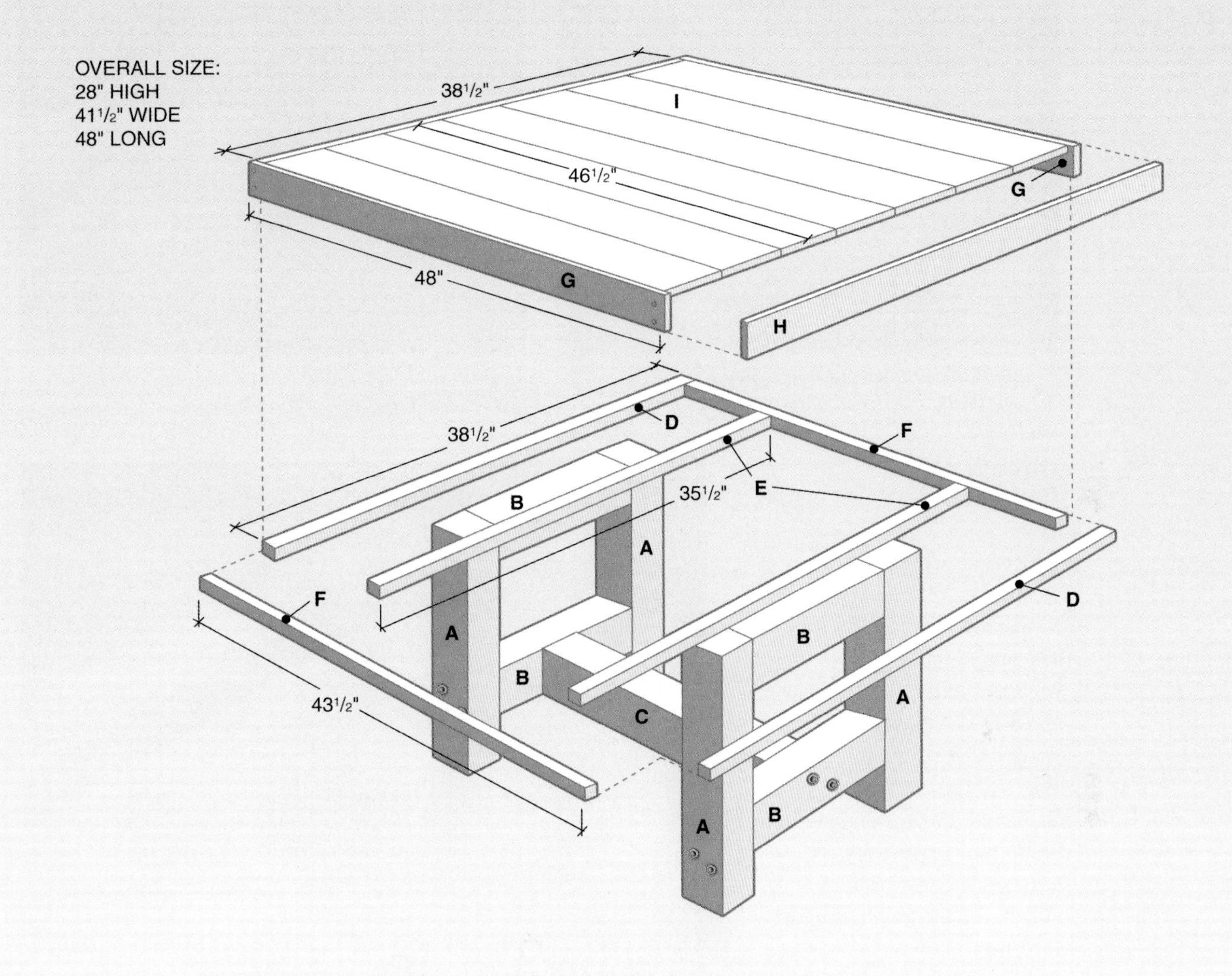

Key	Part	Dimension	Pcs.	Material
A	Leg	3½ × 3½ × 27¼"	4	Cedar
B	Stretcher	3½ × 3½ × 20"	4	Cedar
C	Spreader	3½ × 3½ × 28"	1	Cedar
D	End cleat	1½ × 1½ × 40"	2	Cedar
E	Cross cleat	1½ × 1½ × 37"	2	Cedar

Cutting List

Key	Part	Dimension	Pcs.	Material
F	Side cleat	1½ × 1½ × 43½"	2	Cedar
G	Side rail	¾ × 3½ × 48"	2	Cedar
H	End rail	¾ × 3½ × 40"	2	Cedar
I	Top slat	¾ × 5½ × 46½"	7	Cedar

Materials: Moisture-resistant glue, 2" and 3" deck screws, ⅜ × 6" lag screws with washers (20), finishing materials.

Note: Measurements reflect the actual size of dimension lumber.

Counterbore two sets of holes on each leg to recess the lag bolts when you attach the legs to the stretchers.

Maintain a ¾" distance from the top edge of the rails to the top edge of the cleats.

Directions: Patio Table

PREPARE THE LEG ASSEMBLY.

1. Cut the legs (A), stretchers (B) and spreader (C) to length. Measure and mark 4" up from the bottom edge of each leg to mark the positions of the bottom edges of the lower stretchers.

2. Test-fit the legs and stretchers to make sure they are square. The top stretchers should be flush with the top leg ends.

3. Carefully position the pieces and clamp them together with pipe clamps. The metal jaws on the pipe clamps can damage the wood, so use protective clamping pads.

TIP

Buy or make wood plugs to fill screw holes and conceal screw heads. Building centers and woodworker's stores usually carry a variety of plug types in several sizes and styles. To cut your own wood plug, you can either use a special-purpose plug-cutting tool (sold at woodworker's stores), or a small hole saw that mounts to your power drill (sold at building centers). The diameter of the plug must match the counterbore drilled into the wood.

Use pencils or dowels to set even gaps between top slats. Tape slats in position with masking tape.

BUILD THE LEG ASSEMBLY.

1. Drill ⅞"-× ⅜"-deep counterbores positioned diagonally across the bottom end of each leg and opposite the lower stretchers **(photo A).** Drill ¼" pilot holes through the counterbores and into the stretchers.

2. Unclamp the pieces and drill ⅜" holes for lag screws through the legs, using the pilot holes as center marks.

3. Apply moisture-resistant glue to the ends of the stretchers. Attach the legs to the stretchers by driving lag screws with washers through the legs and into the stretchers. Use the same procedure to attach the spreader to the stretchers.

ATTACH CLEATS AND RAILS.

1. Cut the side rails (G) and end rails (H) to length. Drill two evenly spaced, ⅛" pilot holes through the ends of the side rails. Counterbore the holes ¼" deep, using a counterbore bit. Apply glue and fasten the side rails to the end rails

Fasten cross cleats to the tabletop for strength, and to provide an anchor for the leg assembly.

Keep a firm grip on the tabletop slats when drilling deck screws through the cleats.

Before you stain or treat the patio table, sand the surfaces smooth.

with 2" deck screws.

2. Cut the end cleats (D), cross cleats (E) and side cleats (F) to length. Fasten the end cleats to the end rails ¾" below the top edges of the rails with glue and 2" deck screws **(photo B).** Repeat this procedure with the side cleats and side rails.

CUT AND ATTACH THE TOP SLATS.

1. Cut the top slats (I) to length. Lay the slats into the tabletop frame so they rest on the cleats. Carefully spread the slats apart so they are evenly spaced. Use masking tape to hold the slats in place once you achieve the correct spacing **(photo C).**

2. Stand the tabletop frame on one end and fasten the top slats in place by driving two 2" deck screws through the end cleats and into each slat **(photo D).** Hold or clamp each slat firmly while fastening to prevent the screws from pushing the slats away from the frame.

CONNECT THE LEGS AND TOP.

1. Turn the tabletop over and center the legs on the underside. Make sure the legs are the same distance apart at the top as they are at the bottom.

2. Lay the cross cleats along the insides of the table legs. Fasten the cross cleats to the tabletop with 2" deck screws **(photo E).** Fasten the cross cleats to the legs with 3" deck screws.

APPLY FINISHING TOUCHES.

1. For a more finished appearance, fill exposed screw holes with cedar plugs or buttons (see *Tip,* page 128). Smooth the edges of the table and legs with a sander or router **(photo F).**

2. If you want to fit the table with a patio umbrella, use a 1½"-dia. hole saw to cut a hole into the center of the tabletop. Use a drill and spade bit to cut the 1½"-dia. hole through the spreader.

3. Finish the table as desired—use clear linseed oil for a natural, nontoxic, protective finish.

Gate-leg Picnic Tray

Make outdoor dining on your porch, patio or deck a trouble-free activity with our picnic tray. Built with gate legs, it provides a stable surface for plates or glasses, yet folds up easily for convenient storage.

PROJECT POWER TOOLS

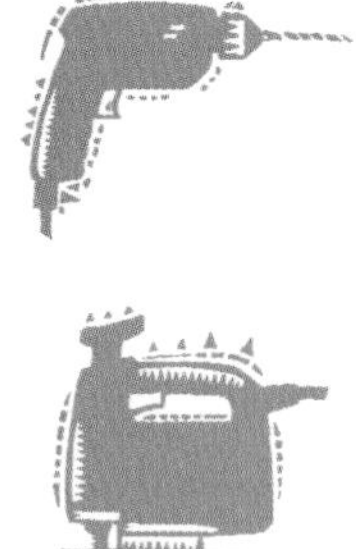

Outdoor dining doesn't need to be a messy, shaky experience. Whether you're on the lawn or patio, you can depend on our gate-leg picnic tray. A hinged leg assembly allows you to fold the tray for easy carrying and storage. Two of the legs are fastened directly to supports beneath the main tray surface with hinges, while the third is attached only to the other legs, allowing it to swing back and forth like a gate and aid compact storage. A small wedge fits under the tray to prevent the swinging leg from moving once the tray is set up.

Our picnic tray also features a hinged bottom shelf that swings down and locks in place with a hook-and-eye clasp to keep the legs in place. Of course, the most conspicuous feature of the project is the tray surface. Plastic tubing is a durable material, and it makes cleaning the tray top easy. The tubing also gives the project an interesting look—it's a companion piece to the patio chair on pages 82-87. Use a portable drill and drill stand to make the holes in the tray sides and insert the plastic tubing. CPVC tubing is a relatively lightweight material, but the strong cedar frame gives our project more stability than you'll get in conventional folding trays.

Even on grass, our gate-leg picnic tray will serve you well, allowing you to enjoy your meal without fear of a messy, dinnertime disaster.

CONSTRUCTION MATERIALS

Quantity	Lumber
1	1 × 2" × 12' cedar
1	1 × 3" × 6' cedar
1	1 × 4" × 8' cedar
1	1 × 12" × 2' cedar
4	½" × 10' CPVC tubing

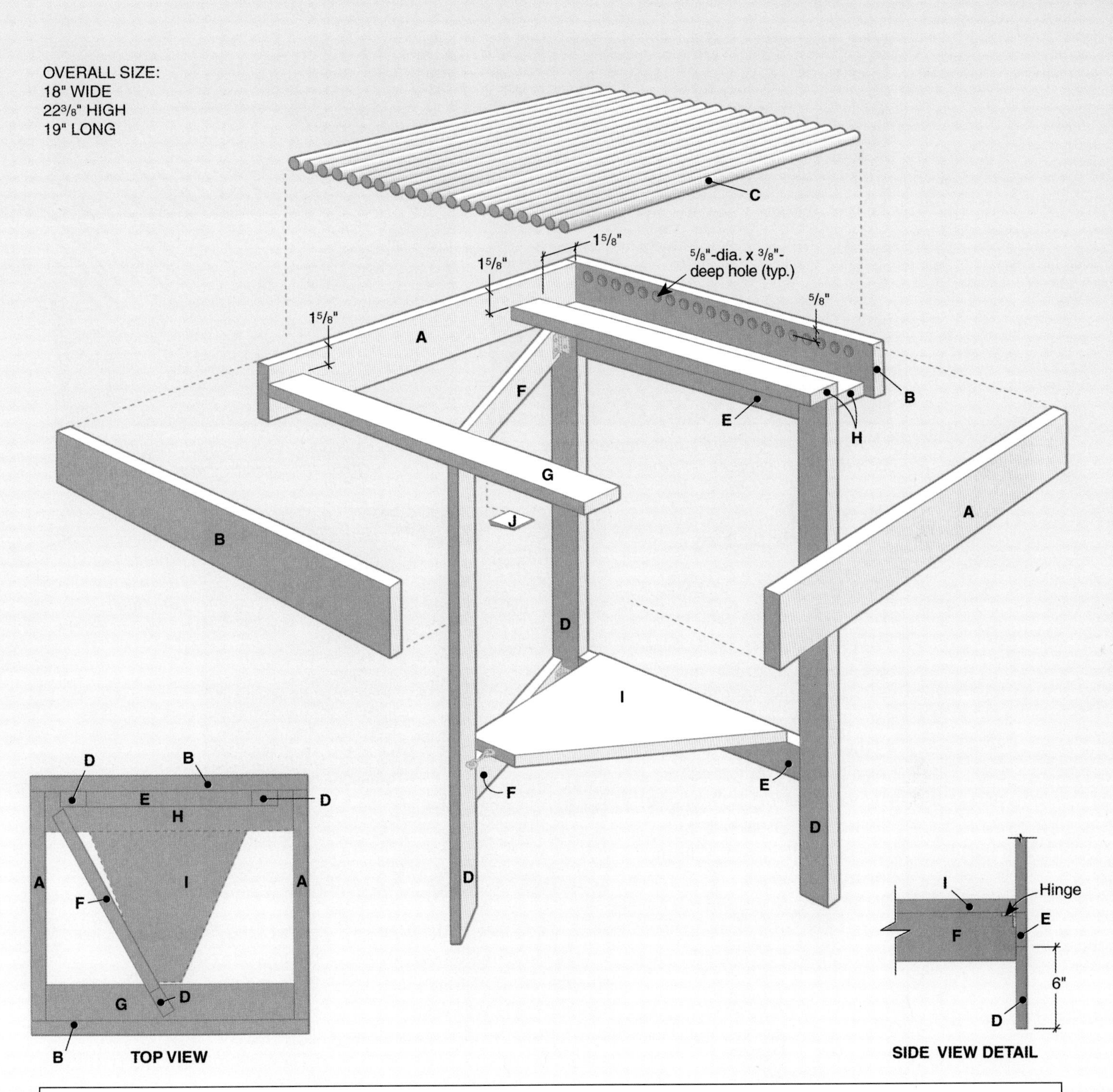

Cutting List

Key	Part	Dimension	Pcs.	Material
A	Side	¾ × 3½ × 16½"	2	Cedar
B	Cap	¾ × 3½ × 19"	2	Cedar
C	Tube	⅝"-dia. × 17¼"	20	CPVC
D	Leg	¾ × 1½ × 20"	3	Cedar
E	Short rail	¾ × 2½ × 13"	2	Cedar

Cutting List

Key	Part	Dimension	Pcs.	Material
F	Long rail	¾ × 1½ × 13½"	2	Cedar
G	Gate support	¾ × 2½ × 17½"	1	Cedar
H	Hinge support	¾ × 1½ × 17½"	2	Cedar
I	Shelf	1 × 10 × 12¾"	1	Cedar
J	Wedge	¼ × 2 × 2"	1	Cedar

Materials: Moisture-resistant glue, deck screws (1¼", 1½", 2", 3"), wire brads, exterior wood putty, hinges, finishing materials.

Note: Measurements reflect the actual size of dimension lumber.

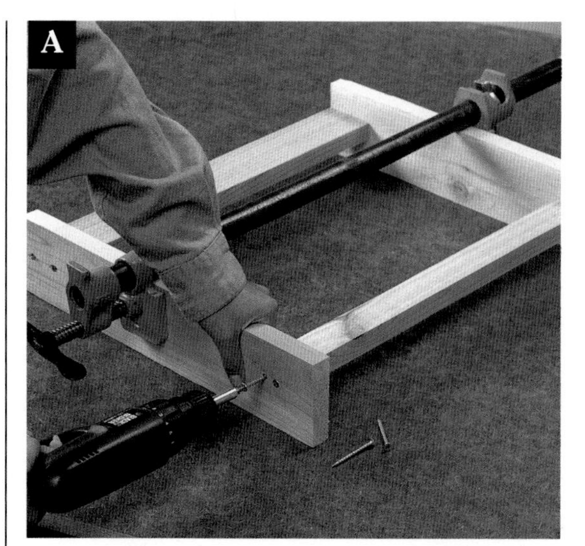

Clamp the pieces to hold them in place, and attach the hinge support 1⅝" from the top side edge.

Once the final cap has been attached with deck screws and glue, the basic tray frame is complete.

Directions: Gate-leg Picnic Tray

Attach the rails to the legs with counterbored deck screws and glue.

MAKE THE TRAY FRAME.

1. Cut the sides (A) and caps (B) to length from 1 × 4 cedar.

2. Use a power drill to make ⅝ × ⅜"-deep holes for CPVC tubing, 1" apart on the inside faces of the caps. We recommend using a portable drill stand for this step (see *Tip*, below). Start the holes 1½" from one end and center the holes ⅝" from the top cap edges.

3. Cut the gate support (G) from 1 × 3 cedar and cut the hinge supports (H) from 1 × 2 cedar.

4. Drill two evenly spaced ¼ × ⅜"-dia. counterbored holes into the outside face of the front end of each of the sides, where the gate support will be attached. Drill the holes 1⅝" down from the top edges of the sides.

5. Apply moisture-resistant glue to the joints and clamp the piece with a bar clamp. Drill 3⁄16"-dia. pilot holes through each center, then fasten the gate support to the sides with 2" deck screws.

6. On the opposite side of the frame, fasten one hinge support between the sides **(photo A).** Make sure the hinge support is fastened 1⅝" from the ends of the sides, and 1⅝" from the top side edges.

7. Fasten one of the caps to the side assembly with deck screws.

CUT & INSTALL THE TUBES.

1. Use a jig saw or compound miter saw to cut 24 pieces of ½"-dia. CPVC tubing (C) to 17¼" in length. For more information on working with plastic tubing, see *Patio Chair,* pages 82-87.

2. Wash the grade stamps from the tubing with lacquer thinner and rinse them with clean water.

3. When the tubes are dry, insert them into their holes in the frame.

TIP

Mount your electric hand drill on a portable drill stand for accurate vertical or angled drilling. Drill guides can keep bits centered on the workpiece. Make sure these tools are securely attached to the worksurface when you use them. Always protect the worksurface with a piece of scrap wood to avoid damaging tearout on the other side of the workpiece. To prevent drilling too far into a piece of wood, some portable drill stands are equipped with depth stops.

Attach the leg frames with high-quality hinges.

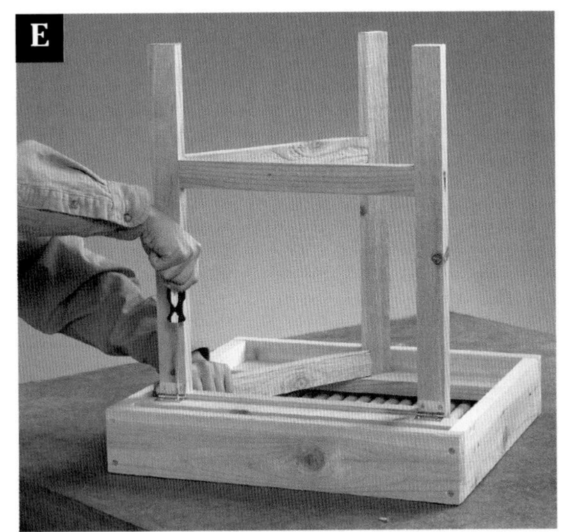

Fasten the stationary leg frame to the lower hinge support on the underside of the tray frame.

Use wire brads and glue to attach the lock wedge, which holds the gate leg assembly in place.

4. Fasten the remaining cap to the frame with glue and deck screws **(photo B).**
5. Attach the remaining hinge support to the sides, starting ⅞" from the end on the outside edge. This hinge support should be flush with the bottom edges of the sides. Use glue and countersunk deck screws to attach the pieces.

BUILD THE LEG ASSEMBLY. The leg assembly consists of two leg frames and a series of rails. One leg frame is stationary and attached to the bottom hinge support with hinges. The other frame has one leg, which swings like a gate and is attached only to the first frame. These braces are attached to the other leg frame.
1. Cut the legs (D), short rails (E), and long rails (F) from 1 × 2 cedar.
2. Fasten the rails to the legs (see *Diagram,* page 131) with counterbored deck screws and glue to form two leg frames **(photo C).** The bottom edge of the rails should be 4" from the bottom of the legs. The top rails should be flush with the top leg edges.
3. Attach the gate leg frame to the stationary frame with hinges **(photo D).**
4. Fasten the stationary legs to the bottom hinge support **(photo E).**
5. To prevent the gate leg from swinging back and collapsing the picnic tray, cut a lock wedge (J) from 1 × 3 cedar.
6. Open the gate leg to the normal standing position, which should be roughly the center of the gate support. Draw a line on the gate support along the edge of the gate leg to locate the wedge position. Use wire brads and glue to attach the lock wedge so that its thin end is against the gate leg **(photo F).** The leg will slide over this wedge slightly and be held fast.
7. Cut the shelf (I) to size and shape.
8. Attach it to the short rails with hinges.
9. Install a hook-and-eye clasp on the gate leg and shelf to secure the open assembly.
10. Sand all sharp edges and finish the project with clear wood sealer.

TIP

Use cedar plugs to fill the counterbores for the screws in the tray frame.

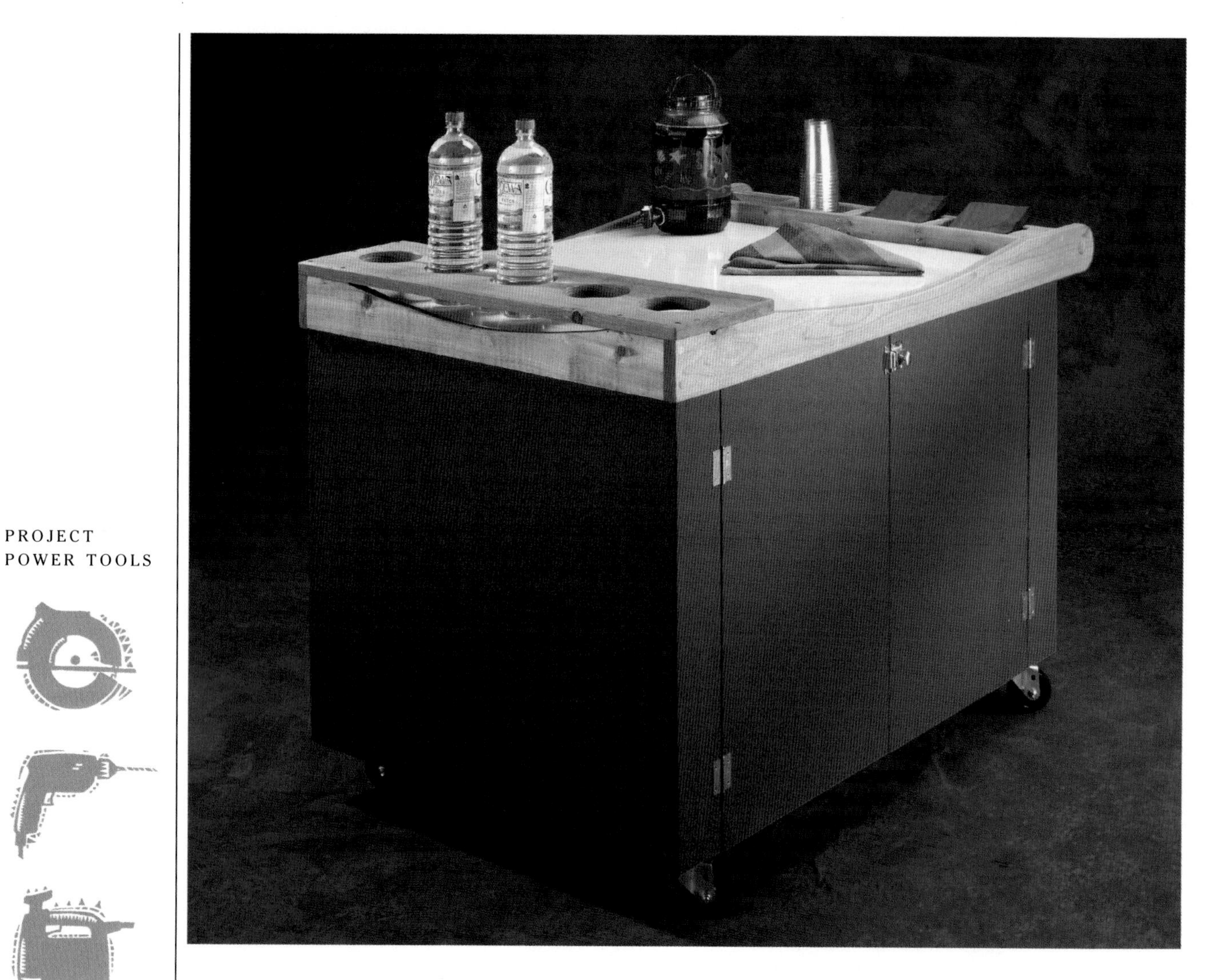

PROJECT
POWER TOOLS

Party Cart

A fully insulated cooler on wheels, our party cart has some clever bells and whistles, as well as distinctive design elements, so it will fit in even at the most formal outdoor gatherings.

CONSTRUCTION MATERIALS

Quantity	Lumber
4	2 × 4" × 8' cedar
2	1 × 4" × 8' cedar
1	1 × 10" × 4' cedar
3	1 × 3" × 8' cedar
1	1 × 2" × 6' cedar
2	4 × 8' × ½"-thick BCX plywood
1	4 × 8' sheet tileboard
1	1"-dia. × 3' oak dowel
1	1½" × 4 × 8' foam insulation

Outdoor entertaining events often turn into an endless parade between the patio and the refrigerator, or a loose huddle around an old foam cooler. With this portable party cart, you can keep your guests refreshed on site and in style. The tileboard top is generously proportioned and easy to clean. With a capacity of 15 cubic feet, the insulated cooler compartment has plenty of room for cans, bottles, even kegs, as well as ice and snacks. You can add accessories to help the cart meet your needs. A few suggestions: attach a bottle opener, paper towel holder, or plastic cup dispenser to the cabinet side; drill a 1"-diameter hole through the side to create a passage for a keg hose, then cover the hole with a plastic grommet; mount a flagpole holder on the back of the cart to hold a beach umbrella.

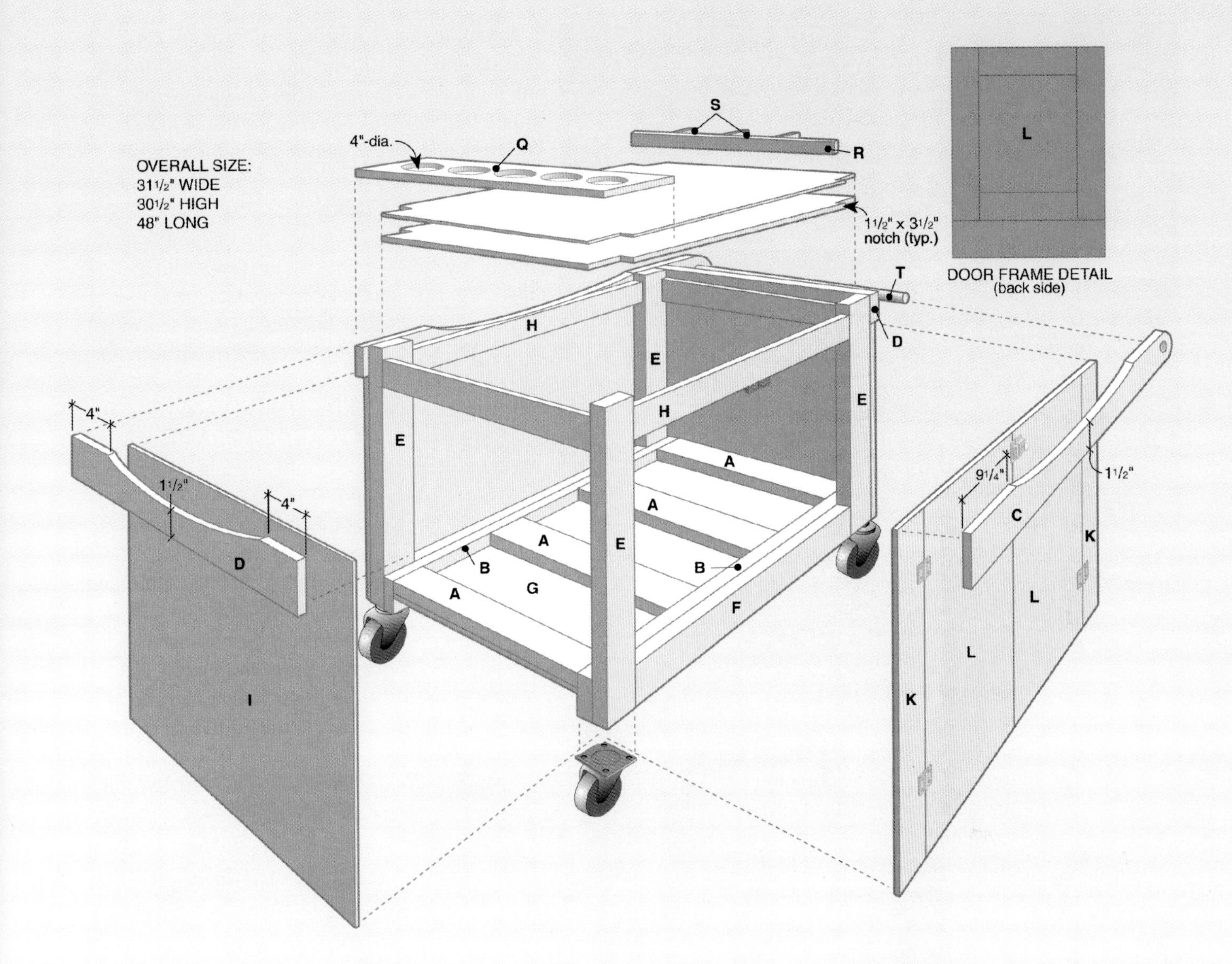

Key	Part	Dimension	Pcs.	Material
A	Bottom stretcher	1½ × 3½ × 24"	4	Cedar
B	Bottom side rail	1½ × 3½ × 42"	2	Cedar
C	Top side rail	⅞ × 3½ × 48"	2	Cedar
D	Top end rail	⅞ × 3½ × 30"	2	Cedar
E	Post	1½ × 3½ × 30"	4	Cedar
F	Rail filler	1½ × 3½ × 35"	2	Cedar
G	Bottom	½ × 24 × 42"	1	BCX plywood
H	Tabletop cleat	⅞ × 2½ × 35"	2	Cedar
I	End panel	½ × 30 × 27"	2	BCX plywood
J	Side panel	½ × 43 × 27"	1	BCX plywood
K	Post cover	½ × 4 × 27"	2	BCX plywood

Cutting List

Key	Part	Dimension	Pcs.	Material
L	Door	½ × 17⅜ × 27"	2	BCX plywood
M	Door stile	⅞ × 2½ × 21⅞"	4	Cedar
N	Door rail	⅞ × 2½ × 10¼"	4	Cedar
O	Tabletop	½ × 30 × 42"	1	BCX plywood
P	Waterproof panel	⅛ × 30 × 42"	2	Tileboard
Q	Bottle caddy	⅞ × 9¼ × 31½"	1	Cedar
R	Bin side	⅞ × 1½ × 30"	1	Cedar
S	Bin divider	⅞ × 1½ × 8¼"	3	Cedar
T	Handle	1"-dia. × 31½"	1	Oak dowel
U	Tabletop end cleat	¾ × 2½ × 27"	2	Cedar

Materials: 1", 1½", 2", 2½" and 3" deck screws, 6d casing nails, 2" brass butt hinges (4), brass clasp, magnetic door catches (2), brass window sash handles (2), heavy-duty locking casters (4), moisture-resistant glue, tileboard adhesive, panel adhesive, ⅜"-dia. cedar plugs.

Note: Measurements reflect the actual size of dimension lumber.

Attach the scooped top frame pieces to the posts with 2" deck screws. Posts are attached to the base frame with 3" deck screws.

Directions: Party Cart

BUILD THE CABINET FRAME.

The main structural element of our party cart is a cabinet frame made from 2 × 4 and 1 × 4 cedar.

1. Cut the bottom stretchers (A) and bottom side rails (B) from 2 × 4s.

2. Lay the rails on edge on a worksurface, then set the stretchers facedown between the rails. One stretcher should be set flush at each end, with the other two spaced evenly between the ends (the assembly should look like a ladder).

3. Drill pilot holes, then fasten the side rails to the stretchers with 3" deck screws.

4. Cut the top side rails (C) and top end rails (D) from 1 × 4s.

5. Cut a gentle, 1½"-deep scoop into each top rail, using a jig saw. The scoops should start 4" from each end of the top end rails. In the top side rails, cut the scoops 9¼" from the front ends, and 11¼" from the back ends.

6. Assemble the top side and top end rails into a square frame. The back end rail should be recessed 6" from the back ends of the side rails.

7. Drill pilot holes with ⅜"-dia. × ¼"-deep counterbores through the side rails and into the end rails.

8. Fasten the rails together with glue and 2" deck screws, then plug the counterbores with ⅜"-dia. cedar plugs.

9. Cut the 2 × 4 posts (E), and arrange them so they fit at the outside corners of the 2 × 4 frame base and the inside corners of the 1 × 4 frame top. Make sure the frame base is positioned with the recesses beneath the stretchers, facing down.

10. Attach the top and bottom frames to the posts with 3" deck screws and glue at the bottom, and 2" deck screws and glue at the top **(photo A).**

ATTACH SIDE PANELS TO CABINET FRAME.

1. Cut and attach a 2 × 4 rail filler (F) to the outer face of each bottom side rail in the frame base with 2½" deck screws. The rail fillers eliminate gaps between the side rails and the sides of the cabinet.

2. Cut the bottom panel (G) from ½"-thick plywood and attach it to the underside of the

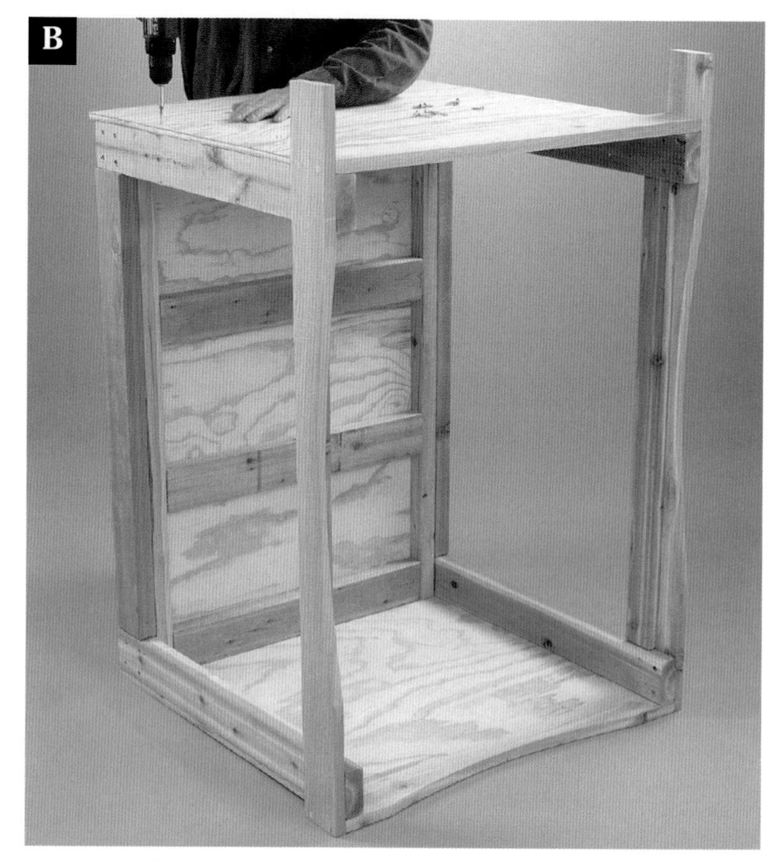

Attach the plywood cabinet sides and the cabinet base to the cabinet frame with 1¼" deck screws

Attach the 1 × 3 door rails and stiles by driving 1" screws through the front faces of the plywood door panels.

Bond tileboard to the plywood tabletop with tileboard adhesive.

frame base with 1½" deck screws.

3. Measure down 2¼" from the tops of the frame pieces and draw reference lines for the cleats that support the top panel.

4. Cut the 1 × 3 tabletop cleats (H) and attach them just below the reference lines, using 1½" deck screws.

5. Cut the end panels (I) and the side panel (J) from plywood, and position them against the cabinet frame so all panels overhang the frame base by ½". Attach the panels to the frame with 1" deck screws **(photo B).**

6. Cut post covers (K) from ½" plywood and attach them to the 2 × 4 posts at the open side.

BUILD THE DOORS.

The cooler compartment doors are made from ½" plywood, with 1 × 3 rails and stiles attached to the back side to stiffen the plywood.

1. Cut the door panels (L), door stiles (M), and door rails (N).

2. Attach the door rails and stiles with glue and 1" screws driven through the fronts of the door panels **(photo C);** attach the upper door rail to the inside (unsanded) edge of each door panel, 1⅝" down from the top edge and centered side to side; attach the lower rail 4⅛" up from the bottom of the panel; attach vertical stiles so they are flush with the outer edges of the rails to complete the doors.

BUILD AND INSTALL THE CART TOP.

The cart top is cut from ½" plywood, then covered with tileboard to create a smooth, water-resistant surface that is easy to clean.

1. Cut the plywood tabletop panel (O).

2. Cut a piece of ¼"-thick tileboard (P) to the same size as the tabletop panel (O). Tileboard is usually sold in 4 × 8' sheets that resemble interior paneling. It is available in a wide variety of textures and finishes. We chose a fairly neutral, biscuit-colored style. If you are willing to spend the extra money, you can substitute fiberglass shower liner panels for a more long-lasting tabletop surface.

3. Attach the tileboard to the sanded side of the top panel with exterior-rated tileboard adhesive **(photo D),** according to the recommended application methods and drying times.

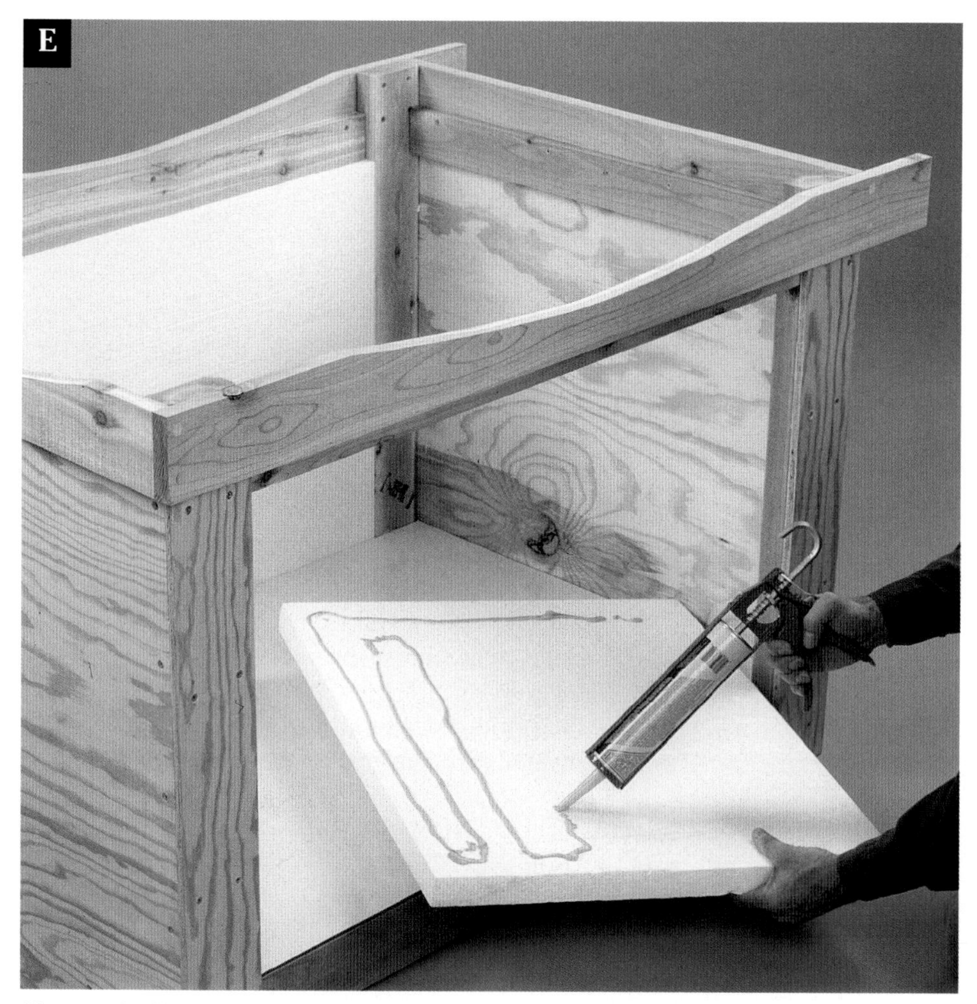

Use panel adhesive to install rigid foam insulation on the interior walls of the cooler compartment.

4. After the adhesive has set, use a jig saw to cut out 1½ × 3½" notches from the corners of the cart top to fit around the cabinet-frame posts. Make sure the cutouts are oriented correctly.

5. Apply a thick bead of panel adhesive to the tops of the 1 × 3 cleats mounted on the inside faces of the frame top, then set the cart top onto the cleats and press down firmly. Do not nail or screw the top in place. Set some heavy weights on the surface while the adhesive dries.

TIP

Read the usage recommendations before purchasing panel adhesive. Some products may dissolve foam or plastic.

LINE THE COOLER COMPARTMENT.

We used 1½"-thick open-cell foam insulation boards to insulate the cooler compartment. For greater durability and better insulation performance, you can substitute closed-cell insulation boards with a puncture-resistant facing.

1. Use a sharp utility knife to cut insulation boards to fit into the gaps between the stretchers on the floor of the cabinet, to fit between the rails and stiles on the doors, and to fit the walls and top of the compartment. Cut insulation slightly oversized, so compression will help hold it in place. Attach all the insulation boards with a panel adhesive **(photo E).**

2. Cut a second piece of tileboard (P) the same size as the top panel.

3. Apply tileboard adhesive to the tops of the exposed stretchers in the frame bottom, and install the tileboard to make a bottom for the compartment.

4. To protect the inside walls and top of the compartment, cut tileboard to fit, and attach it to the insulation boards with panel adhesive (optional).

HANG THE DOORS.

1. Center the doors over the door opening, with the top edges aligned, and hang them from the plywood post covers (K) with 2" brass butt hinges. Leave a gap of about ⅛" between the doors.

ACCESSORIZE THE TOP.

The bottle caddy and napkin/condiment bin are optional features that give the party cart greater versatility.

1. Cut the bottle caddy (Q) from 1 × 10 cedar.

2. Cut evenly spaced 4"-dia. holes in the board, then smooth out the cuts with a drum sander mounted on your electric drill.

3. Set the bottle caddy over the frame top at the front end of the cart, and attach it to the frame with 1" screws.

4. Cut the bin side (R) and the bin dividers (S) from 1 × 2 cedar.

5. Space the dividers evenly along the divider side, and attach with 1" deck screws.

6. Apply panel adhesive to the bottoms of the dividers and divider side, then set the assembly on the tabletop, with the free ends of the dividers flush against the back rail. Secure the rail to the dividers with 1" screws.

INSTALL THE HANDLE.
We used a 1"-dia. oak dowel to make the handle that is mounted between the top frame rails at the back of the cart.
1. Cut the handle (T) to length.
2. Measure in 1" from the back ends of the top frame rails, and mark a point that is centered top-to-bottom. Use this point as a centerpoint to drill 1"-dia. holes through the rails, using a spade bit, to accommodate the handle. For added visual appeal, we used a jig saw to round off the ends of the rails once our handle position was established.
3. Insert the handle through both 1" holes, so the ends are flush with the outside faces of the rails. Secure the handle by drilling ⅛" pilot holes through the rails and into the ends of the handle, then driving a 6d casing nail into each pilot end of the handle.

FINISH THE CART.
1. Fill exposed plywood edges and nail hole with wood putty, then sand smooth with at least 120-grit sandpaper.
2. Apply a clear finish to exposed cedar trim, and paint plywood surfaces with an exterior grade enamel.
3. To keep the doors closed securely, install a brass clasp and magnetic door catches.
4. Attach door pulls. We used brass window sash handles.
5. Because the cart is designed to carry a load in excess of 100 pounds, install heavy-duty, locking casters to the bottom of each post.

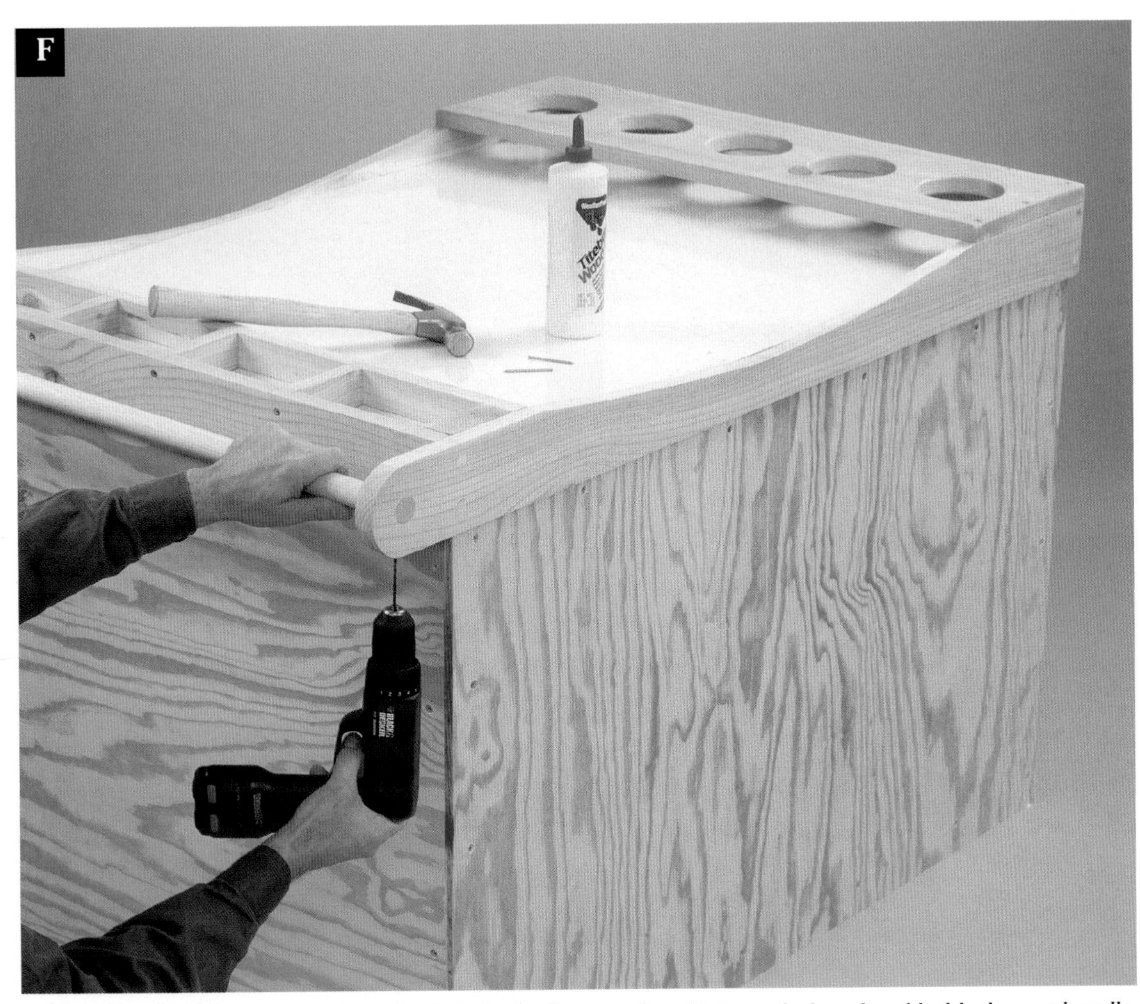

A 6d casing nail driven through a pilot hole in the frame rail and into each dowel end holds the cart handle in position.

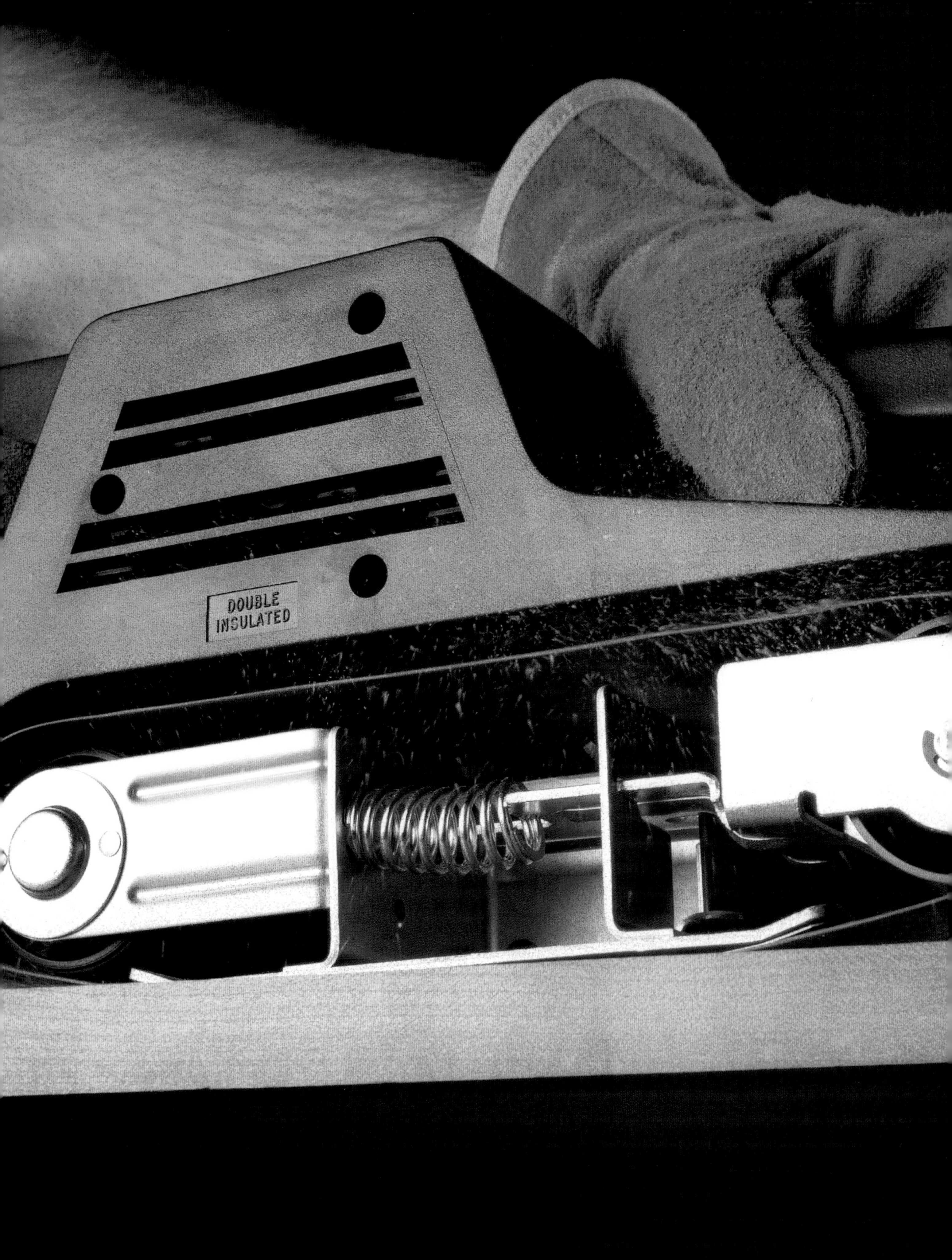
DOUBLE
INSULATED

Children's Play Projects

Children are naturally drawn to the outdoors. It's a great big, wonderful playground filled with bugs, dirt, and challenges. Inspire your youngsters to cultivate their curiosity and expend their energy with kid-sized projects that make bug collecting, gardening, and playing easier for them. The projects shown here bring the big, beautiful world down to kid size. Your little ones thank you with hours of outdoor play.

PROJECT
POWER TOOLS

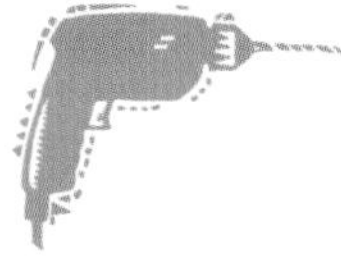

Observation Station

Create a temporary viewing station for bugs and small animals.

CONSTRUCTION MATERIALS

Quantity	Lumber
3	2 × 2" × 12' cedar
1	½" × 4' × 4' exterior plywood
6	¼" × 1⅛" × 8' molding

An observation station allows children to safely observe bugs, insects, and wildlife in their own backyard. If the station will be used primarily for caterpillars, fireflies, grasshoppers, and butterflies, use a dense screen so they won't escape. For snakes, turtles, and frogs, a stronger screen with larger holes is appropriate.

Remember that wild animals don't make good pets, so open the lid and tip the observation station on its side to release critters after watching them. Also, don't place the station in direct sun, as many animals can be injured by the heat.

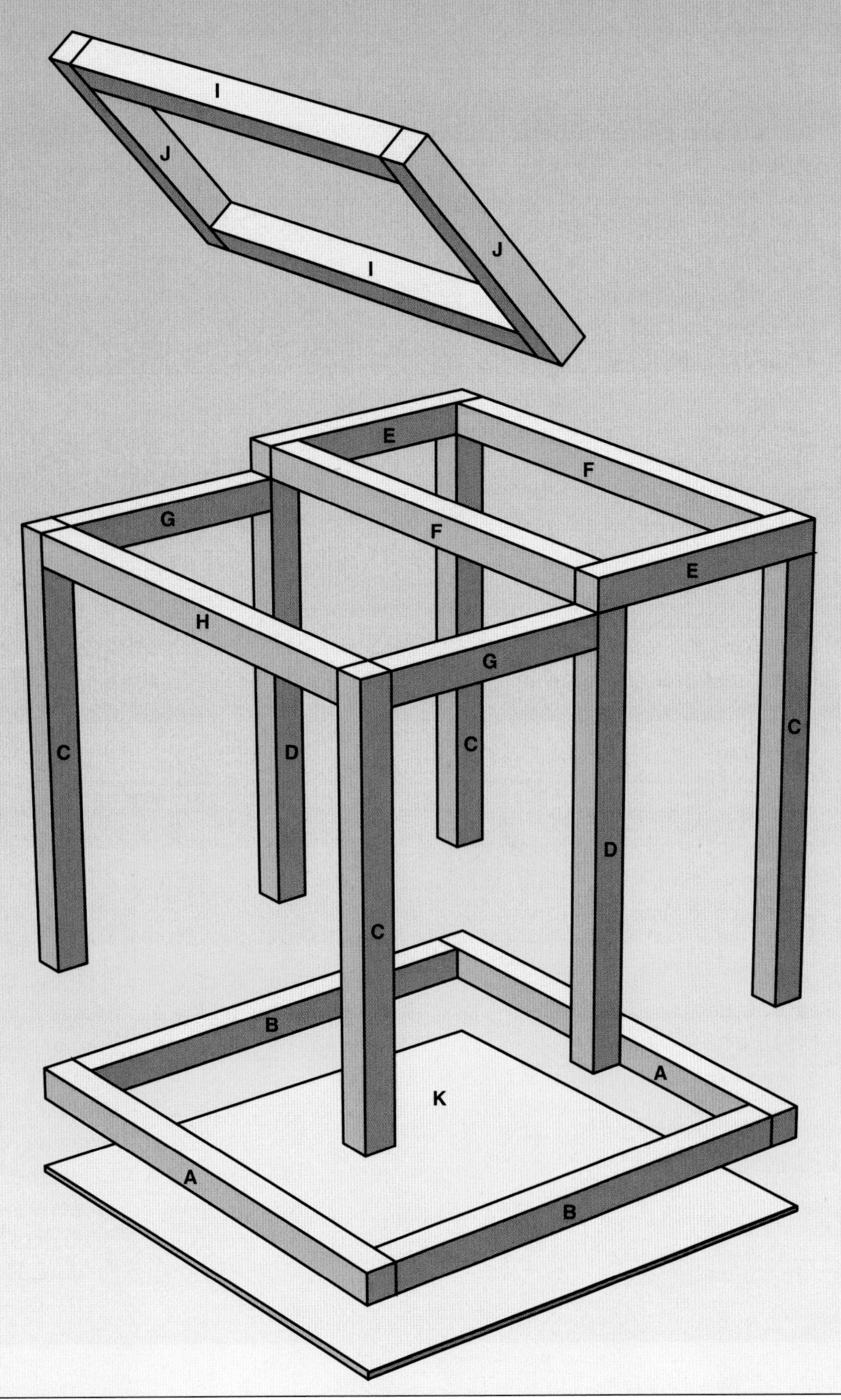

Cutting List

Key	Part	Dimension	Pcs.	Material
A	Base side	1½ × 1½ × 24"	2	Cedar
B	Base rail	1½ × 1½ × 21"	2	Cedar
C	Corner post	1½ × 1½ × 20½"	4	Cedar
D	Center post	1½ × 1½ × 20½"	2	Cedar
E	Top side	1½ × 1½ × 12"	2	Cedar
F	Top rail	1½ × 1½ × 21"	2	Cedar

Cutting List

Key	Part	Dimension	Pcs.	Material
G	Lid support	1½ × 1½ × 10½"	2	Cedar
H	Lid support rail	1½ × 1½ × 21"	1	Cedar
I	Lid rail	1½ × 1½ × 21"	2	Cedar
J	Lid side	1½ × 1½ × 12"	2	Cedar
K	Base	½ × 24 × 24"	1	Plywood
L	Trim	¼ × 1⅛ × *	24	Molding

Materials: 2½" and 1¼" deck screws, brad nails, aluminum screen, staples, butt hinges (2), handle, sandpaper.

Note: Measurements reflect actual size of dimension lumber.

* cut to fit.

Directions: Observation Station

ASSEMBLE THE FRAME.

1. Cut the base sides (A) and base rails (B) to length.

2. Butt the end of a base rail against the inside edge of a side rail. With the top and the ends of the rails flush and square, drill two 3/32" pilot holes through the side and into the end of the rail. Attach the rails using 2½" deck screws. Fasten the second base side on the outside of the opposite end of the base rail by drilling pilot holes and using deck screws. Fasten the second base rail between the base sides at the open end to complete the base.

3. Cut the corner posts (C) and side posts (D) to length. Place the base assembly on its side and fasten a corner post at each corner, making sure the edges are flush. Drill pilot holes and use deck screws to fasten the corners in place.

4. Measure 9" from the inside of a corner on the base and mark the base side. Attach a side post outside the mark, making sure the outside edges are flush. Attach the other side post on the opposite side of the frame at the same location.

5. Cut the top sides (E), top rails (F), lid supports (G), and lid support rail (H) to length. Return the frame to an upright position. Place a top side piece over a side and corner post, making sure it's flush on the outside and end. Drill pilot holes and insert 2½" deck screws through the top of the top side piece into the ends of the corner and side posts. Repeat with the other top side piece.

6. Place a top side rail between the ends of the side pieces. Align the rail with the top and outside corners of the sides, drill pilot holes, and insert deck screws. Repeat with second rail at the other end of the top sides.

7. Place the lid rail between the remaining corner support posts, flush with the top of the posts. Drill pilot holes, and fasten with 2½" deck screws.

8. Place a lid support between a corner and side support post, flush with the top of the posts. Drill pilot holes, and fasten with 2½" deck screws. Repeat with the other lid support **(photo A).** Note: the lid supports are 1½" lower than the top sides and rails.

Fasten the rails together by drilling pilot holes and inserting 2½" deck screws.

Attach the hinges to the lid and the frame.

Place the screen flat over the framework, then staple it every 2".

ATTACH THE LID AND THE BASE.

1. Cut the lid rails (I) and sides (J) to length. Butt the lid sides against the ends of the lid rails. Align the edges. Drill pilot holes and fasten with 2½" deck screws.

2. Attach two hinges to the back of the lid using the screws that came with the hinges.

3. Place the lid on the observation station. Attach the other end of the hinges to the adjacent rail **(photo B).** When closed, the lid should sit securely on the lid supports.

4. Cut the base (K) to size. Turn the station on its side and fasten the plywood to the base rails using 1¼" deck screws. The plywood should be flush with all four corners of the station. If it isn't, adjust the box so it's square before fastening the screws.

5. Lightly sand any rough areas using 220-grit sandpaper. Wipe away any dust, then stain or paint the wood. Wait until the stain or paint is completely dry before continuing.

Fasten the screen to the box.

1. Cut the aluminum screen to size using wire cutters. Cut four screens at 22½ × 22½" for the sides and two screens at 10½ × 22½" for the top and lid.

2. Place each section of screen on the outside of the rails, overlapping each rail by ¾". Make sure the screen is flat against the rails. Staple the screen to the rails every 2" **(photo C).**

ATTACH THE HANDLE AND TRIM.

1. Center a handle on the face of the front rail of the lid and attach it using the screws that came with the handle.

2. Cut a piece of ¼" × 1⅛" molding at 24" and place it along a rail on one side of the observation station. Align the molding with the end and outside edge of the rail, covering the screen, and fasten it to the rail using brad nails.

3. Measure the distance between the inside edge of the molding and the opposite corner. Cut a piece of trim to that size, then nail it in place. Do the same for the remaining two sides **(photo D).** Repeat to fasten trim to all sides of the observation station.

4. Paint or stain the molding to match the box.

Cut pieces of trim to size, then nail them over the edges of the screen.

PROJECT
POWER TOOLS

Rolling Garden

Why not take the garden with you? This rolling garden allows children to have their own garden spot and still be close to mom or dad at all times.

CONSTRUCTION MATERIALS

Quantity	Lumber
1	1 × 10" × 4' cedar
1	1 × 8" × 6' cedar
3	1 × 2" × 8' cedar

This garden on wheels is sure to be a favorite with your children. The unique design features dividers in the cart so kids can plant different flowers and vegetables or leave one section empty for carrying their watering can or other treasures. The back of the cart contains a tool caddy so children can take their gardening tools with them wherever they go. The 3" caster wheels make the cart easy to pull, and the swivel wheels in the front allow the cart to turn smoothly without tipping over.

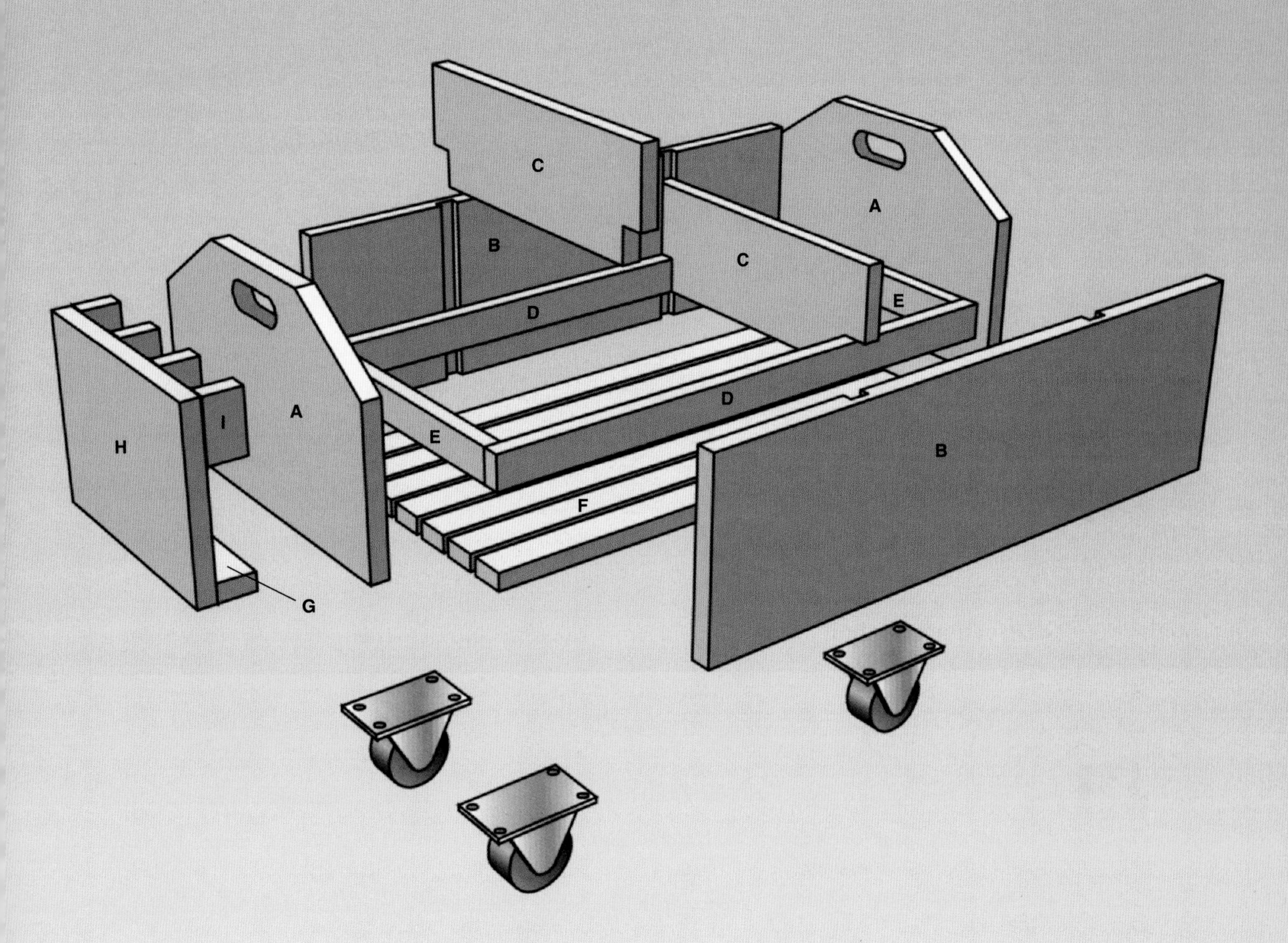

Cutting List

Key	Part	Dimension	Pcs.	Material
A	End panel	¾ × 9½ × 14"	2	Cedar
B	Side panel	¾ × 7½ × 24"	2	Cedar
C	Divider	¾ × 4½ × 12½"	2	Cedar
D	Side cleats	¾ × 1½ × 24"	2	Cedar
E	End cleats	¾ × 1½ × 10½"	2	Cedar

Cutting List

Key	Part	Dimension	Pcs.	Material
F	Floor slats	¾ × 1½ × 24"	7	Cedar
G	Caddy floor	¾ × 1½ × 10"	1	Cedar
H	Caddy wall	¾ × 7½ × 10"	1	Cedar
I	Caddy dividers	¾ × 1½ × 2½"	4	Cedar

Materials: 1½" galvanized screws, 3" swivel caster wheels (2), 3" rigid caster wheels (2), ¼ × 1" lag screws, landscape fabric, rope, exterior wood glue, sandpaper.

Note: Measurements reflect actual size of dimension lumber.

Mark the angled cut-offs. Drill holes for the handles, then cut away the remaining material.

Cut dado grooves in the side panels for the removable divider walls.

Directions: Rolling Garden

PREPARE THE END PANELS.

1. Cut the end panels (A) using a circular saw.

2. Make a mark 2¾" from the top along an edge of an end panel for the corner angle. Make another mark 4" from that edge along the top of the panel. Connect the two points using a straightedge. Repeat for the other corner.

3. Mark a horizontal line across the panel 1½" from the top edge. Measure 6" from each side and mark cross points on the horizontal line.

4. Set the two end panels together, making sure the edges are flush, and clamp them to your work surface. Drill holes for the handles using a 1⅛" spade bit centered on the cross points **(photo A).** Cut away the remaining material from the handle using a jig saw. Cut the corner angles using a jig saw. Sand all edges smooth using 220-grit sandpaper.

BUILD THE CART FRAME AND DIVIDERS.

1. Cut the side panels (B) to size.

2. To make grooves for the removable dividers, clamp the two side panels side by side on your work surface. Cut two ¾"-wide × ⅜"-deep dadoes in the side panels, 6" from each end, using a router and a straightedge guide **(photo B).**

3. Butt the panels together with the end panels over the side panels. Drill ⅛" pilot holes through the end into the side panels. Attach the side and end panels using glue and 1½" galvanized screws.

4. Cut the dividers (C) by cutting a 1 × 10 to length and use a circular saw to rip this piece in half to form two dividers. Use a jig saw to cut a 1"-wide × 1½"-long notch in the bottom corners of each divider to fit around the support cleats.

BUILD THE CART FLOOR.

1. Cut the side cleats (D), end cleats (E) and floor slats (F).

2. Make two marks on the inside of each side panel 1½" from the bottom edge. Align the bottom edge of a side cleat along the marks and attach one to each side panel using 1½" galvanized screws spaced every 6-8".

3. Position the end cleats between the side cleats, making sure the edges are flush. Attach the cleats to the end panels with 1½" galvanized screws.

4. Turn the cart upside down and install the floor slats, leaving about ¼" of space between each slat to allow for drainage. Drill pilot holes through the

Attach the floor slats to the support cleats with galvanized screws.

Assemble the tool caddy and attach it to the back end panel.

ends of the slats and attach them to the end cleats using 1½" galvanized screws **(photo C).**

INSTALL THE TOOL CADDY.

1. Cut the tool caddy floor (G), wall (H) and dividers (I) to size.

2. Assemble the pieces of the tool caddy, then fasten the pieces together by drilling pilot holes and inserting 1½" screws **(photo D).**

3. Position the caddy on the back end panel so the base is flush with the bottom edge of the cart. From inside the cart, drill pilot holes and insert screws into the two outer caddy dividers. Insert two additional screws from the underside of the cart into the caddy base.

FINISH THE CART.

1. Attach caster wheels to the floor slats, using ¼ × 1" lag screws **(photo E).** The wheels should be recessed inside the frame. For easy maneuvering, install the swivel casters at the front of the cart.

2. Drill two ½" holes for the pull rope in the front panel, 3" from the side and 2" from the top. Thread a 36" piece of rope through the holes and knot the ends.

3. Line the bottom of the cart with landscape fabric and staple it along the panels. Insert dividers into the grooves to make separate growing or storage areas. Fill the cart with 6" of potting soil, then add seeds or plants.

Attach caster wheels so they are recessed inside the frame.

PROJECT POWER TOOLS

Children's Picnic Table

Grown-ups get a picnic table their size, why not build a sturdy, beautiful table sized just for kids?

CONSTRUCTION MATERIALS

Quantity	Lumber
2	2 × 4" × 6' cedar
5	2 × 6" × 6' cedar
3	2 × 8" × 8' cedar

A picnic table is a wonderful addition to any backyard. Like other projects in this section, the children's picnic table is built at the right size for kids. Its light weight allows you to move the table around the yard for impromptu tea parties on the deck or dinner under the trees.

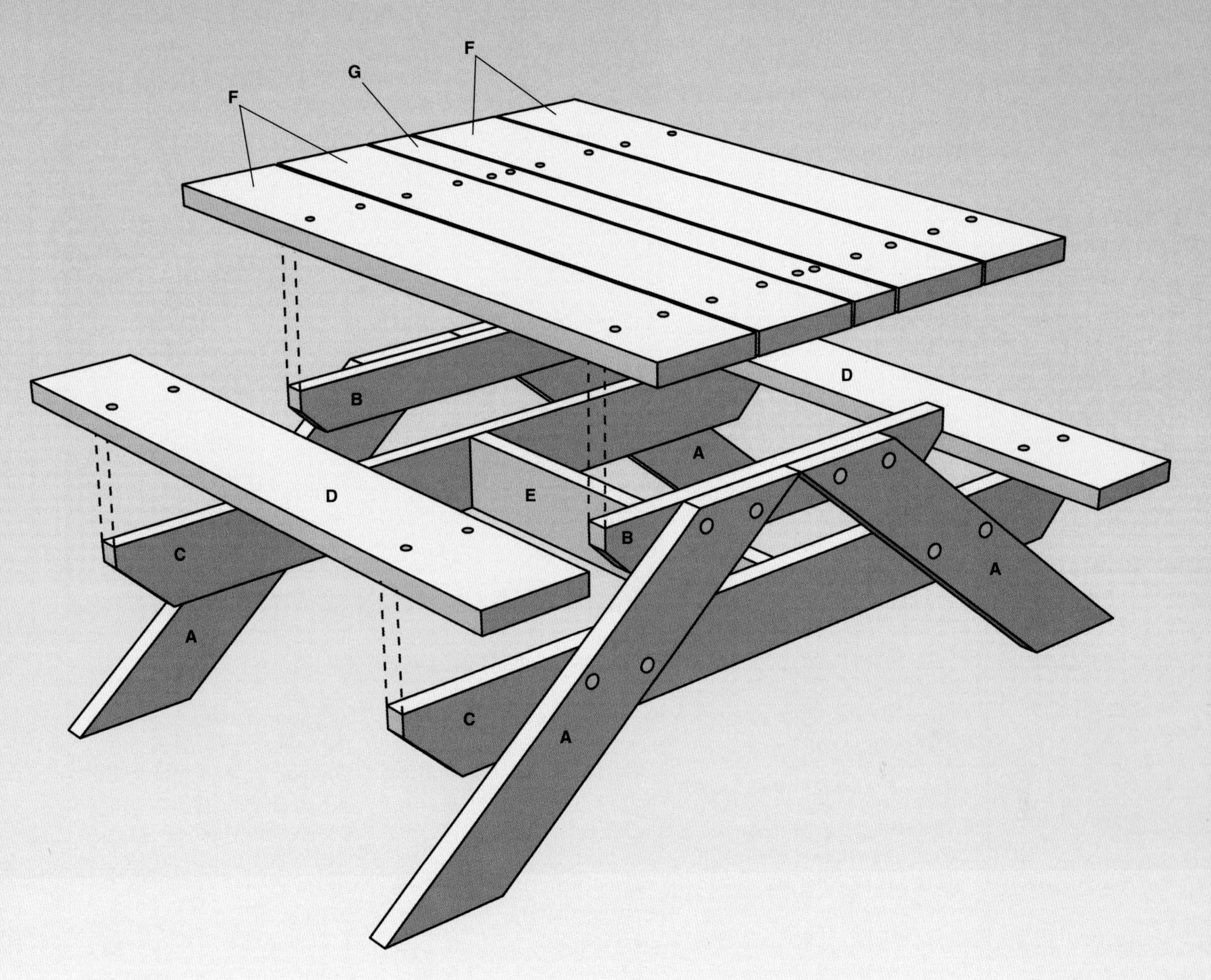

Cutting List

Key	Part	Dimension	Pcs.	Material
A	Leg	1½ × 5½ × 32"	4	Cedar
B	Table support	1½ × 3½ × 29¾"	2	Cedar
C	Seat support	1½ × 5½ × 60"	2	Cedar
D	Seat	1½ × 7½ × 48"	2	Cedar

Cutting List

Key	Part	Dimension	Pcs.	Material
E	Brace	1½ × 5½ × 30"	1	Cedar
F	Table top	1½ × 7½ × 48"	4	Cedar
G	Table top	1½ × 3½ × 48"	1	Cedar

Materials: 2½" deck screws, ⅜ × 3" carriage bolts with nuts (16) and washers (32).

Note: Measurements reflect the actual size of dimension lumber.

Directions: Children's Picnic Table

CUT THE ANGLED LEGS AND SUPPORTS.

1. To make the angled legs (A), use a saw protractor to mark a 50° angle on one end of a 2 × 6 **(photo A).** Cut the angle using a circular saw. Measure 32" from the tip of the angle, then mark and cut another 50° angle parallel to the first. Do this for all four legs, cutting two legs from one piece of lumber.

3. Cut the tabletop supports (B) to length. Measure 1½" in from each end of both supports and make a mark. Make a 45° angle starting at the mark and going in the direction of the board end. This relieves the sharp end of the board to prevent injuries and also looks more pleasing.

4. Cut the seat supports (C) to length. Measure 2½" from the ends of both supports, make a mark, and cut a 45° angle to relieve the sharp ends.

Use a saw protractor to mark a 50° angle on the end of the table leg, then cut the angle using a circular saw.

ASSEMBLE THE A-FRAMES.

1. Place one of the legs against the tabletop support so the inside edge of the leg is at the centerpoint of the support. Align the top of the leg with the top of the support. Clamp the pieces together.

2. Drill two ⅜" holes through the leg and support. Stagger the holes. To keep the bolts from causing scrapes, recess both the bolt head and the nut. Countersink 1" holes about ¼" deep into the leg and the tabletop support using the ⅜" holes as a guide. Insert a ⅜ × 3" carriage bolt and washer into each hole. Tighten a washer and nut on the end of the bolt using a ratchet wrench. Repeat these steps to fasten the second leg in place. NOTE: If your washers are larger than 1", you'll need a larger recess.

3. Measure along the inside edge of each leg and make a mark 12½" from the bottom. Center the seat support over the leg assembly, on the same side of the legs as the tabletop support, with the 45° cuts facing down and the bottom flush with the 12½" marks. Drill ⅜" holes, countersink 1" holes, and fasten the seat support to the legs using carriage bolts, nuts and washers **(photo B).**

4. Repeat this step to assemble the second A-frame.

ATTACH THE TABLE TOP AND SEATS.

1. Cut the seats (D) to length.

2. Stand one of the A-frames upright. Place a seat on the seat support so the seat overhangs the outside of the support by 7½". Align the back edge of the seat with the end of the support. Drill two 3/32" pilot holes through the seat into the support, then insert 2½" deck screws. Attach the seat to the second A-frame the same way. Fasten the seat on the other side of the table using the same method.

2. Cut the brace. Center the brace between the seat supports, making sure they're flush at the bottom. Drill two $\frac{3}{32}$" pilot holes through the supports on each side, then fasten the brace to the supports using 2½" deck screws.

3. Cut the table top boards (F & G) to length. Place the 2 × 4 table top across the center of the table top supports, overhanging the supports by 7½". Drill two $\frac{3}{32}$" pilot holes on both ends of the top board where it crosses the supports. Attach it to the supports with 2½" deck screws.

4. Place a 2 × 8 table top board across the supports, keeping a ¼" gap from the 2 × 4. Drill pilot holes in the end of the board, then insert 2½" deck screws **(photo C).** Install the remaining top boards the same way, spacing them evenly with a ¼" gap. Allow the outside boards to overhang the end of the tabletop supports.

5. Sand any rough surfaces and splinters, and round over edges on the seat and tabletop, using 220-grit sandpaper.

6. Apply a stain, sealer, or paint following the manufacturer's instructions.

Fasten the table top and seat supports to the legs with carriage bolts. The nuts are recessed to prevent injury.

Install the table top boards by drilling pilot holes and inserting deck screws.

PROJECT
POWER TOOLS

Children's Balance Beam & Portable Putting Green

These two easy-to-build projects will provide hours of fun for your children.

Construction Materials

BALANCE BEAM

Quantity	Lumber
1	4 × 4" × 8' cedar
2	2 × 6" × 4' cedar

PORTABLE PUTTING GREEN

Quantity	Lumber
1	1 × 2" × 10' cedar
1	1 × 4" × 8' cedar
1	½" × 4 × 4' plywood

Large play structures are great for building children's strength, but children also need practice building balance and hand-eye coordination. This two projects are fun to play with and develop both of these skills.

The safe balance beam has rounded corners and sturdy supports to keep it stable. Just because it sits on the ground doesn't mean it won't challenge your child's sense of balance. Try it yourself!

The portable putting green is a great toy for developing the next generation of golfers. Use it with plastic putters for young kids, then graduate to the real thing as they grow. You can practice your technique right alongside them.

Both of these durable projects will continue to challenge your children as they grow.

OVERALL SIZE:
5" HIGH
24" WIDE
96" LONG

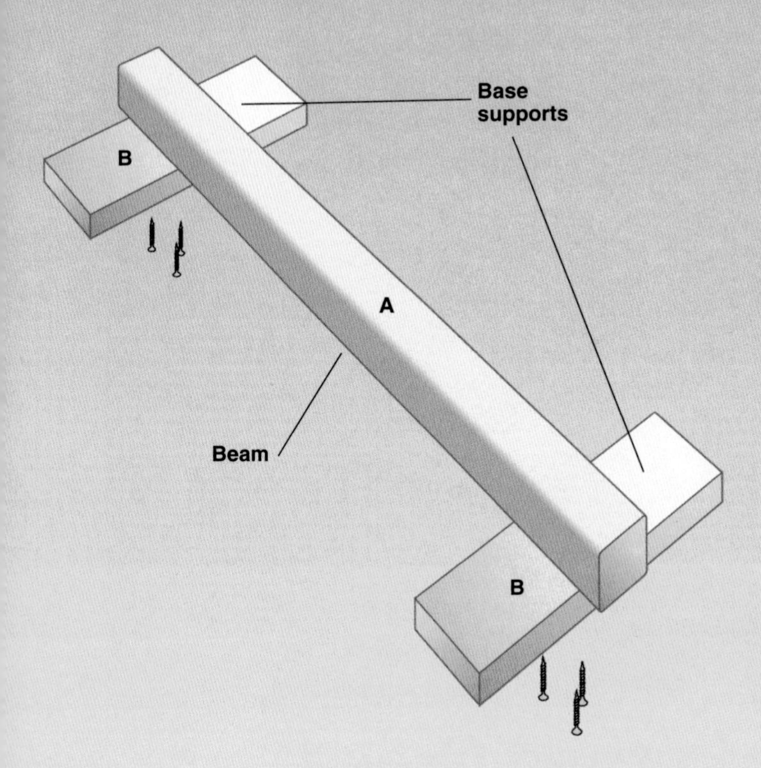

OVERALL SIZE:
3½" HIGH
31½" WIDE
72" LONG

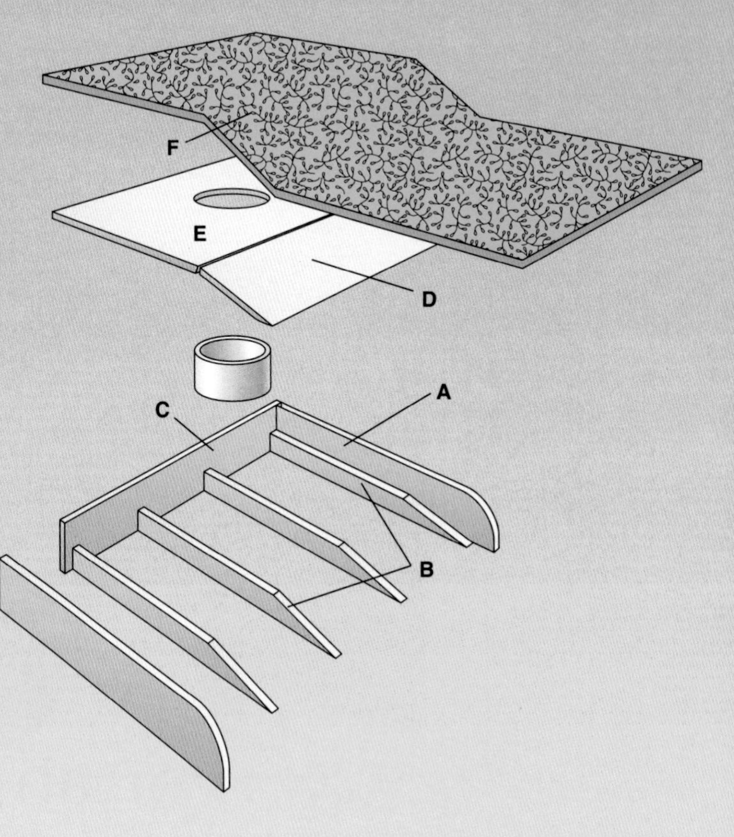

Cutting List–Balance Beam

Key	Part	Dimension	Pcs.	Material
A	Beam	3½ × 3½ × 96"	1	Cedar
B	Base	1½ × 5½ × 24"	2	Cedar

Materials: 2½" galvanized screws, sandpaper, finishing materials.

Note: Measurements reflect the actual size of dimension lumber.

Cutting List–Portable Putting Green

Key	Part	Dimension	Pcs.	Material
A	Side panel	¾ × 3½ × 30½"	2	Cedar
B	Base support	¾ × 1½ × 29"	4	Cedar
C	Back panel	¾ × 3½ × 31½"	1	Cedar
D	Ramp	½ × 11 × 30"	1	½" plywood
E	Base	½ × 20 × 30"	1	½" plywood
F	Green	30 × 72"	1	Carpet

Materials: 1¼" galvanized screws, sandpaper, 4" pvc end cap, staples, indoor/outdoor carpet, doublestick carpet tape.

Note: Measurements reflect actual size of dimension lumber.

Round the top edges and ends of the balance beam with a router.

Directions: Balance Beam

ROUND THE BEAM AND SUPPORTS.

1. Round the top edges of the beam (A), using a router with a $\frac{3}{8}$" rounding bit, and round both ends of the beam, using a router.
2. Lightly sand the rounded edges using a finish sander and 220-grit sandpaper.
3. Sand the beam to remove any splinters.
4. Cut the base supports (B) using a circular saw.
5. Round the edges and ends of the two base supports, using a router.
6. Lightly sand the base supports.

ASSEMBLE THE BEAM.

1. Position the beam on your work surface so the bottom is facing up.
2. Center the base supports on the beam. Keep the edge of the supports 2" from the end of the beam. Use a framing square to make sure the supports are perpendicular to the beam. Clamp the supports to the beam.
3. Drill three $\frac{3}{32}$" pilot holes in each base support. Stagger the holes to prevent splitting.
4. Fasten the supports to the beam using $2\frac{1}{2}$" galvanized screws.
5. Sand out any rough spots using 220-grit sandpaper. Stain or paint the balance beam.

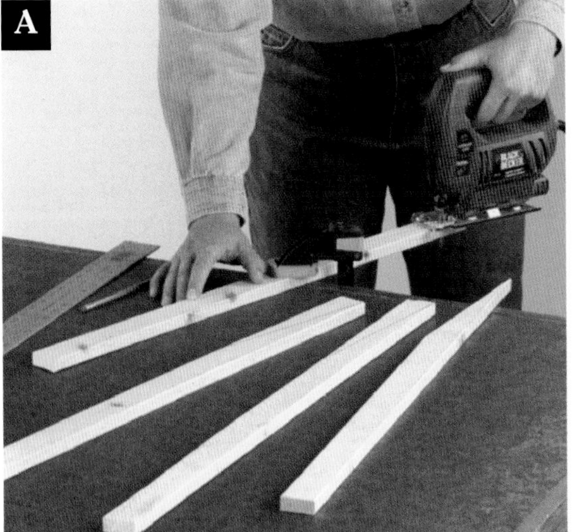

Cut the slope on the first base support, then use it as a template for the others.

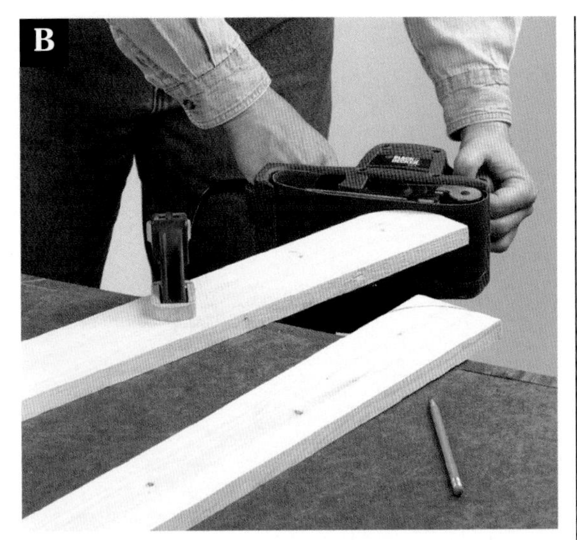

Round one corner at each side panel, using a belt sander.

Directions: Putting Green

CUT THE SLOPED BASE SUPPORTS

1. Cut four 1 × 2 base supports (B) at 29".

2. Cut the slope on the first base support by first marking a point 9" from one end. Then draw a line from that point to the opposite corner at the end of the board. Cut the line using a jig saw. Use this board as a template to cut the other supports.

BUILD THE PLYWOOD BASE.

1. Cut the base (E) at 20 × 30" and the ramp (D) at 11 × 30" from ½" plywood, using a circular saw.

2. Position the supports (B) across the bottom of the base. Align the two outside supports with the outside and back edges of the base, and evenly space the other two supports about 9" apart. Drill countersunk pilot holes in the base, then attach the base to the supports, using glue and 1¼" galvanized screws.

3. Sand one long edge of the ramp (D) to a gentle bevel (approximately 15°), using a belt sander. Apply glue to the beveled edge of the ramp and the sloped edge of the supports. Butt the ramp against the edge of the base, then attach it to the supports using 1¼" screws driven through countersunk pilot holes. Don't worry if there is a slight gap between the base and ramp since the carpet will cover it.

4. Use a belt sander to bevel the leading edge of the ramp until it lies flat.

5. Cut two 1 × 4 side panels (A) at 30½" and one back panel (C) at 31½". Round one corner of each side panel using a belt sander. Attach the back and side panels to the base, using 1¼" screws driven through pilot holes.

INSTALL CUP & CARPET.

1. Center the cup hole about 8" from the back wall. Trace the PVC end cap onto the plywood base (E) at the hole location. Drill a starter hole to insert the blade of a jig saw, then cut out the hole. Attach the cup using hot glue.

2. Cut a 30" wide piece of indoor/outdoor carpeting (F) using a utility knife. Attach the carpet to the base using double stick carpet tape and staples. Cut away the carpet over the hole.

Drill a starter hole to insert the blade of a jig saw, then cut out the cup hole.

PROJECT
POWER TOOLS

Timberframe Sandbox

A playground just isn't complete without a sandbox, and this version gives an old favorite a new look.

CONSTRUCTION MATERIALS

Quantity	Lumber
14	4 × 4" × 8' cedar
1	1 × 8" × 12' cedar
2	1 × 6" × 8' cedar
2	2 × 2" × 6' cedar

This sandbox is much more refined than nailing four boards together and hoping for the best. The timber construction is not only charming, it's solid. A storage box at one end gives kids a convenient place to keep their toys. The other end has built-in seats, allowing children to sit above the sand as they play. The gravel bed and plastic sheathing provide a nice base for the sandbox, allowing water to drain while keeping weeds from sprouting in the sand. The structure is set into the ground for stability, and to keep the top of the pavers at ground level so you can easily mow around them. When your children outgrow the sandbox, turn it into a garden bed.

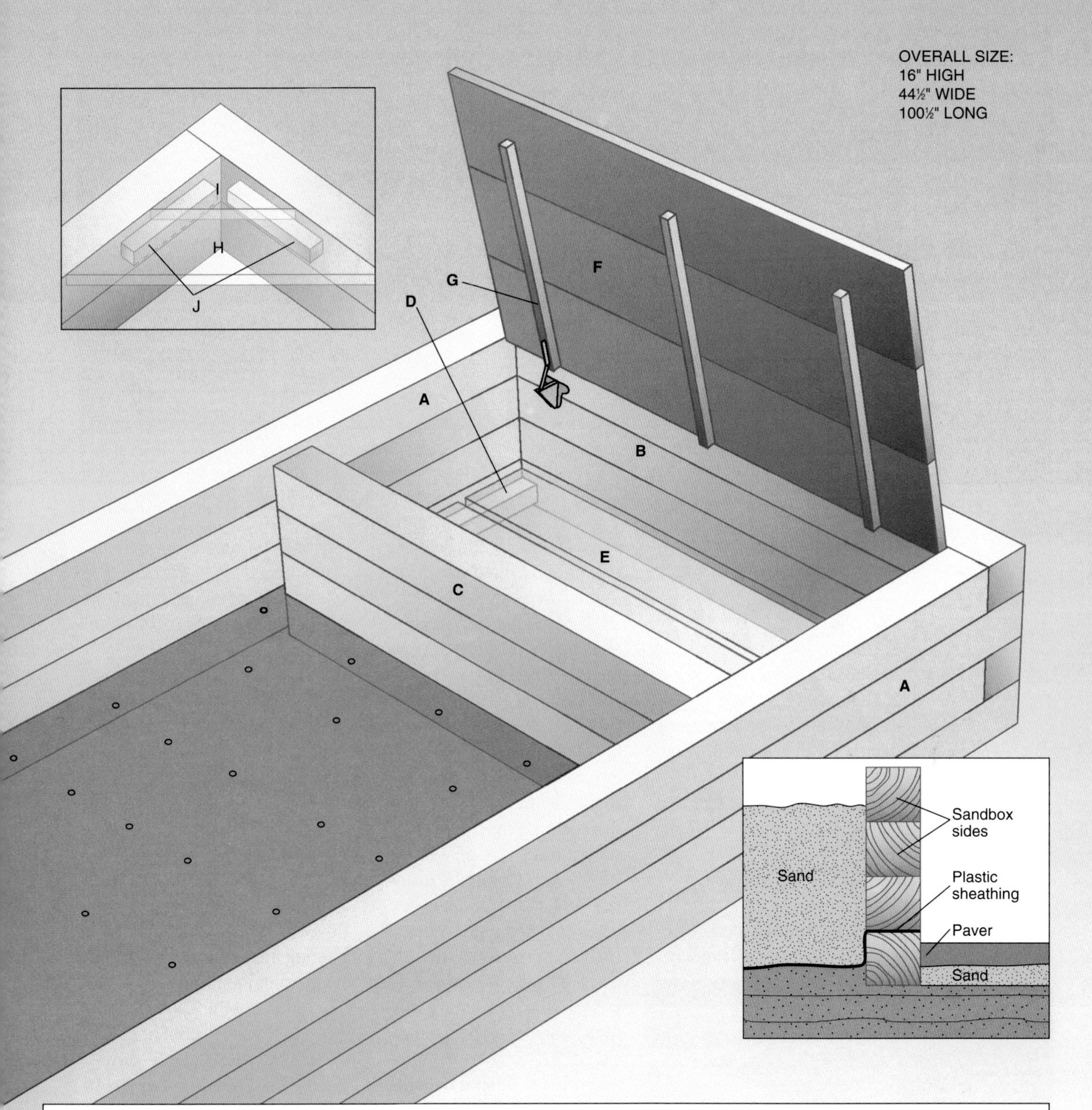

Cutting List

Key	Part	Dimension	Pcs.	Material
A	Sandbox sides	3½ × 3½ × 92½"	8	Cedar
B	Sandbox ends	3½ × 3½ × 44½"	8	Cedar
C	Storage box wall	3½ × 3½ × 41"	4	Cedar
D	Floor cleats	1½ × 1½ × 18"	2	Cedar
E	Floor boards	¾ × 5½ × 43"	3	Cedar

Cutting List

Key	Part	Dimension	Pcs.	Material
F	Lid boards	¾ × 7½ × 43½"	3	Cedar
G	Lid cleats	1½ × 1½ × 18"	3	Cedar
H	Bench boards	¾ × 5½ × 18"	2	Cedar
I	Corner bench boards	¾ × 5½ × 7"	2	Cedar
J	Bench cleats	1½ × 1½ × 10"	4	Cedar

Materials: Coarse gravel, sand, wood sealer/protectant, heavy duty plastic sheathing, 2" galvanized screws, 6" barn nails, pavers, hinges.

Note: Measurements reflect the actual size of dimension lumber.

Use a shovel to remove the grass in the sandbox location, then dig a trench for the first row of timbers.

Lay the first row of timbers, including the wall for the storage box. Fill the sandbox area with a 2" layer of gravel and cover with plastic sheathing.

Directions: Timberframe Sandbox

PREPARE THE SITE.

1. Outline a 48 × 96" area using stakes and strings.

2. Use a shovel to remove all of the grass inside the area. Dig a flat trench 2" deep by 4" wide around the perimeter of the area, just inside the stakes and string **(photo A).**

LAY THE FIRST ROW OF TIMBERS.

1. Cut the side (A), end (B) and storage box wall timbers, using a reciprocating saw. Coat the timbers with a wood sealer and let dry completely.

2. Place the first tier of sides and ends in the trench so the corners alternate (see opening photo). Place a level across a corner, then add or remove soil to level it. Level the other three corners the same way. Drill two 3/16" pilot holes through the timbers at the corners, then drive 6" barn nails through the pilot holes.

4. Measuring from the inside of one end, mark for the inside edge of the storage box at 18" on both sides. Align the storage box wall with the marks, making sure the corners are square, then score the soil on either side of it. Remove the timber and dig a 3" deep trench at the score marks.

5. Replace the storage box timber in the trench. Its top edge must be 3/4" lower than the top edge of the first tier of the sandbox wall. Add or remove dirt until the storage box timber is at the proper height.

6. Drill 3/16" pilot holes through the sandbox sides into the ends of the storage box timber, then drive 6" barn nails through the pilot holes.

7. Pour 2" of coarse gravel into the sandbox section. Rake the gravel smooth.

8. Cover the gravel bed section with heavy duty plastic sheathing **(photo B).** Pierce the plastic with an awl or screwdriver at 12" intervals for drainage.

BUILD THE SANDBOX FRAME.

1. Set the second tier of timbers in place over the first tier and over the plastic sheathing, staggering the joints with the joint pattern in the first tier.

2. Starting at the ends of the timbers, drill 3/16" pilot holes every 24", then drive 6" galvanized barn nails through the pilot holes. Repeat for the remaining tiers of timbers, staggering the joints.

3. Stack the remaining storage box timbers over the first one. Drill 3/16" pilot holes through the sandbox sides into the ends of the storage box timbers, then drive 6" barn nails into the pilot holes **(photo C).**

4. Cut the excess plastic from

Build the rest of the sandbox frame, staggering the corner joints. Drill holes and drive barn nails through the holes.

Attach the bench lid using heavy-duty hinges. Install a child-safe lid support to prevent the lid from falling shut.

Install 2 × 2 support cleats ¾" from the top of the sandbox. Attach the corner bench boards using galvanized screws.

around the outside of the sandbox timbers, using a utility knife.

BUILD THE STORAGE BOX FLOOR AND LID.

1. Cut the floor cleats (D) and position one against each side wall along the bottom of the storage box and attach them using 2" galvanized screws.

2. Cut the floor boards (E) and place over the cleats with a ½" gap between each board to allow for drainage. Fasten the floor boards to the cleats using 2" screws.

3. Cut the lid boards (F) and lay out side by side with the ends flush. Cut the lid cleats (G) and place across the lid, one at each end and one in the middle, making sure the end of each cleat is flush with the back edge of the lid. Drill pilot holes and attach the cleats using 2" galvanized screws.

4. Attach the lid to the sandbox frame using heavy-duty child safe friction hinges **(photo D).**

BUILD CORNER BENCHES.

1. Cut the bench cleats (J). Mark ¾" down from the top edge of the sandbox at two corners. Align the top edge of the bench cleats with the mark and fasten using 2" galvanized screws.

2. Cut the corner bench board (I) to length with a 45° angle at each end. Place it in the corner and attach it to the cleats using 2" screws **(photo E).** Cut the bench board (H) to length with a 45° angle at each end. Butt it against the corner bench board, then attach it to the cleats. Repeat this step to install the second corner bench.

FILL SANDBOX AND INSTALL BORDER.

1. Fill the sandbox section with sand.

2. Mark an area the width of your pavers around the perimeter of the sand box. Remove the grass and soil in the paver area to the depth of your pavers plus another 2", using a spade.

3. Spread a 2" layer of sand into the paver trench. Smooth the sand level using a flat board.

4. Place the pavers on top of the sand base, beginning at a corner of the sandbox **(photo F).** Use a level or a straightedge to make sure the pavers are even and flush with the surrounding soil. If necessary, add or remove sand to level the pavers. Set the pavers in the sand by tapping them with a rubber mallet.

5. Fill the gaps between the pavers with sand. Wet the sand lightly to help it settle. Add new sand as necessary until the gaps are filled.

Place the pavers into the sand base. Use a rubber mallet to set them in place.

PROJECT
POWER TOOLS

Kid-sized Porch Swing

Your children will cherish the memories of spending pleasant summer days on their very own porch swing.

Construction Materials

Quantity	Lumber
1	2 × 4" × 8' cedar (1)
1	2 × 6" × 8' cedar (1)
10	1 × 2" × 10' cedar (10)

This swing is about three-quarters of the size of a full-sized porch swing and is designed specifically for children. The gentle curves of the slatted seat and back are built for their comfort.

You can hang the swing from your porch, a tree branch, or under a deck. Be sure to balance the swing by properly adjusting the links on the chain. This particular swing was built with cedar lumber, which is long-lasting and attractive.

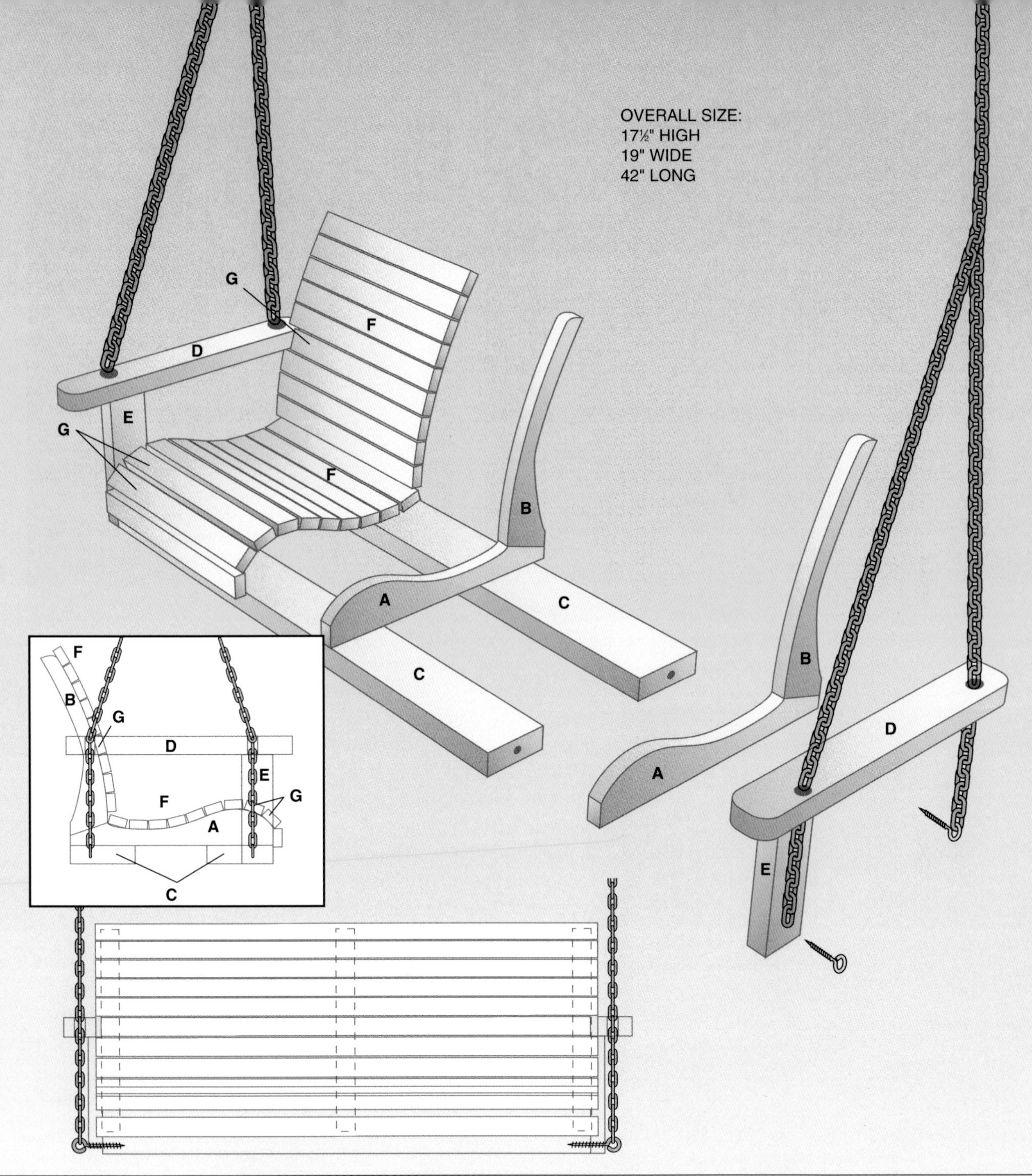

Key	Part	Dimension	Pcs.	Material
A	Seat Support	1½ × * × 17½"	3	Cedar
B	Back Support	1½ × * × 14½"	3	Cedar
C	Support Rail	1½ × 3½ × 39"	2	Cedar
D	Arm	1½ × 2⅝ × 19"	2	Cedar

Key	Part	Dimension	Pcs.	Material
E	Arm Upright	1½ × 2⅝" × 8¾	2	Cedar
F	Long Slats	¾ × 1½ × 40"	16	Cedar
G	Short Slats	¾ × 1½ × 39"	3	Cedar

Materials: Exterior wood glue, 2½" galvanized screws, ¼ × 3" lag screws (2), 5⁄16 × 6" lag eye screw (4), heavy chains (2), 2" chain connectors (6)

Note: Measurements reflect the actual size of dimension lumber.

* cut to size using template

Make sure the inside corner is flush, then attach each seat support to a back support with glue and galvanized screws.

Attach support rails to the seat supports using galvanized screws.

Drill a hole through the back support into the arm and attach it with a lag screw.

Directions: Kid-sized Porch Swing

BUILD THE SEAT AND BACK SUPPORTS.

1. Enlarge the seat support and back support templates on page 165 using the grid system or a photocopier.
2. Trace the patterns for three sets of seat (A) and back supports (B) onto a 2 × 6, then cut out the pieces using a jig saw. Sand the edges smooth using a belt sander.
3. Apply wood glue to the flat section on the top of the seat support. Position the back support on top of the seat support, making sure the two pieces are flush at the inside corner, and clamp. Drill two 3⁄16" pilot holes, then attach the two pieces using 2½" galvanized screws **(photo A).** Repeat this process for each seat-back support set.

ATTACH THE SUPPORT RAILS.

1. Cut the support rails (C) to size.
2. Set the joined seat and back supports face down on the edge of your work surface so the back support sections hang off the edge. Place a support rail across the seat supports. Align the end of the rail with the outside and back edges of a side seat support. Drill two pilot holes in the support rail, then attach it using 2½" galvanized screws. Attach the rail to the other side seat support the same way.
3. Align the second support rail along the front edge of the seat supports, making sure the edges are flush. Drill pilot holes and attach the rail to the two outside seat supports using 2½" galvanized screws **(photo B).**
4. Center the third seat support between the side seat supports and attach it to the front and rear rails using two 2½" galvanized screws through each rail.

BUILD AND ATTACH ARMS.

1. Cut the arms (D) and arm uprights (E) by ripping a 2 × 6 down the middle, then cutting the pieces to length to create two sets of arms and uprights about 2⅝" wide. Round the corners of the arms using a jig saw or a belt sander. Sand all the edges smooth.
2. Align the edge of an arm upright with the front edge of the

Install the finished slats, started at the crook and covering the seat and back. Space the slats evenly.

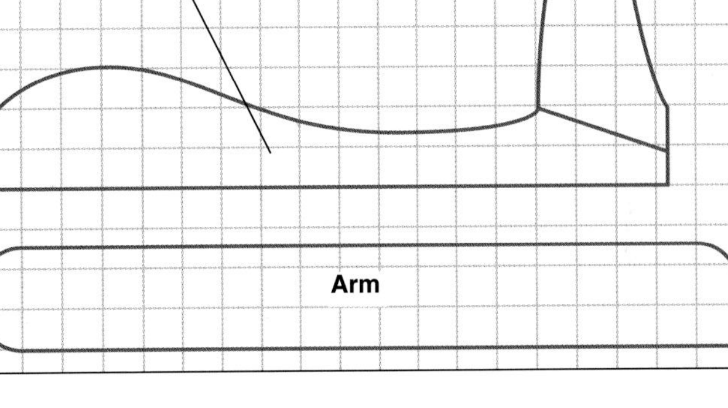

front support rail. Make sure the upright is plumb, then drill two pilot holes and attach it to the support rail using 2½" galvanized screws.

3. Position the arm on top of the upright so it is flush with the inside of the upright and overhangs the front by 1½". Drill two pilot holes, then attach the arm to the upright using 2½" galvanized screws.

4. Make sure the arm is level, then drill a ⅛" pilot hole through the back support into the arm. Attach the arm to the back support using a ¼ × 3" lag screw **(photo C).** Repeat steps 2 through 4 for the other arm and upright.

CUT AND ATTACH SLATS.

1. Cut the long slats (F) and short slats.

2. Paint or stain the swing and the slats as desired. Allow the paint 24 hours to dry before continuing.

3. Install the long slats, beginning at the crook of the seat, spacing them evenly. Drill pilot holes, then attach each slat using a 1¼" galvanized screw driven into each seat support.

4. Install the short slats between the arms by driving a 1¼" galvanized screw into each seat support.

HANG THE SWING.

1. Measure and make a mark on top of the arm 3" from the front edge and ¾" from the outside edge for the front chain hole. Make a mark for the rear chain hole at 1½" from the back edge and centered from side to side. Drill holes through the marked points using a ¾" spade bit.

2. Drill pilot holes and insert screw eyes into the ends of the rear support rail and ¾" from the bottom of the arm uprights.

3. Insert the chain through the arm holes and hook the chain to the screw eyes using chain connectors. Suspend the swing and adjust the length of the chain and the position of the connectors until the swing balances properly.

4. Hang the swing using heavy screw eyes inserted into ceiling joists or into a 2 × 4 lag-screwed across the ceiling joists.

Suspend the swing and adjust the chain length and the position of the connectors until the swing balances.

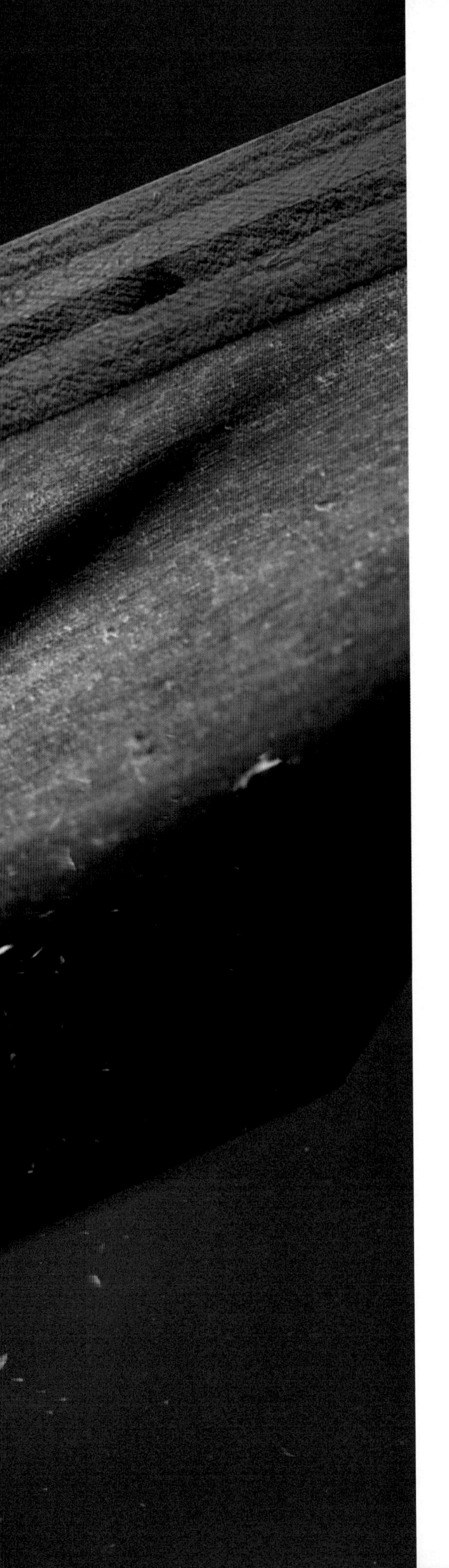

Bird & Pet Projects

Birds bring motion and sound to the yard and garden. Invite them to linger a while longer by creating a stationary or hanging feeder designed just for them. For those animals that hold a permanent place in your yard and heart, we've included plans for cozy homes that provide security for them and style for your outdoor home.

PROJECT POWER TOOLS

Birdhouse

Give your local birds a dry, clean shelter with these easy-to-build birdhouses.

CONSTRUCTION MATERIALS

Quantity	Lumber
1	1 × 6" × 4' cedar

Birds are always looking for nesting areas. Why not help out with this simple house? It's also a great project for children.

Our version is constructed with a swing-out door to make annual fall cleaning easy. There is no perch because it is not necessary.

You can embellish this basic birdhouse many ways, as in the examples below. There are, however, a few important things to keep in mind: don't paint or apply preservatives to the inside of the house, the inside edge of the entrance hole, or within ¼" of the face of the entrance hole or it will keep away the birds. The birdhouse can be hung with simple eyescrews and a chain, mounted on a post, or vertically mounted to a tree or other structure.

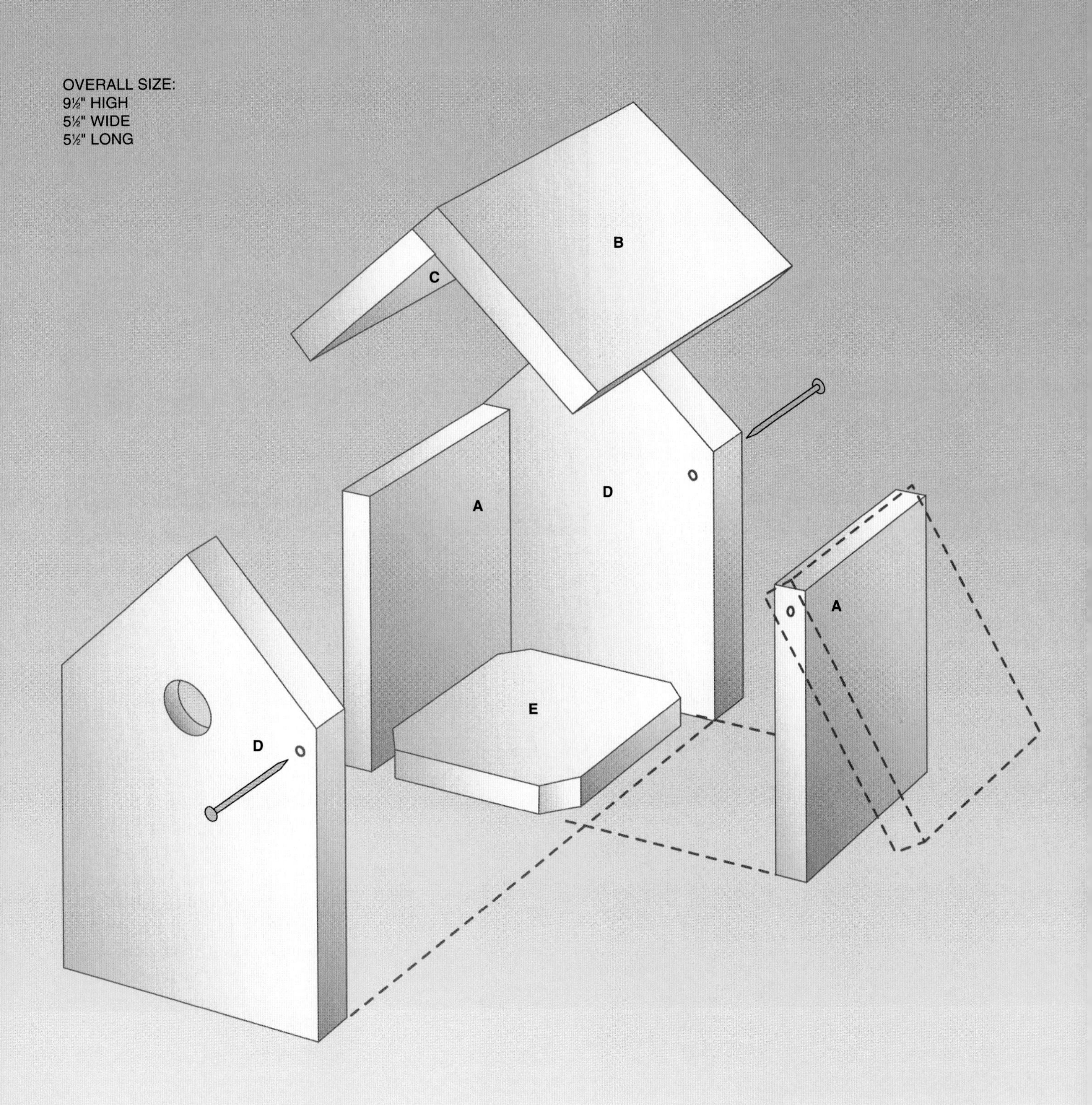

Key	Part	Dimension	Pcs.	Material
A	Side	¾ × 4 × 5½"	2	Cedar
B	Roof	¾ × 5½ × 6½"	1	Cedar
C	Roof	¾ × 4¾ × 6½"	1	Cedar

Key	Part	Dimension	Pcs.	Material
D	Front/Back	¾ × 5½ × 8¾"	2	Cedar
E	Bottom	¾ × 4 × 4"	1	Cedar

Cutting List

Materials: 4d galvanized finish nails, exterior wood glue, shoulder hook.

Note: Measurements reflect the actual size of dimension lumber.

Directions: Birdhouse

PREPARE THE PARTS.

1. Cut the sides (A) and bottom (E) by first ripping a 15¼" piece of 1 × 6 to a width of 4". Cut the pieces to length. On the bottom piece, make a diagonal cut across each corner, ½" from the end, to allow for drainage.

2. Cut the front and back (D) to length. To make the peaks, make a mark on each side of these pieces, 2¾" from the top. Mark the center point at the top. Mark lines from the center point to each side, then cut along them **(photo A).**

3. Mark a point on the front piece 6¾" from the base, centering the mark from side to side. Use an appropriately-sized spade bit to drill an entrance hole, usually 1¼ to 1½" (see Tip).

4. Use a wood screw or awl to make several deep horizontal scratches on the inside of the front piece, starting 1" below the entrance hole. (These grip lines help young birds hold on as they climb up to the entrance hole.)

Use a speed square as a cutting guide and gang-cut the table parts when possible for uniform results.

ASSEMBLE THE BASE AND SIDES.

1. Apply wood glue to one edge of the bottom piece. Butt a side piece against the bottom piece so the bottoms of the two pieces are flush.

2. Drill ⅙" pilot holes and attach the pieces using 4d galvanized finish nails. Repeat this process for the front and back pieces, aligning the edges with the side piece **(photo B).**

3. Set the remaining side piece in place, but do not glue it. To attach the side to the front and back pieces, drill ⅙" pilot holes and drive a 4d nail through the front wall and another through the back wall, each positioned about ⅝" from the top edge. This arrangement allows the piece to pivot.

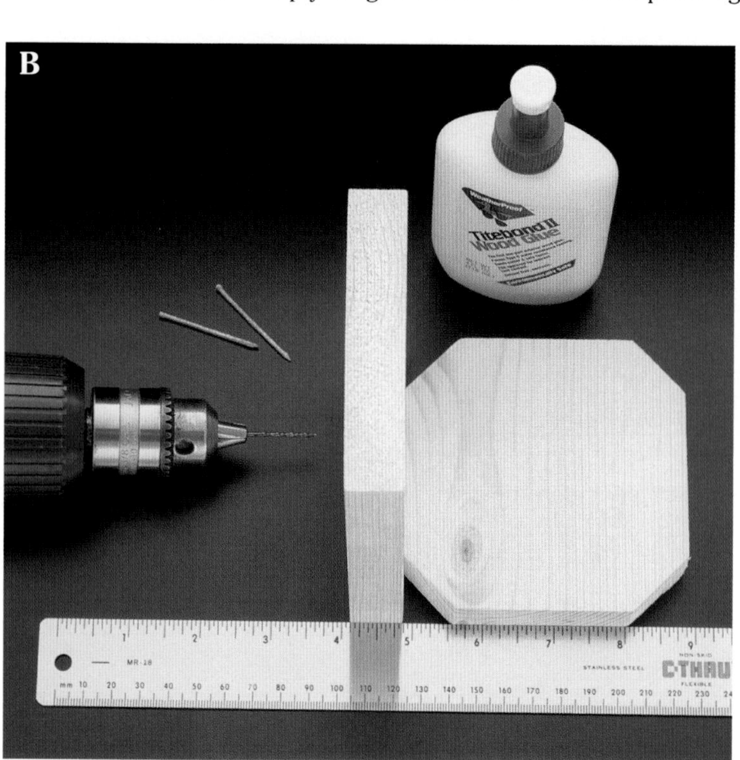

Mark the ends of the tapers on the leg sides, then connect the marks to make taper cutting lines.

ADD THE ROOF AND FINISH.

1. Cut the roof (B & C) pieces to size.

2. Apply glue to the top edges of one side of the front and back pieces. Set the smaller roof piece on the house so its upper edge is aligned with the peak of the house.

3. Apply glue to the top edges

Use a jig saw or circular saw to cut the leg tapers.

TIP

Different bird species prefer different sized nesting boxes. Some species, like robins, will not nest in boxes, but prefer platforms on which to build their nests. Many publications give even more specific information on how to attract nesting birds to your yard.
Keeping predators and invasives species like sparrows from invading nesting boxes is important. Drilling the proper size entrance hole protects your house from becoming a home to sparrows or squirrels. Do not use perches, as these allow predatory birds to sit and wait for adults and nestlings to emerge.
The following chart shows nesting box dimensions for common bird species.

NEST BOX DIMENSIONS

Species	**Box floor**	**Box height**	**Hole height**	**Hole diameter**	**Box placement**
Eastern Bluebird	*5 × 5"*	*8 to 12"*	*6 to 10"*	*1½"*	*4 to 6 ft.*
Chickadees	*4 × 4"*	*8 to 10"*	*6 to 8"*	*1⅛"*	*4 to 15 ft.*
Titmice	*4 × 4"*	*10 to 12"*	*6 to 10"*	*1¼"*	*5 to 15 ft.*
Red-breasted Nuthatch	*4 × 4"*	*8 to 10"*	*6 to 8"*	*1¼"*	*5 to 15 ft.*
White-breasted Nuthatch	*4 × 4"*	*8 to 10"*	*6 to 8"*	*1⅜"*	*5 to 15 ft.*
Northern Flicker	*7 × 7"*	*16 to 18"*	*14 to 16"*	*2½"*	*6 to 20 ft.*
Yellow-bellied Sapsucker	*5 × 5"*	*12 to 15"*	*9 to 12"*	*1½"*	*10 to 20 ft.*
House wrens	*4 × 4"*	*6 to 8"*	*4 to 6"*	*1¼"*	*5 to 10 ft.*
Carolina Wren	*4 × 4"*	*6 to 8"*	*4 to 6"*	*1½"*	*5 to 10 ft.*
Wood Ducks	*10 × 18"*	*10 to 24"*	*12 to 16"*	*4"*	*10 to 20 ft.*

on the opposite side of the front and back pieces. Place the larger roof piece in position. Drill pilot holes and drive 4d nails through the roof into the front piece and then the back **(photo C).**

4. Drill a pilot hole in the edge of the front piece on the pivot wall side, placed about 1" from the bottom edge of the house. Screw in a shoulder hook, positioning it to hold the side piece closed.

5. Sand the birdhouse smooth, then paint or decorate it as desired. We hot-glued sticks and straw to the bird houses on the opposite page to create a "Three Little Pigs" theme. The brick exterior on the third bird house is contact paper and the roof is hot-glued into place. These items can be purchased at craft stores.

Bird Feeder

A leftover piece of cedar lap siding is put to good use in this rustic bird feeder.

Construction Materials

Quantity	Lumber
1	¾ × 16 × 16" plywood scrap
1	¾" × 6' cedar stop molding
1	8" × 10' cedar lap siding
1	1 × 2" × 8' cedar
1	1"-dia. × 3' dowel

Watching birds feeding in your backyard can be a very relaxing pastime. In this bird feeder project, you will use a piece of 8"-wide cedar lap siding to build a decorative feeder box and then mount it on a piece of scrap plywood. The birds won't mind the leftover building materials. And you'll like the bird feeder because it costs almost nothing to build. Even the plastic viewing window covers that you place inside the feeder box can be made with clear acrylic scrap left over from another project. To fill this cleverly designed bird feeder with seed, turn the threaded rod that serves as a hook so it is aligned with the slot in the roof. Then, simply lift up the roof and add the bird food.

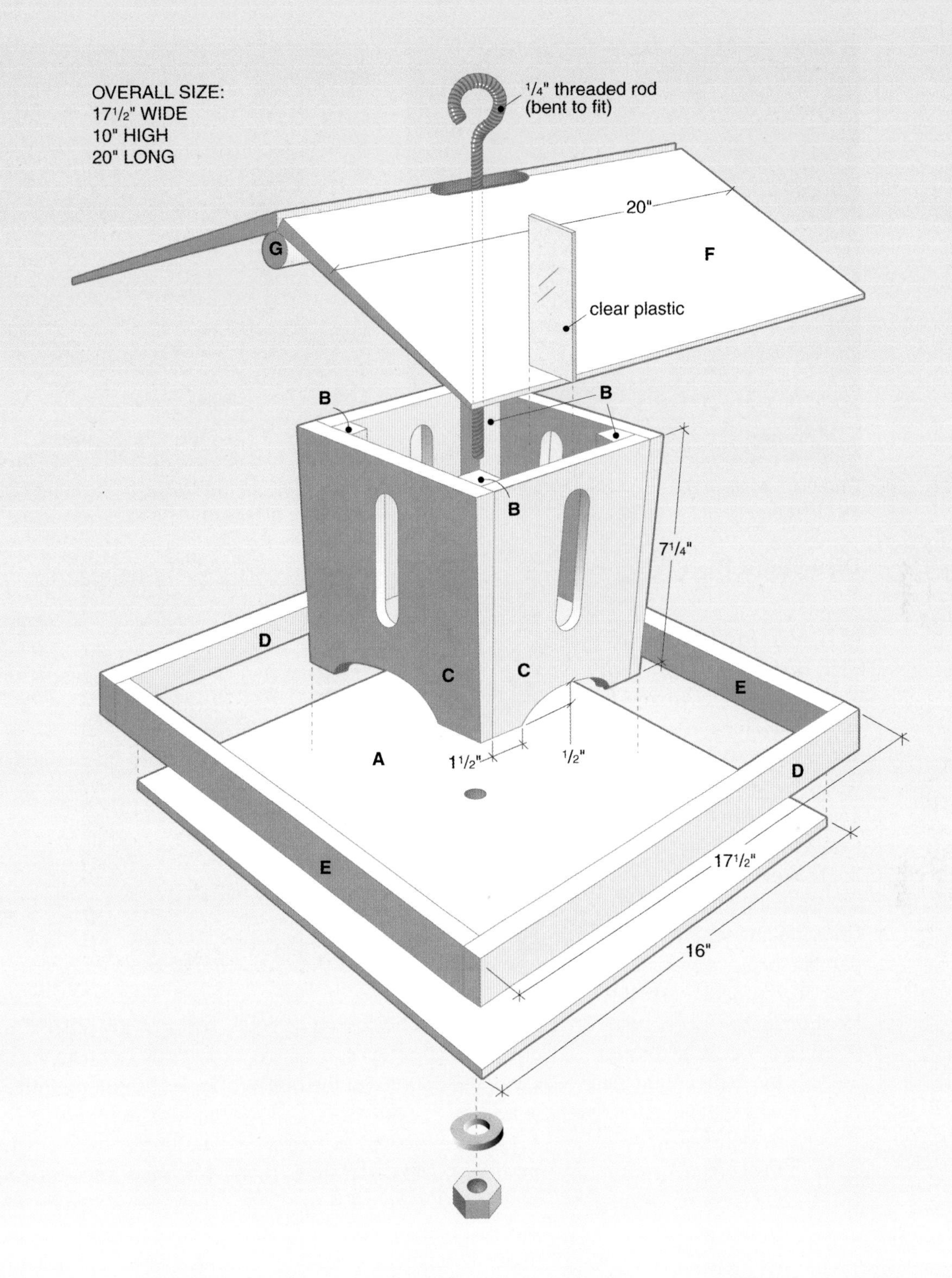

Cutting List

Key	Part	Dimension	Pcs.	Material
A	Base	¾ × 16 × 16"	1	Plywood
B	Post	¾ × ¾ × 7¼"	4	Cedar
C	Box side	5⁄16 × 6 × 7¼"	4	Cedar siding
D	Ledge side	¾ × 1½ × 17½"	2	Cedar

Cutting List

Key	Part	Dimension	Pcs.	Material
E	Ledge end	¾ × 1½ × 16"	2	Cedar
F	Roof panel	5⁄16 × 7¼ × 20"	2	Cedar siding
G	Ridge pole	1"-dia. × 20"	1	Dowel

Materials: ¼"-dia. threaded rod with matching nut and washer, hot glue, 4d common nails, rigid acrylic or plastic.

Note: Measurements reflect the actual size of dimension lumber.

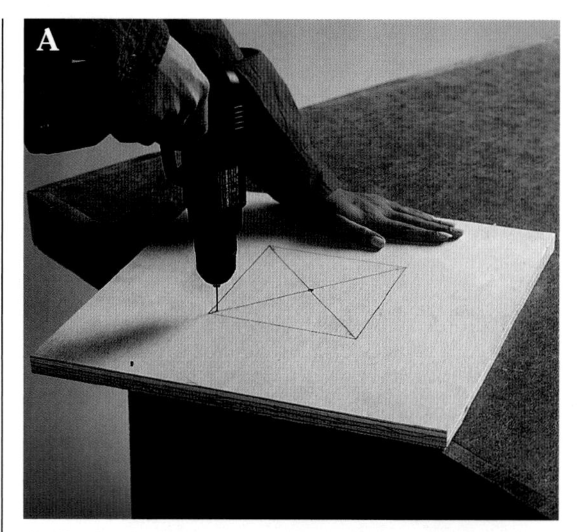

Drill pilot holes in the corners of the feeder box location that is laid out on the plywood base.

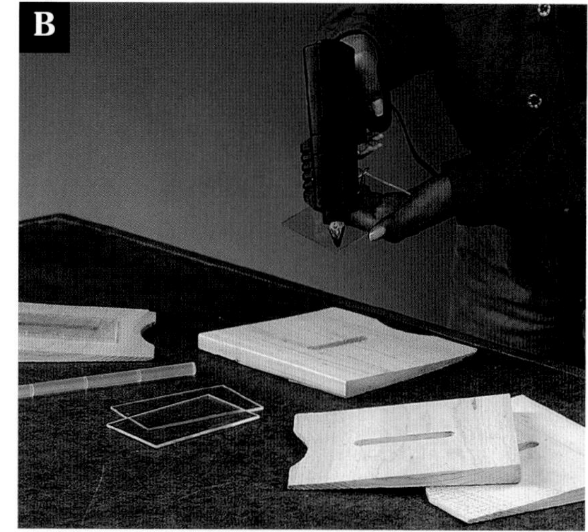

Cover the viewing slots by hot-gluing clear plastic or acrylic pieces to the inside face of each panel.

Directions: Bird Feeder

Mark the profile of the bevel of the siding onto two of the box sides for trimming.

CUT AND PREPARE THE BASE.

The base provides room for several feeding birds and seed.

1. Cut the base (A) from ¾" plywood. Draw straight diagonal lines from corner to corner to locate the center of the base.

2. Measure and mark a 6" square in the middle of the base, making sure the lines are parallel to the edges of the base. This square marks the location for the feeder box.

3. Drill a ¼"-dia. hole through the center of the base where the lines cross.

4. Measure in toward the center ⅜" from each corner of the 6" square and mark points. Drill 1/16" pilot holes all the way through at these points **(photo A).**

PREPARE THE FEEDER BOX PARTS.

The posts and box sides form the walls of the feeder box. Vertical grooves in the box sides let you check seed levels. Seed flows through small arcs cut in the bottoms of the box sides.

1. Cut the posts (B) to length from ¾"-square cedar stop molding. (Or, rip a 3'-long piece of ¾"-thick cedar to ¾" in width to make the posts.)

2. From 8" cedar lap siding (actual dimension is 7¼") cut two 6"-wide box sides (C). Then, cut two more panels to about 7" in width to be trimmed later to follow the lap-siding bevels.

3. Cut viewing slots. First, drill two ½" starter holes for a jig saw blade along the center of each box side—one hole 2" from the top, and the other 2" from the bottom. Connect the starter holes by cutting with a jig saw to form the slots.

4. Cut a ½"-deep arc into the bottom of each box side, using the jig saw. Start the cuts 1½" from each end. Smooth out the arcs with a drum sander on a power drill.

5. Cut strips of clear acrylic or plastic slightly larger than the viewing slots. Hot-glue them over the slots on the inside of the box sides **(photo B).**

6. To mark cutting lines for trimming two of the box sides to follow the siding bevel, tape the box sides together into a

TIPS

Hot glue is often thought of as primarily a product for craftmaking and indoor patch-up jobs. But for lightweight exterior projects, it is a very effective adhesive. The hot glue penetrates deeply into the pores of cedar, and creates a strong, durable bond between wood parts.

Drive 4d common nails through pilot holes to fasten the feeder box to the base.

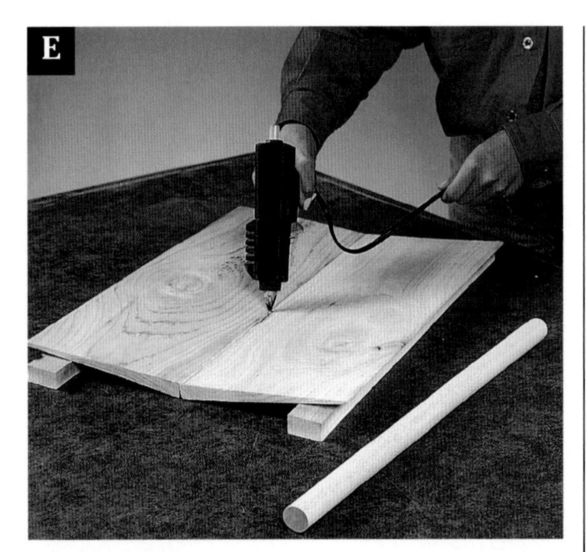

Insert spacers 2" in from the "eaves" of the roof to set the pitch before applying glue to the seam.

box shape. The wide ends of the beveled siding should all be flush. Trace the siding profile onto the inside faces of the two box ends **(photo C).** Disassemble the box. Cut along the profile lines with a jig saw.

ASSEMBLE THE FEEDER BOX.

1. Hot-glue the posts flush with the inside edges on the box sides that were trimmed in Step 6 (above).

2. Hot-glue the untrimmed box sides to the posts.

ATTACH THE BASE.

1. Align the assembled feeder box with the 6" square outline on the base. Hot-glue the box to the base on these lines. Turn the assembly upside down.

2. Attach the base to the feeder box by driving 4d galvanized common nails through the predrilled pilot holes in the base, and into the posts on the feeder box **(photo D).**

3. Cut the ledge sides (D) and ledge ends (E) to length. Next, build a frame around the base that prevents seed spills. Using hot glue, attach the ledge pieces so the bottoms are flush with the bottom of the base. Reinforce the joint with 4d common nails.

MAKE THE ROOF.

1. Cut the ridge pole (G) from a 1"-dia. dowel. Cut the roof panels (F) from 8" siding.

2. To create the roof pitch, lay the panels on your work surface so the wide ends butt together. Place a 1"-thick spacer under each of the narrow ends, 2" in from each end.

3. Apply a heavy bead of hot glue into the seam between the panels **(photo E).** Quickly press the ridge pole into the seam. Let the glue harden for at least 15 minutes.

4. Set the roof right-side-up, and rest each end of the ridge pole on a 2 × 4 block. Drill ⅜" starter holes down through the roof and the ridge pole, 1" to either side of the ridge's midpoint. Connect the starter holes by cutting a slot between them, using a jig saw. Widen the slot until the ¼"-dia. threaded rod passes through with minimal resistance.

5. Cut the threaded rod to 16" in length. Use pliers to bend a 1½"-dia. loop in one end of the rod. Place the roof on the feeder box. Then, thread the unbent end of the rod through the roof and the hole in the base **(photo F).** Spin the rod loop so it is perpendicular to the roof ridge.

6. Tighten a washer and nut onto the end of the rod, loosely enough so the loop can be spun with moderate effort. For a rustic look, don't apply a finish to your bird feeder.

The bird feeder is held together by a looped, threaded rod that runs through the roof and is secured with a washer and nut on the underside of the base.

Bird Feeder Stand

Send an invitation to flocks of colorful backyard guests by hanging bird feeders from this sturdy cedar stand.

PROJECT POWER TOOLS

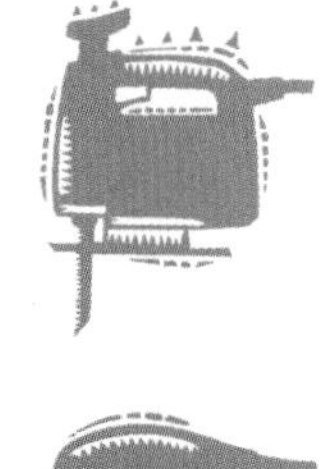

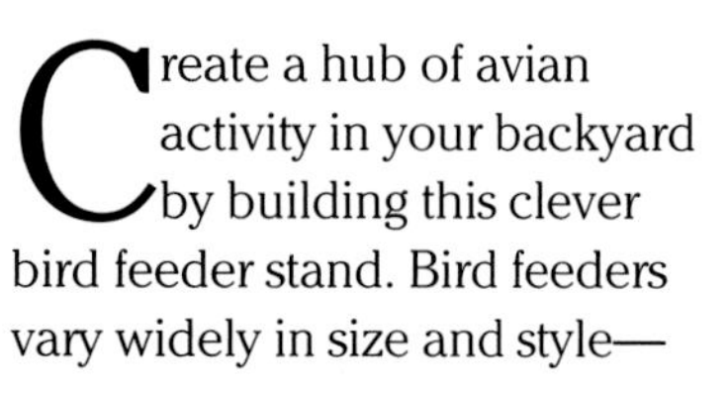

Create a hub of avian activity in your backyard by building this clever bird feeder stand. Bird feeders vary widely in size and style—from small and plain to large and fanciful. This stand can support more than one kind of bird feeder at a time, letting you show off your favorite types. If you want to attract different species of birds to your feeding area, hang feeders that contain different foods. Then sit and enjoy the sight of a variety of birds fluttering and roosting in one central area.

One important benefit of this cedar bird feeder stand is that it has a freestanding, open design. Birds are always in full view as they eat.

The heavy stand base, made from cedar frames, provides ample support for the post and hanging arms. To simplify cleanup of any spilled food (and to make it accessible to hungry birds), you can attach a layer of window screening over the slats in the top of the base. Cleaning the bird feeder stand is easy—just remove the feeders, tip the stand on its side and spray it down with a hose.

CONSTRUCTION MATERIALS

Quantity	Lumber
1	1 × 4" × 8' cedar
1	1 × 4" × 10' cedar
2	1 × 4" × 12' cedar
2	1 × 6" × 12' cedar
2	2 × 4" × 12' cedar
1	2 × 6" × 6' cedar

OVERALL SIZE:
72" HIGH
35¼" WIDE
35¼" LONG

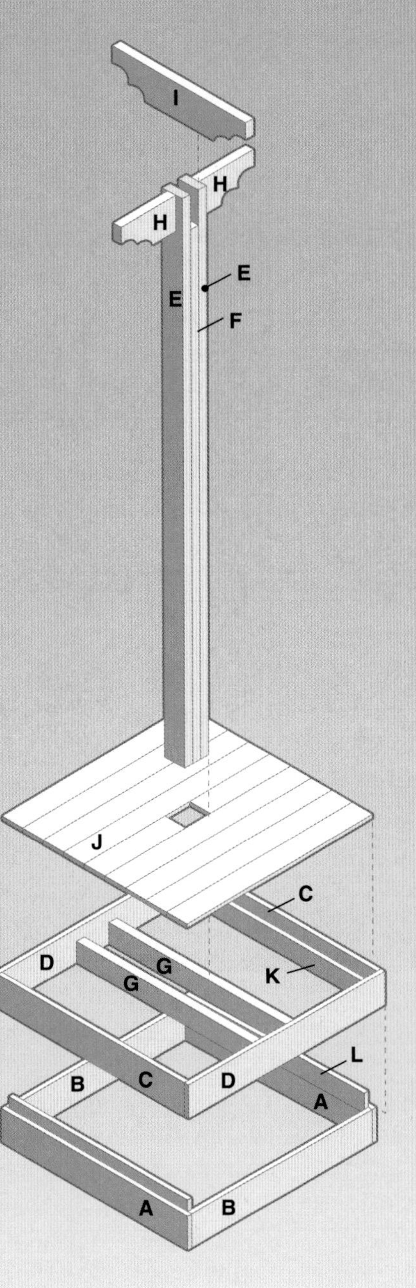

1" squares

PART H DETAIL

Cutting List

Key	Part	Dimension	Pcs.	Material
A	Bottom end	⅞ × 5½ × 33½"	2	Cedar
B	Bottom side	⅞ × 5½ × 31¾"	2	Cedar
C	Top end	⅞ × 5½ × 33½"	2	Cedar
D	Top side	⅞ × 5½ × 35¼"	2	Cedar
E	Post board	1½ × 3½ × 72"	2	Cedar
F	Center board	1½ × 3½ × 66½"	1	Cedar

Cutting List

Key	Part	Dimension	Pcs.	Material
G	Post support	1½ × 3½ × 33½"	2	Cedar
H	Outside arm	1½ × 5½ × 10¼"	2	Cedar
I	Inside arm	1½ × 5½ × 36"	1	Cedar
J	Floor board	⅞ × 3½ × 33½"	9	Cedar
K	Floor support	⅞ × 3½ × 33½"	2	Cedar
L	Bottom cleat	⅞ × 3½ × 31¾"	2	Cedar

Materials: 1½" and 2½" deck screws, 18 × 36" window screening (2), eye hooks, finishing materials.

Note: Measurements reflect the actual size of dimension lumber.

Join the top base frame to the bottom base frame by driving screws through the frame cleats.

Use a square to make sure the inside arm is perpendicular to the post before you secure it into the gap at the top of the post.

Directions: Bird Feeder Stand

BUILD THE BASE FRAMES.
1. Cut the bottom ends (A), bottom sides (B), top ends (C) and top sides (D) to length. Sand the parts smooth. Drill ⅛" pilot holes near the ends of the bottom ends and counterbore the holes to a ¼" depth with a counterbore bit. Fasten the bottom sides between the bottom ends by driving 1½" deck screws through the pilot holes. Repeat this procedure with the top sides and top ends to complete the second base frame.
2. Cut the floor supports (K) to length. Fasten them to the inside faces of the top ends so the bottoms of the supports are flush with the bottoms of the ends. Cut the bottom cleats (L) to length. Attach them with 1½" deck screws to the inside faces of the bottom ends. Make sure the top edge of each bottom cleat is 1½" above the top edge of each bottom end.
3. Set the top frame over the bottom frame. Fasten the top and bottom frames together by driving deck screws through the bottom cleats and into the top frame **(photo A).**

TIP

There is a real art to making and stocking bird feeders, identifying species and enjoying bird watching. If you are just a budding ornithologist, make a visit to your local library—the more knowledge you acquire, the more enjoyment you will experience.

INSTALL POST SUPPORTS.
1. Mark the centerpoints of the top sides on their inside faces. Draw reference lines, 2¼" to each side of the centerpoints. These lines mark the locations for the post supports (G).
2. Cut the post supports to length. Place them in the top frame so their bottom edges rest on the tops of the bottom sides. Position the post supports with their inside faces just outside the reference lines. Drill pilot holes through the frame and counterbore the holes. Fasten the post supports to the top frame by driving 2½" deck screws through the frame and into the supports.

BUILD THE ARMS.
1. Cut the two outside arms (H) and the inside arm (I) to length. Use a pencil to draw a 1"-square grid pattern on one of the arms. Using the grid patterns as a reference (see *Diagram,* page 177), lay out the decorative scallops at the end of the arm.
2. Cut along the layout lines with a jig saw. With a 1"-dia. drum sander mounted in an electric drill, smooth the insides of the curves. Use the arm as a template to draw identical scallops on the other arms. Then, cut and sand the other arms to match.

MAKE THE POST.
The post is constructed by sandwiching the center board (F) between two post boards (E). It's easiest to attach the outside arms before you assemble the post.
1. Cut the post boards (E) to length and draw 5½"-long center lines on one face of each post board, starting at the top. Then, draw a 5½"-long line, ¾"

to each side of the center line, to mark the outlines for the outside arms on the post. On the center line, drill pilot holes for the deck screws, 1½" and 4½" down from the top edge. Counterbore the holes.

2. Attach the outside arms to the side posts by driving 2½" deck screws through the posts and into the straight ends of the outside arms. Sandwich the center board between the side post boards, with the bottom and side edges flush.

3. Drive pairs of 2½" deck screws at 8" to 12" intervals, screwing through the face of one post board. Then, flip the assembly over and drive screws through the other post board. Make sure to stagger them so you don't hit screws driven from the other side.

4. Center the inside arm in the gap at the top of the post **(photo B).** Then, drive 2½" deck screws through the post boards and into the inside arm.

5. Install the post assembly by standing the post up between the post supports in the base frame. Be sure the post is centered between the top frame sides and is perpendicular to the post supports. Drive 2½" deck screws through the post supports and into the post to secure the parts.

MAKE THE FEEDING FLOOR.

Floor boards are attached to the floor supports within the top base frame.

1. Cut the floor boards (J) to length. One floor board should be cut into two 14½"-long pieces to fit between the post and frame.

2. Arrange the floor boards across the post supports and floor supports, using ¼"-wide scraps to set ¼"-wide gaps between the boards.

3. To fasten the floor boards to the floor supports and post supports, first drill pilot holes in the floor boards and counterbore the holes. Then, drive 1½" deck screws through the pilot holes and into the floor supports **(photo C).**

Attach the floor boards by driving deck screws through the floor boards and into the post and floor supports.

APPLY FINISHING TOUCHES.

1. Apply exterior wood stain to the bird feeder stand. After it dries, staple two 18 × 36" strips of window screening to the floor to keep food from falling through the gaps **(photo D).**

2. Insert brass screw eyes or other hardware at the ends of the arms to hang your bird feeders. Set the stand in a semisheltered area in clear view of your favorite window or deck.

Staple window screening over the tops of the floor boards to keep bird food from falling through the gaps.

PROJECT
POWER TOOLS

Doghouse

Add a contemporary twist to a traditional backyard project with this cedar-trimmed, arched-entry doghouse.

CONSTRUCTION MATERIALS

Quantity	Lumber
2	1 × 2" × 8' cedar
3	2 × 2" × 8' pine
2	2 × 4" × 8' cedar
2	⅝" × 4 × 8' siding
1	¾" × 4 × 8' ABX plywood

Close your eyes and picture the first image that comes to mind when you think of a doghouse. More than likely it's a boxy, boring little structure. Now consider this updated doghouse, with its sheltered breezeway and contemporary styling. What dog wouldn't want to call this distinctive dwelling home? The sturdy 2 × 4 frame provides a stable foundation for the wall panels and roof. The main area has plenty of room to house an average-sized dog comfortably, and the porch area shelters the entry, while providing an open, shady area for your pet to relax. The rounded feet keep the inside of the house dry by raising the base up off the ground.

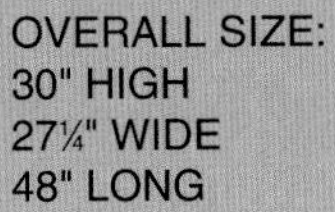

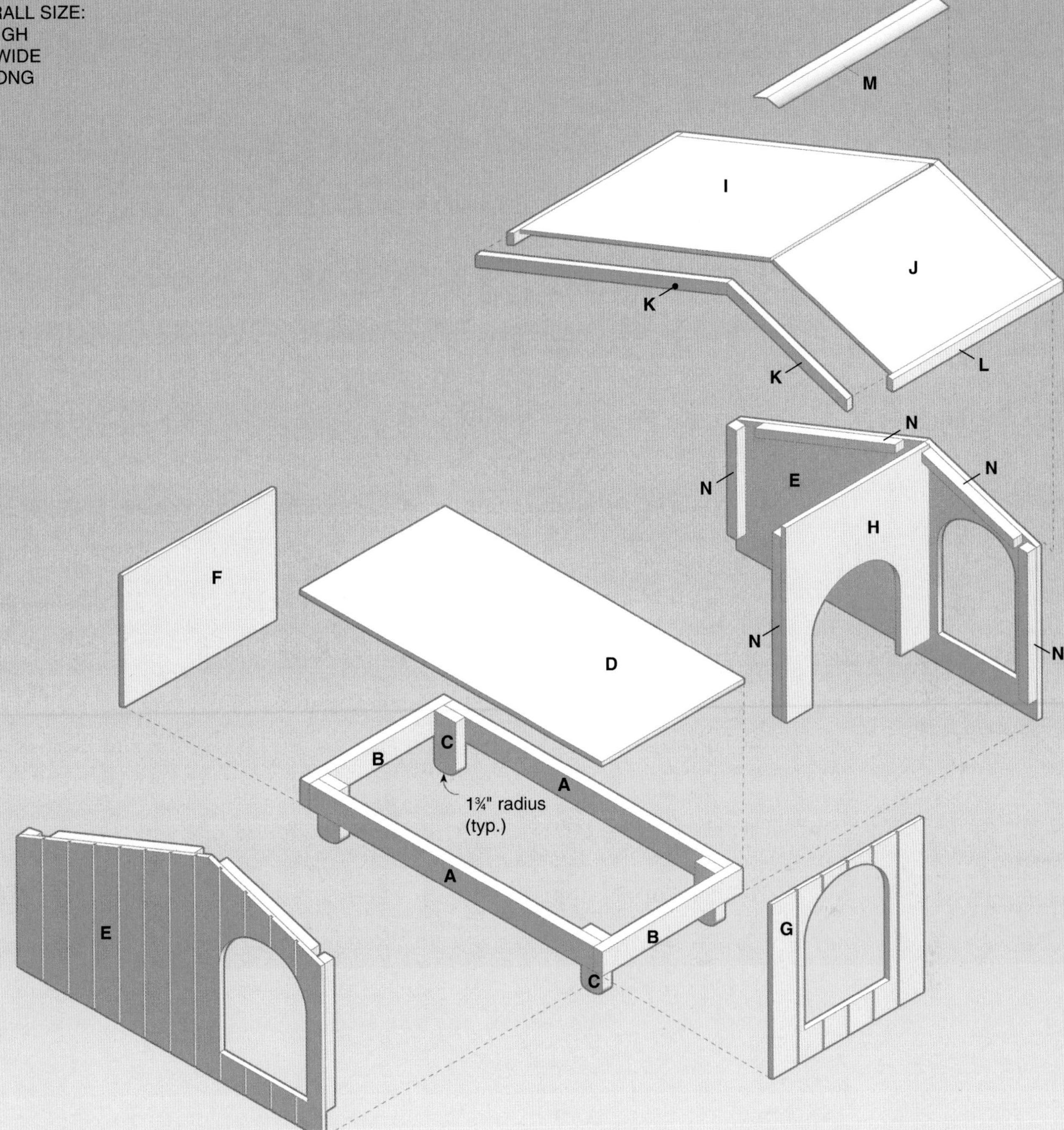

Cutting List

Key	Part	Dimension	Pcs.	Material
A	Frame side	1½ × 3½ × 45"	2	Cedar
B	Frame end	1½ × 3½ × 22⅞"	2	Cedar
C	Feet	1½ × 3½ × 7½"	4	Cedar
D	Floor	¾ × 22⅞ × 48"	1	ABX Plywood
E	Side panel	⅝ × 30 × 48"	2	Siding
F	House end panel	⅝ × 18 × 24"	1	Siding
G	Porch end panel	⅝ × 24 × 24"	1	Siding

Cutting List

Key	Part	Dimension	Pcs.	Material
H	Center panel	⅝ × 22⅞ × 29¾"	1	Siding
I	House roof	¾ × 25½ × 35"	1	ABX Plywood
J	Porch roof	¾ × 25½ × 23"	1	ABX Plywood
K	Side roof trim	⅞ × 1½ × *	4	Cedar
L	End roof trim	⅞ × 1½ × 27¼"	2	Cedar
M	Flashing	¹⁄₁₆ × 4 × 27¼"	1	Galv. flashing
N	Cleat	1½ × 1½ × *"	10	Pine

Materials: 2" and 3" deck screws, 6d galvanized finish nails, 2d galvanized common nails, silicone caulk, roofing nails with rubber washers, finishing materials.

*Cut to fit **Note:** Measurements reflect the actual size of dimension lumber.

Directions: Doghouse

BUILD THE FRAME & FLOOR.

The frame of the doghouse is the foundation for the floor, sides and roof. It is built from 2 × 4 cedar lumber.

1. Cut the frame sides (A) and frame ends (B) to length. Place the frame sides between the frame ends to form a rectangle, then fasten together with 3" deck screws. Make sure to keep the outside edges flush.

2. Cut the feet (C) to length. Use a compass to lay out a 1¾"-radius roundover curve on one end of each foot, then cut with a jig saw to form the roundover. Smooth out the jig-saw cuts with a power sander.

3. Fasten a foot in each corner of the frame with 3" deck screws **(photo A).** Be sure to keep the top edges of the feet flush with the top edges of the frame.

4. Cut the floor (D) to size from ¾"-thick exterior plywood, and fasten it to the top of the frame with 2" deck screws. The edges of the floor should be flush with the outside edges of the frame.

Fasten the 2 × 4 cedar feet to the inside frame corners with 3" galvanized deck screws.

MAKE THE WALLS.

The walls for the doghouse are cut from ⅝"-thick siding

Lay out the roof angle on the side panels using a straightedge.

panels—we chose panels with grooves cut every 4" for a more decorative effect.

1. Cut the side panels (E) to the full size listed in the *Cutting List* on page 181.

2. Create the roof line by cutting peaks on the top of the panels. To make the cuts, first mark points 18" up from the bottom on one end, and 24" up from the bottom on the other end. Measure in along the top edge 30" out from the end with the 24" mark, and mark a point to indicate the peak of the roof. Connect the peak mark to the marks on the ends with straight lines to create the cutting lines **(photo B).** Lay the side panels on top of one another, fastening them with a screw or two in the waste area. Then cut both panels at the same time, using a circular saw or jig saw and straightedge cutting guide.

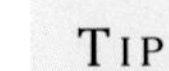

TIP

With most siding products, whether they are sheet goods or lap siding boards, there is a definite front and back face. In some cases, it is very easy to tell which face is meant to be exposed, but you always should be careful not to confuse the two.

Lay out the opening archway on the side panels using a ruler and pencil.

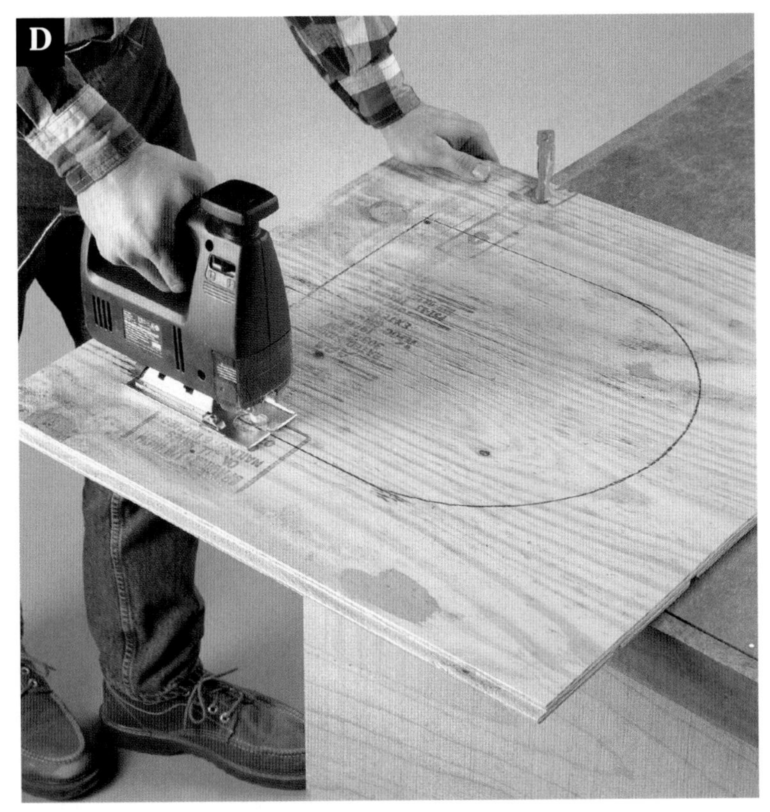

Cut out the openings in the panels with a jig saw.

3. Make the arched cutouts in the front (taller) sections of the side panels, by first measuring and marking a point 2" and 16" in from the 24"-tall end of one panel, then drawing lines from the bottom to the top of the panel, through the points. Measure up 4¼" and 15¾" from the bottom edge and draw horizontal lines to complete the square. Find the centerpoint between the sides of the square cutout outline, and measure down 7" from the top of the cutout at that point. Press down on the end of a ruler so it pivots at that point, and use the ruler and a pencil like a compass to draw a curve with a 7" radius across the top of the cutout **(photo C).** Drill a starter hole at a corner of the cutout outline, then cut the opening with a jig saw **(photo D).** Trace the cutout onto the other side panel, then make that cutout.

4. Cut the center panel (H) and porch end panel (G) to full size. Use one of the side panel cutouts to trace an arched

Fasten the center panel by driving screws through the side panels into the cleats. Use a combination square to keep the panel even.

cutout outline onto the porch end panel so the sides are 4½" from each side edge and the top is 15¾" up from the bottom. Mark an arched cutout outline on the center panel, 3⅞" from each side edge and 15¾" up from the bottom.

5. Make the cutouts with a jig saw, then sand all cut edges smooth.

TIP

If plan dimensions do not meet your needs, you can recalculate them to a different scale. The doghouse shown here is designed for an average dog (about 15" tall). If you own a larger dog, add 1" to the size of the entry cutouts and panels for every inch that your dog is taller than 15".

ATTACH THE WALLS & FRAME.

1. Cut the house end panel (F).

2. Fasten the side panels (E) to the frame with 2" deck screws, so the bottoms of the panels are flush with the bottoms of the frame, and the ends of the panels are flush with the frame ends.

3. Fasten the house end panel (F) and the porch end panel (G) to the frame so the bottoms of the panels are flush with the bottom of the frame (the sides of the end panels will overlap the side panels by ⅝" on each side).

4. Cut the 10 cleats (N) long enough to fit in the positions shown in the *Diagram* on page 181—there should be a little space between the ends of the cleats, so exact cutting is not important. Just make sure the edges are flush with the edges of the panel they are attached to.

5. Fasten four cleats along the perimeter of each side panel (E), using 2" deck screws.

6. Fasten the remaining two cleats at the edges of the back side of the center panel (H).

7. Set the center panel between the side panels so the front is aligned with the peak in the roof. Make sure the center panel is perpendicular, then attach it with 2" deck screws driven through the side panels and into the cleats at the edges of the center panel **(photo E).**

ATTACH THE ROOF & TRIM.

The roof and trim are the final structural elements to be fastened to the doghouse.

1. Cut the house roof (I) and porch roof (J) to size from ¾"-thick exterior plywood.

2. Fasten the roof panels to the cleats at the tops of the side walls, making sure the edges of the panels butt together to form the roof peak.

3. Cut the trim pieces to frame the roof (K, L) from 1 × 2 cedar. The end roof trim pieces are square-cut at the ends, but

the ends of the side roof trim pieces (K) need to be miter-cut to form clean joints at the peak and at the ends, where they meet the end trim. To mark the side trim pieces for cutting, first cut the side trim pieces so they are an inch or two longer than the space between the end of the roof panel and the roof peak. Lay each rough trim piece in position, flush with the top of the roof panel. On each trim piece, mark a vertical cutoff line that is aligned with the end of the roof panel. Then, mark a cutoff line at the peak, making sure the line is perpendicular to the peak. Cut the trim pieces with a power miter saw or miter box and backsaw.
4. Attach the trim pieces to the side panels with 6d galvanized finish nails **(photo F).**

Cut each side roof trim piece to fit between the peak and the end of the roof panel, mitering the ends so they will be perpendicular when installed. Attach all the roof trim pieces with galvanized finish nails.

APPLY FINISHING TOUCHES.

1. Sand all the wood surfaces smooth, paying special attention to any sharp edges, then prime and paint the doghouse. Use a good-quality exterior primer and at least two coats of paint, or you can do as we did and simply apply two or three coats of nontoxic sealant to preserve the natural wood tones. We used linseed oil.
2. Cut a strip of galvanized steel flashing (M) to cover the roof peak (or you can use aluminum flashing, if you prefer). Use tin snips or aviator snips to cut the flashing, and buff the edges with emery paper to help smooth out any sharp points.
3. Lay the flashing lengthwise on a wood scrap, so the flashing overhangs by 2". Bend the flashing over the edge of the board to create a nice, crisp peak, then attach the flashing with roofing nails with neoprene (rubber) washers driven at 4" intervals **(photo G).**

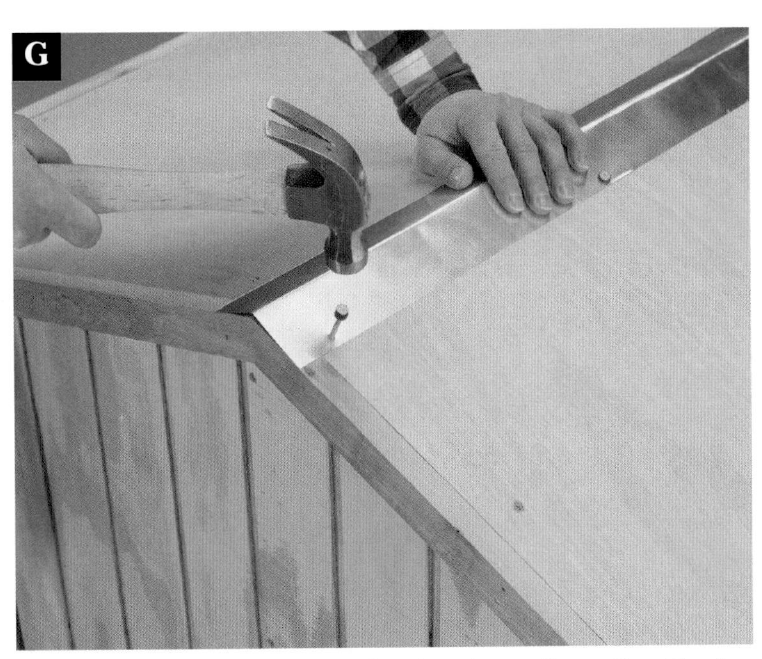

Install metal flashing over the roof peak, using roofing nails with rubber washers.

PROJECT
POWER TOOLS

Rabbit Hutch

With its two compartments, this rabbit hutch provides both a breezy and cozy home for your bunny.

CONSTRUCTION MATERIALS

Quantity	Lumber
7	2 × 2 × 6' cedar
7	2 × 4" × 6' cedar
1	⅝" × 4 × 8' grooved cedar plywood siding

This rabbit hutch is an easy to build outdoor shelter for your bunny. The floor is made of hardware cloth which allows droppings to fall through but is easy on the rabbit's feet. A large airy compartment is enclosed with hardware cloth and a cozy smaller compartment is sided. Each compartment has a door to make feeding and cage cleaning an easier task.

Place straw or wood shavings in the compartment to make comfortable bedding for bunny.

Finish the rabbit hutch with an animal safe exterior stain. Place the hutch in a protected area out of direct sun.

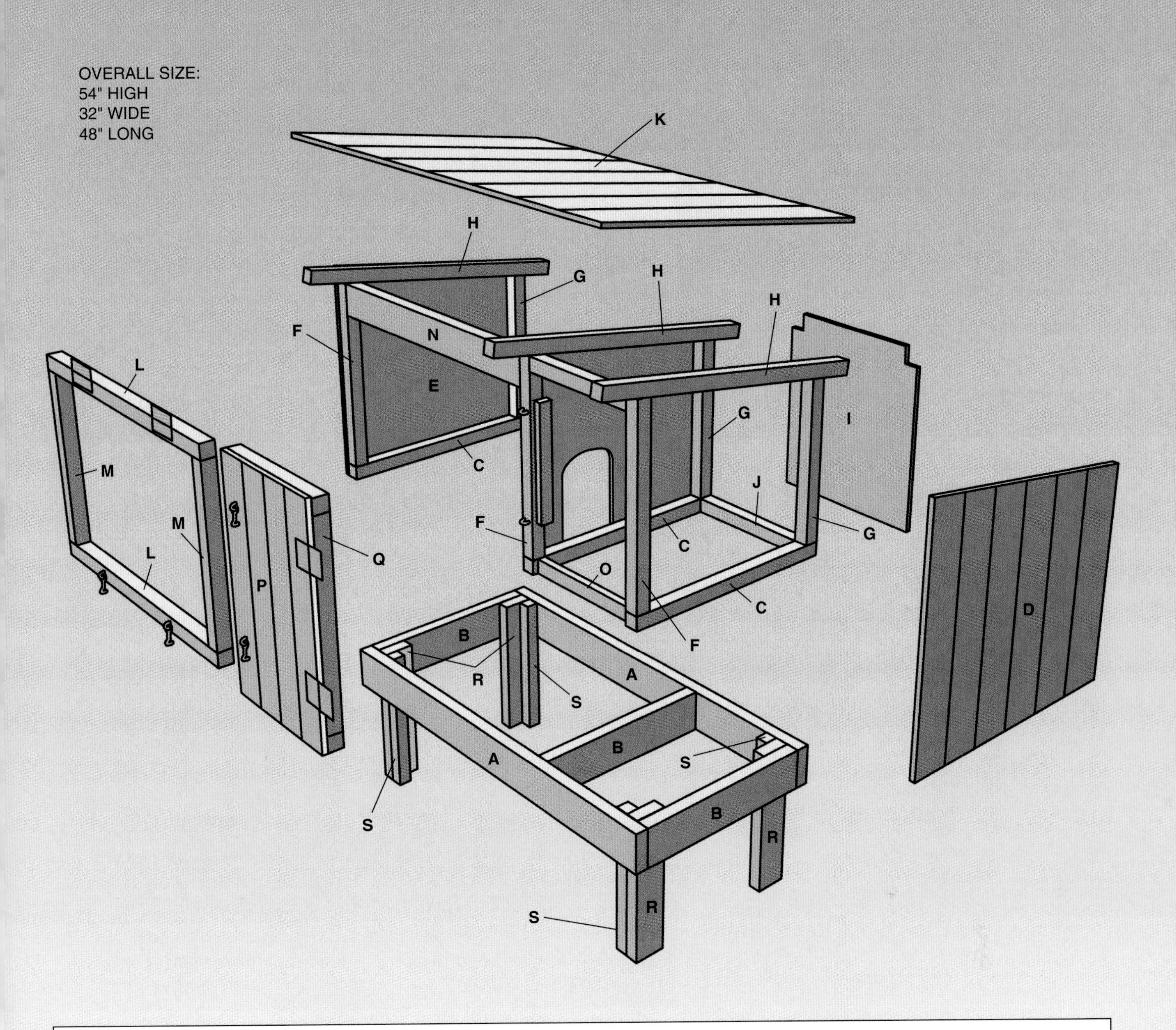

Cutting List

Key	Part	Dimension	Pcs.	Material
A	Floor side	1½ × 3½ × 47½"	2	Cedar
B	Floor crosspiece	1½ × 3½ × 21"	3	Cedar
C	Frame base	1½ × 1½ × 24"	3	Cedar
D	Right side wall	½ × 24 × 24"	1	Siding
E	Left side wall	½ × 24 × 24"	1	Siding
F	Frame front	1½ × 1½ × 21"*	3	Cedar
G	Frame back	1½ × 1½ × 17½"*	3	Cedar
H	Frame top	1½ × 1½ × 32"	3	Cedar
I	Back wall	½ × 17¼ × 20"	1	Siding
J	Back wall stop	1½ × 1½ × 13¼"	1	Siding

Cutting List

Key	Part	Dimension	Pcs.	Material
K	Roof	½ × 32 x 48"	1	Siding
L	Door crosspiece	1½ × 1½ × 29½"*	2	Cedar
M	Door side	1½ × 1½ × 17¾"*	2	Cedar
N	Hinge support	1½ × 3½ × 29½"*	1	Cedar
O	Door jamb	1½ × 1½ × 13¼"*	1	Cedar
P	Compartment door	½ × 13 × 22½"*	1	Siding
Q	Door supports	1½ × 1½ × *	4	Cedar
R	Legs	1½ × 3½ × *	4	Cedar
S	Legs	1½ × 1½ × *	4	Cedar

Materials: ½" × 4' × 8' hardware cloth, ¾" fence staples, 1¼" and 2½" deck screws, 3 × 3" hinges (4), hook and eye fasteners (4).
Note: Measurements reflect the actual size of dimension lumber.
*cut to fit.

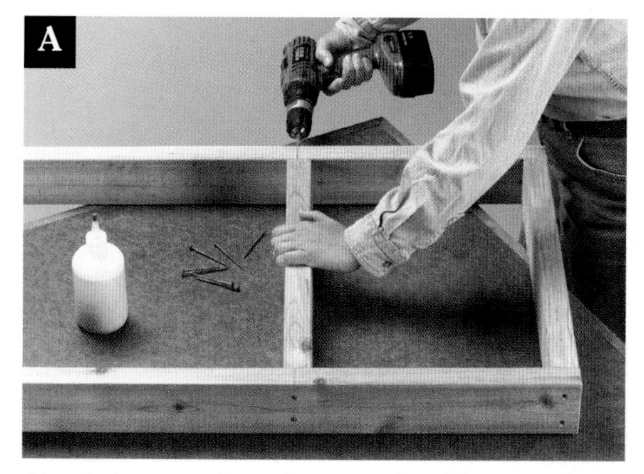

Attach the crosspieces between the sides to make the floor.

Center the frame top and mark the ends so they are parallel to the wall sides.

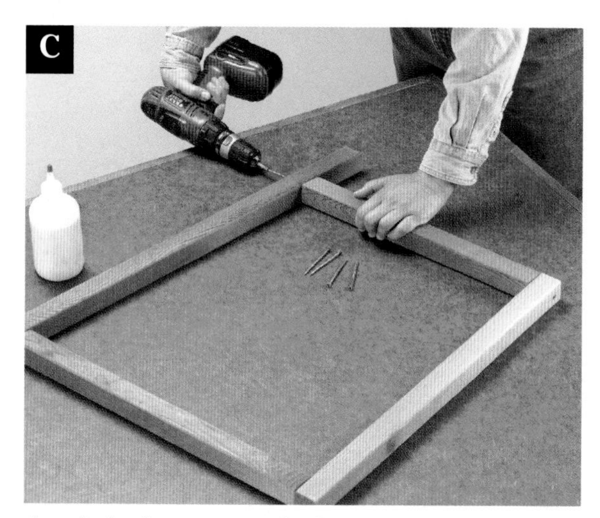

Attach the frame top to the frame back, sides, and base to create the cage side.

Directions: Rabbit Hutch

BUILD THE FLOOR.

For all screws in this project, drill a 9/64" pilot hole and a ⅛" deep counterbore.

1. Cut the floor sides (A) and crosspieces (B). Mark a point 15¾" from the right ends of the side pieces.

2. Set the pieces on edge and apply exterior glue to the crosspiece ends. Drive 2½" deck screws through the sides into the ends. Center the third crosspiece at the mark and attach **(photo A).**

3. Cut a six foot section of hardware cloth. Align one corner of the cloth with the right front corner of the floor. Attach it with ¾" fence staples every 4".

BUILD THE COMPARTMENT SIDES.

1. Cut the frame bases (C) and side walls (D and E). Place the walls together with the smooth sides facing in. Make a mark at 20" on a lengthwise side. Draw a line from the mark to the nearest opposite corner. Cut on the line to create the left and right peaked walls.

2. Make the door cutout on the inside wall by marking a 5 × 5" square 1½" up from the bottom and 4" from the front (longer) edge. Use a compass to draw an arch on top of the square. Drill a starter hole and use a jig saw to cut along the lines.

3. Align a frame base with the inside bottom of a wall. Attach the wall to the frame with 1¼" deck screws. Repeat with the second wall and base.

4. Draw a line across the inside of the walls, 1½" down from the peaked edge. Cut three sets of frame fronts (F) and backs (G) to fit between the frame base and the angled line. Attach using 1¼" deck screws through the siding into the frames.

5. Cut the frame tops (H). Center a top against each wall. Mark the ends so they are parallel with the sides and cut **(photo B).** Using one of these frame tops, cut the third frame top to match.

BUILD THE CAGE SIDE.

1. Assemble the third set of frame pieces cut in the previous step. Drive 2½" deck screws through the frame base into the square ends of the frame front and frame back.

2. Center the frame top across the front and back. Make sure it matches the extension of the two compartment sides. Use 2½" deck screws to attach the top to the mitered ends of the front and back **(photo C).**

ATTACH THE FRAMES.

1. Place the cage frame over the left floor crosspiece and attach using 2½" deck screws.

2. Place the sided frames on the middle and right end crosspiece, and attach using 2½" deck screws **(photo D).**

MAKE THE CAGE.

1. Fold the hardware cloth against the back of the frames. Attach it to the left side frame and middle frame, using ¾" fencing staples every 4". Cut the hardware cloth along the

Attach the frames to the floor crosspieces.

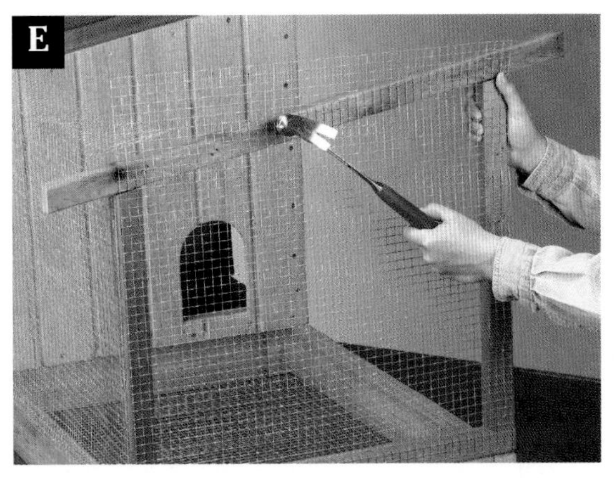

After folding and attaching the hardware cloth to the back, cut off excess cloth.

edges of the middle and left end frame. Cut along the floor by the enclosed compartment and discard this piece.
2. Fold the hardware cloth up against the left frame and attach using ¾" fencing staples every 4" **(photo E).** Cut off the excess hardware cloth.

MAKE THE ROOF AND BACK WALL.
1. Cut the back wall (I) and back wall stop (J). Notch the wall upper corners to fit around the top frames.
2. Attach the stop to the floor between the side walls using 2½" deck screws. Attach the back to the back frames and stop with 1¼" deck screws.
3. Cut the roof (K), with the siding grooves oriented vertically. Attach the roof to the top frames with 1¼" deck screws **(photo F).**

MAKE THE DOORS.
1. Measure the openings for the doors to make sure sizing is correct, then cut the door crosspieces (L), sides (M) hinge support (N) and door jamb (O).
2. Place the sides between the crosspieces, apply exterior glue and attach with 2½" deck screws. Cut hardware cloth to fit the frame and attach with ¾" fencing staples.
3. Position the door jamb between the compartment sides and attach with 2½" deck screws. Position the hinge support between the cage sides and attach with 2½" deck screws.
4. Cut the compartment door (P) and door supports (Q). Attach the door supports to the back of the door, using 1¼" deck screws.
5. Mount the doors with two 3" hinges each **(photo G).** Attach two hook and eye fasteners to secure each door.

ATTACH THE LEGS.
Cut the legs (R and S) to the desired length. Align a 2 × 2 against the wide side of a 2 × 4 to make an L. Use 2½" deck screws to attach. Attach the legs to the inside corners of the base with 2½" deck screws.

Attach the roof to the frames using 1¼" deck screws.

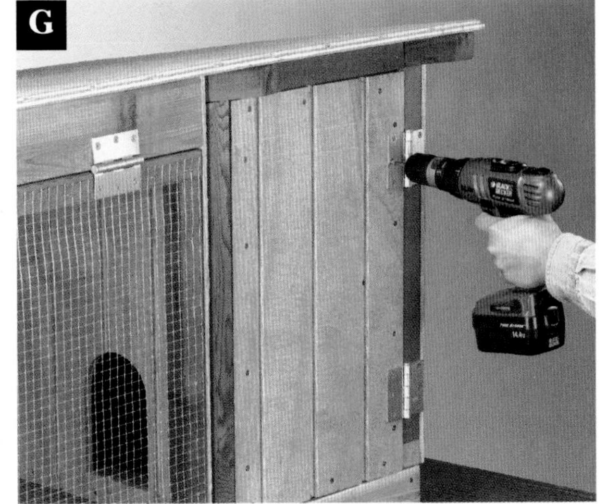

Mount the doors with 3" hinges.

ON
BLACK&DECKER
Finishing Sander
1/4 Sheet All Ball Bearing
Orbital Action

Garden Accessories

Outdoor living comes with its own list of accessories. For gardeners, it's tools, supplies, and compost bins. For boaters, it's life vests, swimsuits, and sporting equipment. For those who like to get away from it all at the cabin, it's food, beverages, and weekend supplies. Hauling all that stuff around is back-breaking work. The projects on the following pages will make moving and organizing your outdoor life quicker and easier, so there'll be more time for enjoying your outdoor activity.

PROJECT POWER TOOLS

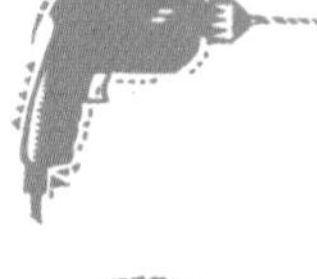

Gardener's Tote

Organize and transport your essential gardening supplies with this handy cedar tote box.

Construction Materials

Quantity	Lumber
1	1 × 10" × 6' cedar
1	1 × 6" × 6' cedar
1	1 × 4" × 6' cedar
1	1 × 2" × 6' cedar

This compact carrying tote has plenty of room and is ideal for gardeners. With special compartments sized for seed packages, spray cans and hand tools, it is a quick and easy way to keep your most needed supplies organized and ready to go. The bottom shelf is well suited to storing kneeling pads or towels. The gentle curves cut into the sides of the storage compartment make for easy access and provide a decorative touch. The sturdy cedar handle has a comfortable hand-grip cutout. You'll find this tote to be an indispensible gardening companion, whether you're tending a small flower patch or a sprawling vegetable garden.

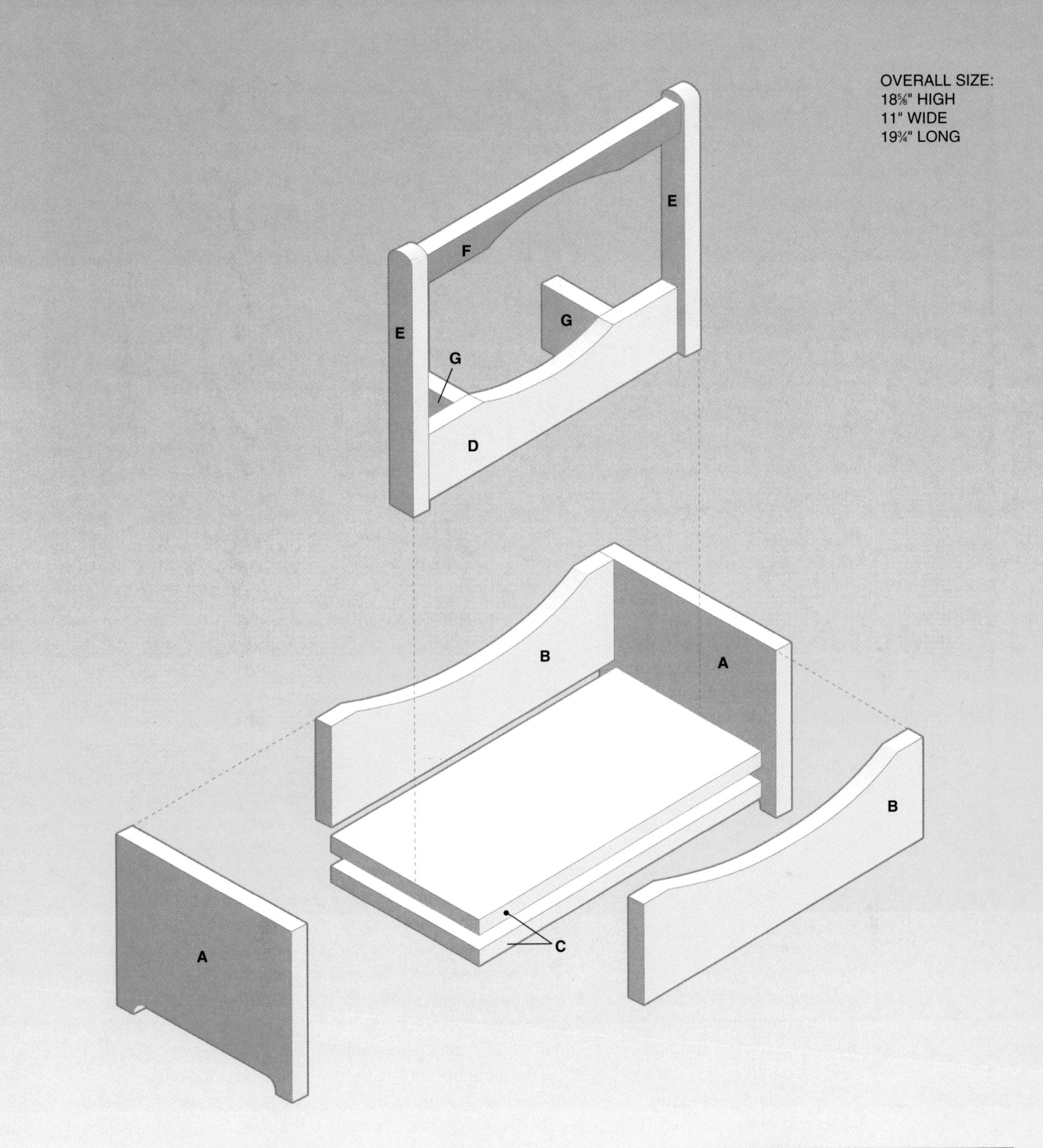

Cutting List

Key	Part	Dimension	Pcs.	Material
A	End	⅞ × 9¼ × 11"	2	Cedar
B	Side	⅞ × 5½ × 18"	2	Cedar
C	Shelf	⅞ × 9¼ × 18"	2	Cedar
D	Divider	⅞ × 3½ × 16¼"	1	Cedar

Cutting List

Key	Part	Dimension	Pcs.	Material
E	Post	⅞ × 1½ × 14"	2	Cedar
F	Handle	⅞ × 1½ × 16¼"	1	Cedar
G	Partition	⅞ × 3½ × 3⅞"	2	Cedar

Materials: Moisture-resistant glue, 1¼" and 2" deck screws, finishing materials.

Note: Measurements reflect the actual size of dimension lumber.

Use a jig saw to cut the curves on the bottom edge of each end, forming feet for the box.

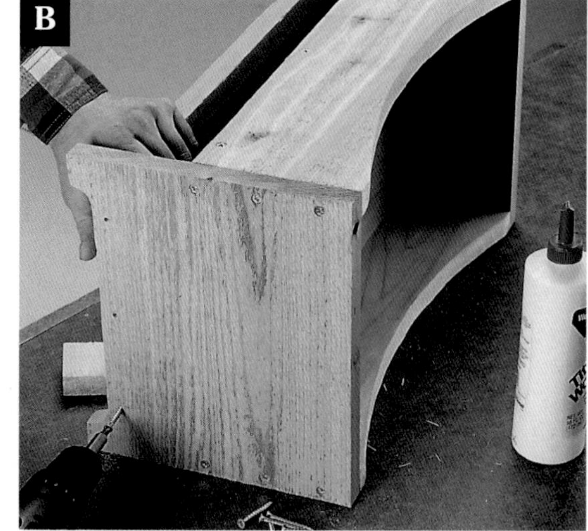

Attach the shelves by driving deck screws through the end pieces and into the ends of the shelves.

Directions: Gardener's Tote

BUILD THE BOX.

The gardener's tote has curved cutouts to improve access and scalloped ends to create feet. All screws are counterbored to ¼" depth for a smooth appearance. A counterbore bit will help you avoid drilling too deep.

1. Cut the ends (A), sides (B) and shelves (C) to size. Sand all parts smooth with medium-grit sandpaper.

2. On one side, mark points on one long edge, 1½" in and 1½" down. Draw a graceful curve between the points to form the cutting line for the curve. Cut the curve with a jig saw and sand it smooth.

3. Position the sides so the edges and ends are flush. Then, trace the curve onto the uncut side and cut it to match. Clamp the sides together, and gang-sand both curves until smooth.

4. Use a compass to draw ¾"-radius semicircles on the bottom edge of the end pieces, with centerpoints 1¾" from each end.

5. Using a straightedge, draw a line connecting the tops of the semicircles to complete the cutout shape. Cut the curves with a jig saw **(photo A),** and sand the ends smooth.

6. To attach the end and side pieces, drill ⅛" pilot holes at each end, 7⁄16" in from the edges. Position the pilot holes 1", 3" and 5" down from the tops of the ends. Counterbore the holes.

7. Apply glue to the ends of the side pieces—making sure the top and outside edges are flush—and fasten them to the end pieces with 2" deck screws, driven through the end pieces and into the side pieces.

8. Mark the shelf locations on the inside faces of the ends. The bottom of the lower shelf is ¾" up from the bottoms of the ends, and the bottom of the upper shelf is 3¾" up from the bottoms of the ends.

9. Drill pilot holes 7⁄16" up from the lines. Apply glue to the shelf ends, and position the shelves flush with the lines marked on the end pieces. Drive 2" deck screws through the pilot holes in the end pieces and into the shelves **(photo B).**

BUILD THE DIVIDER ASSEMBLY.

The divider and partitions are assembled first, and then inserted into the box.

1. Cut the divider (D), posts (E), handle (F) and partitions (G) to size.

2. Draw a ⅜"-radius semicircle, using a compass, to mark the cutting line for a roundover at one end of each post. Use a sander to make the roundover.

3. The divider and handle have shallow arcs cut on one long edge. To draw the arcs, mark points 4" in from each end. Then, mark a centered point, ⅝" in from one long edge on the handle. On the divider, mark a centered point, ⅝" in from one long edge.

4. Draw a graceful curve to connect the points, and cut along the lines with a jig saw. Sand the parts smooth.

5. Drill two pilot holes on each end of the divider, 7⁄16" out from the start of the curve. Counterbore the holes. Attach the divider to the partitions, using glue and 2" deck screws, driven through the divider and

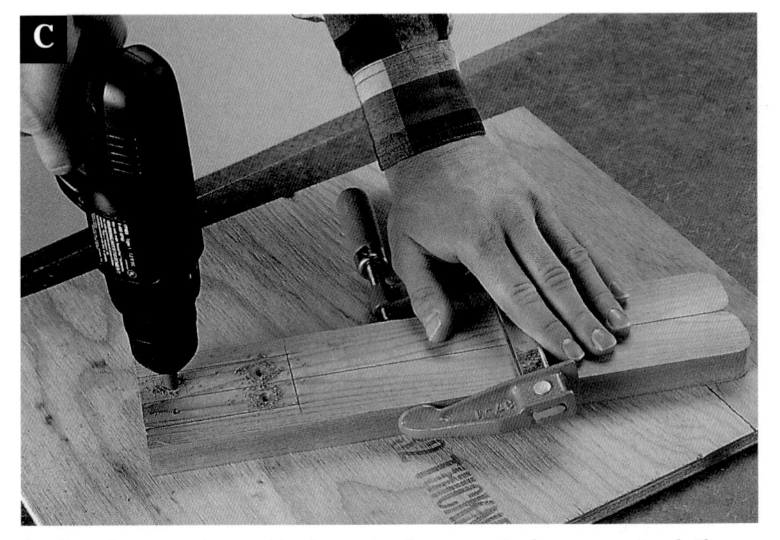

Drill and counterbore pilot holes in the posts before you attach them to the handle and divider.

TIP

Many seed types, soil additives and other common gardening supplies should not be stored outdoors in subfreezing temperatures. If you live in a colder climate, load up your tote with these items in the fall, and store the tote in a warm spot for the winter.

into the edges of the partitions.
6. To mark the positions of the divider ends, clamp the posts together with their edges flush, and mark a 3½"-long reference line on each post, ⅞" from the meeting point between the two posts **(photo C).** Start the reference lines at the square post ends. Connect the lines at the tops to indicate the position of the divider ends.
7. Drill two pilot holes through the posts, centered between each reference line and the inside edge **(photo C).** Counterbore the holes. Drill and counterbore two more pilot holes in each post, centered ½" and 1" down from the tops.
8. Position the handle and divider between the posts, aligned with the pilot holes. One face of the divider should be flush with a post edge. Fasten the handle and divider between the posts with moisture-resistant glue and 2" deck screws, driven through the post holes. Set the assembly in the box. Make sure the partitions fit square with the side.

INSTALL THE DIVIDER ASSEMBLY.

1. Trace position lines for the posts on the end pieces **(photo D).** Apply glue where the posts will be fastened. Drill pilot holes through the posts and counterbore the holes. Then, attach the posts with 1¼" deck screws, driven through the pilot holes and into the ends.
2. Drill two evenly spaced pilot holes in the side adjacent to the partitions. Counterbore the holes. Then, drive 2" deck screws through the holes and into the edges of the partitions.

APPLY THE FINISHING TOUCHES.

Sand all surfaces smooth with medium (100- or 120-grit) sandpaper. Then finish-sand with fine (150- or 180-grit) sandpaper. If you want to preserve the cedar tones, apply exterior wood stain to all surfaces. Or, you can leave the wood uncoated for a more rustic appearance. As you use the tote, it will slowly turn gray.

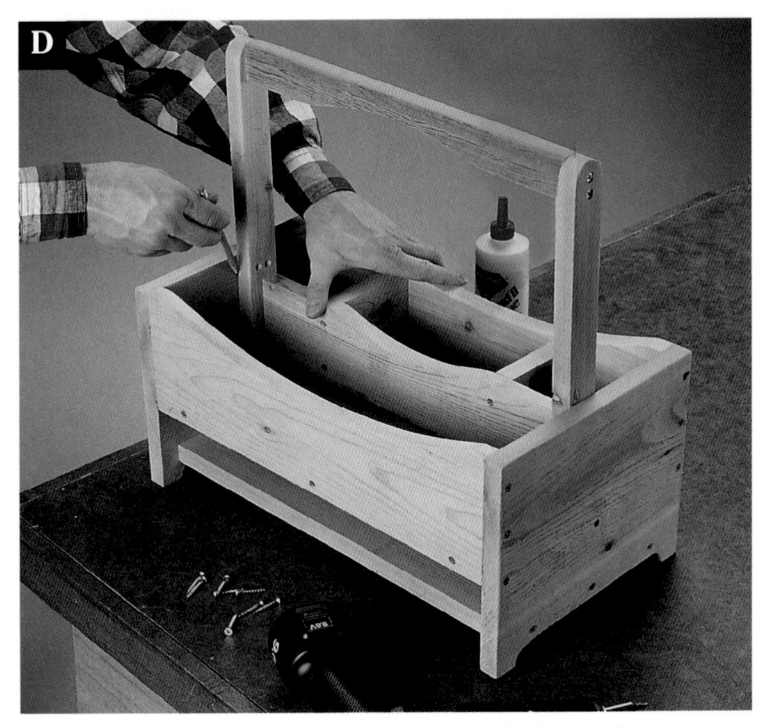

Draw reference lines for the post position on the box ends.

PROJECT
POWER TOOLS

Yard & Garden Cart

With a 4-cubic-foot bin and a built-in rack for long-handled tools, this sleek utility cart is hardworking and versatile.

CONSTRUCTION MATERIALS

Quantity	Lumber
1	2 × 6" × 8' cedar
4	2 × 4" × 8' cedar
2	1 × 6" × 8' cedar
2	1 × 4" × 8' cedar
1	1"-dia. × 3' dowel

This sturdy yard-and-garden cart picks up where a plain wheelbarrow leaves off. It includes many clever features that help make doing yard work more efficient, without sacrificing hauling capacity. And because it's made of wood, this cart will never dent or rust. The notches in the handle frame keep long-handled tools from being jostled about as the cart rolls across your yard. The handle itself folds down and locks in place like a kickstand when the cart is parked. When you're pushing the cart, the handle flips up to form an extra-long handle that takes advantage of simple physics to make the cart easier to push and steer.

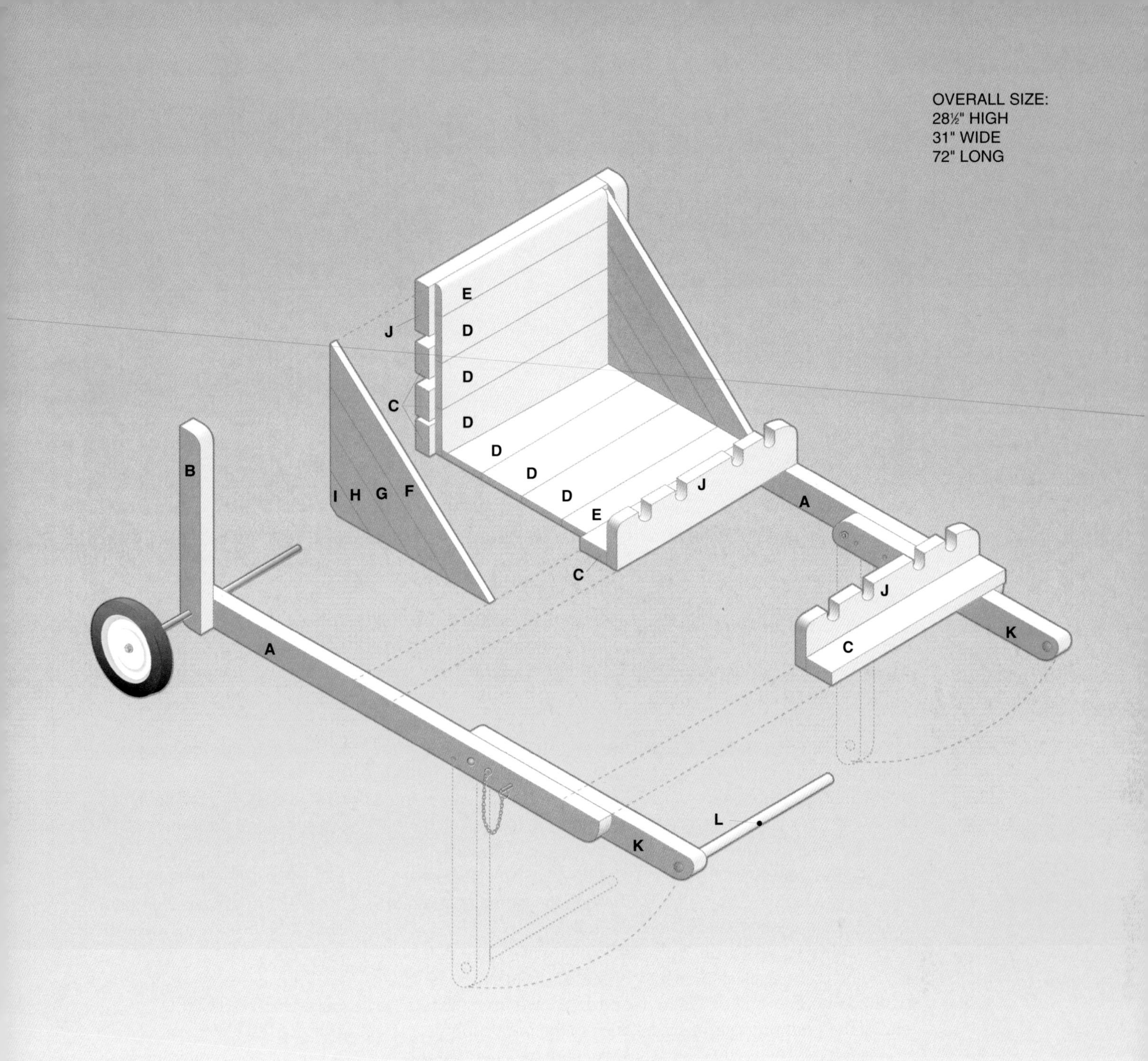

Key	Part	Dimension	Pcs.	Material
A	Back support	1½ × 3½ × 57"	2	Cedar
B	Front support	1½ × 3½ × 23½"	2	Cedar
C	Cross rail	1½ × 3½ × 24"	5	Cedar
D	Bin slat	⅞ × 5½ × 22¼"	6	Cedar
E	End slat	⅞ × 3½ × 22¼"	2	Cedar
F	Bin side	⅞ × 3½ × 28"	2	Cedar
G	Bin side	⅞ × 3½ × 21"	2	Cedar
H	Bin side	⅞ × 3½ × 14"	2	Cedar
I	Bin side	⅞ × 3½ × 7"	2	Cedar
J	Top rail	1½ × 5½ × 24"	3	Cedar
K	Arm	1½ × 3½ × 32"	2	Cedar
L	Handle	1"-dia. × 20⅞"	1	Dowel

Materials: 2" and 2½" deck screws, 4d finish nails (2), 10" utility wheels (2), 30" steel axle rod, 3⁄16"-dia. cotter pins, ⅜"-dia. hitch pins and chain (2), ⅜ × 4" carriage bolts (2) with lock nuts (2) and washers (4), finishing materials.

Note: Measurements reflect the actual size of dimension lumber.

Test with a square to make sure the front supports and back supports are joined at right angles.

Make straight cuts from the edge of each rail to the sides of the holes to make the tool notches.

Directions: Yard & Garden Cart

BUILD THE CART FRAME.
Counterbore all pilot holes in this project, using a counterbore bit, so the screw heads are recessed for improved safety and visual appeal.

1. Cut the back supports (A), front supports (B), three cross rails (C) and one of the top rails (J) to length.

2. Use a compass to draw a curve with a 3½" radius on each end of the back supports on the same side, and on each end of the front supports on opposite sides. When the curves are cut, the ends of these parts will have one rounded corner and one square corner. Cut the curves with a jig saw and sand out any rough spots or saw marks.

3. Position a top rail between the ends of the front supports, flush with the square corners of the front supports. Drill ⅛" pilot holes in the supports. Counterbore the holes ¼" deep, using a counterbore bit. Fasten the rail between the supports with glue and drive 2½" deck screws through the supports and into the rail.

4. Position two cross rails between the front supports, 7½" and 13" down from the top ends of the front supports. Make sure the cross rails are aligned with the top rail. Attach them with glue and 2½" deck screws. Fasten another cross rail between the bottom ends of the front supports. The bottom edge of this cross rail should be 3½" up from the bottoms of the front supports and aligned with the other rails.

5. Attach the front supports to the back supports with glue and 2½" deck screws, using a square to make sure the parts are joined at right angles **(photo A).** The unshaped ends of the back supports should be flush with the front and bottom edges of the front supports, and the back supports should be attached to the inside faces of the front supports.

6. Drill centered, ½"-dia. holes for the wheel axles through the bottoms of the front supports and back supports. Position the holes 1¾" from both the bottom ends and the sides of the front supports.

CUT THE NOTCHED TOP RAILS.

1. Cut the two remaining top rails (J) to length. These rails contain notches that are aligned to create a tool rack. Before cutting the notches, use a compass to draw 1½"-radius roundover curves at each end along one side of each rail. Cut the roundovers with a jig saw.

2. To make the tool notches in the top rails, first draw a reference line 1½" in from the rail edge between the roundovers. Mark four drilling points on the line, 3¾" and 8¼" in from each

---

TIP

If you need to round over the end of a board, one easy solution that gets good results is to use your belt sander like a bench grinder. Simply mount the belt sander to a work surface sideways, so the belt is perpendicular to the work surface and has room to spin. Turn on the sander, lay your workpiece on the work surface, and grind away.

Attach the bin slats to the front supports, leaving a ⅞"-wide gap at both ends of each slat.

end. Use a drill and a spade bit to drill 1½"-dia. holes through the drilling points on each rail.
3. Use a square to draw cutting lines from the sides of the holes to the near edge of each rail. Cut along the lines with a jig saw to complete the tool notches **(photo B).**

ATTACH RAILS BETWEEN THE BACK SUPPORTS.
1. Cut two cross rails (C) to length and lay them flat on your work surface. Attach a top rail to one edge of each cross rail, so the ends are flush and the notched edges of the top rails are facing up. Drive 2½" deck screws at 4" intervals through the top rails and into the edges of the cross rails.
2. Set one of the assemblies on the free ends of the back supports, flush with the edges. The free edge of the cross rail should be flush with the ends of the back supports. Attach the cross rail with 2½" deck screws driven down into the back support.
3. Attach the other rail assembly to the top edges of the back supports so the top rail faces the other rail assembly, and the free edge of the cross rail is 22⅜" from the front ends of the back supports.

ATTACH THE BIN SLATS.
1. Cut the bin slats (D) and end slats (E) to length. Position one end slat and three bin slats between the front supports, with the edge of the end slat flush with the edge of the top rail and the last bin slat butted against the back supports. There should be a ⅞" gap between the ends of each slat and the front supports. Attach the slats with glue and 2" deck screws driven down through the slats and into the cross rails and top rail **(photo C).**
2. Fasten the rest of the bin slats to the top edges of the back supports, with a ⅞" recess at each end. Start at the bottom of the bin, and work your way up, driving 2½" deck screws through the slats and into the tops of the back supports. Fasten the end slat between the last bin slat and the lower cross rail on the back supports.
3. Use a grinder or belt sander with a coarse belt to round the front edges of the front end slat **(photo D).**

ATTACH THE BIN SIDES.
1. Square-cut the bin sides (F, G, H, I) to the lengths shown in the *Cutting List,* page 197. Draw a 45° miter-cutting line at each end of each bin side. Make the miter cuts with a circular saw and straightedge, or with a power miter saw.
2. Fit the short, V-shaped sides into the recesses at the sides of the bin, and attach them to the front supports with glue and 2"

TIP

Cut pieces of sheet aluminum or galvanized metal to line the cart bin for easy cleaning after hauling. Simply cut the pieces to fit inside the bin, then attach them with special roofing nails that have rubber gaskets under the nail heads. Make sure that no sharp metal edges are sticking out from the bin.

Round the tips of the front supports and the front edge of the end slat, using a belt sander.

Fasten the bin sides in a V-shape with glue and deck screws.

Drill a pilot hole through each arm and into the ends of the handle, then drive 4d finish nails into the holes to secure the handle.

deck screws. Install the rest of the bin sides **(photo E).**

MAKE THE ARMS.

The arms serve a dual purpose. First, they support the handles when you wheel the cart. Second, they drop down and lock in place to support the cart in an upright position.

1. Cut the arms (K) to length. Mark the center of each end of each arm, measured from side to side. Measure down 3½" from each end, and mark a point. Set the point of a compass at each of these points, and draw a 1¾"-radius semicircle at each end on both arms. Cut the curves with a jig saw.

2. Drill a 1"-dia. hole for the handle dowel at one of the centerpoints at the end of each arm. At the other centerpoint, drill a ⅜"-dia. guide hole for a carriage bolt.

ATTACH THE ARMS.

1. Drill ⅜"-dia. holes for carriage bolts through each back support, 19" from the handle end, and centered between the top and bottom edges of the supports.

2. Insert a ⅜"-dia. × 4"-long carriage bolt through the outside of each ⅜"-dia. hole in the back supports. Slip a washer over each bolt, then slip the arms over the carriage bolts. Slip another washer over the end of each bolt, then secure the arms to the supports by tightening a lock nut onto each bolt. Do not overtighten the lock nut—the arms need to be loose enough to pivot freely.

3. Cut the handle (L) to length from a 1"-dia. dowel (preferably hardwood). Slide it into the 1"-dia. holes in the ends of the arms. Secure the handle by drilling pilot holes for 4d finish nails through each arm and into the dowel **(photo F).** Then, drive a finish nail into the dowel at each end.

Secure the wheels by inserting a cotter pin into a hole at the end of each axle, then bending the ends of the pin down with pliers.

ATTACH THE WHEELS.
Make sure to buy a steel axle rod that fits the holes in the hubs of the 10" wheels.
1. Cut the axle rod to 30" in length with a hacksaw. Remove any burrs with a file or bench grinder. (Rough-grit sandpaper also works, but it takes longer and is harder on the hands.) Secure the axle rod in a vise, or clamp it to your work surface, and use a steel twist bit to drill a $\frac{3}{16}$"-dia. hole through the rod $\frac{1}{8}$" in from each end of the axle.
2. Slip the axle through the $\frac{1}{2}$"-dia. holes drilled at the joints between the front and back supports. Slide two washers over each end of the axle.
3. Slip a wheel over each axle end, add two washers and insert $\frac{3}{16}$"-dia. cotter pins into the holes drilled at the ends of the axle. Secure the wheels by bending down the ends of the cotter pins with a pair of pliers **(photo G).**

LOCK THE ARMS IN PLACE.
1. On a flat surface, fold down the arm/handle assembly so the arms are perpendicular to the ground. Drill a $\frac{3}{8}$"-dia. guide hole through each back support, 1" below the carriage bolt that attaches the arms to the supports. Extend the holes all the way through the arms **(photo H).** Insert a $\frac{3}{8}$"-dia. hitch pin (or hinge pin) into each hole to secure the arms.
2. To avoid losing the pins when you remove them, attach them to the back supports with a chain or a piece of cord. Now, remove the pins and lift the arms so they are level with the tops of the back supports. Drill $\frac{3}{8}$"-dia. holes through the arms and back supports, about 12" behind the first pin holes, for locking the arms in the cart-pushing position.

APPLY FINISHING TOUCHES.
Smooth out all the sharp edges on the cart with a sander. Also sand the surfaces slightly. Apply two coats of exterior wood stain to the wood for protection. Squirt some penetrating/lubricating oil or synthetic lubricant on the axle on each side of the wheels to reduce friction.

Drill $\frac{3}{8}$"-dia. holes through the back supports and into the arms for inserting the hitch pins that lock the arms in position.

PROJECT
POWER TOOLS

Cabin Porter

Shuttle heavy supplies from car to cabin or down to your dock with this smooth-riding cedar cart.

CONSTRUCTION MATERIALS

Quantity	Lumber
3	2 × 4" × 8' cedar
11	1 × 4" × 8' cedar

Transporting luggage and supplies doesn't need to be an awkward, back-breaking exercise. Simply roll this cabin porter to your car when you arrive, load it up and wheel your gear to your cabin door or down to the dock. The porter is spacious enough to hold coolers, laundry baskets or grocery bags, all in one easy, convenient trip. Both end gates are removable, so you can transport longer items like skis, ladders or lumber for improvement projects. The cabin porter is also handy for moving heavy objects around your yard. The 10" wheels ensure a stable ride, and the porter is designed to minimize the chances of tipping. The wheels, axle and mounting hardware generally can be purchased as a set from a well-stocked hardware store. For winter use, you might try adding short skis or sled runners, allowing the cabin porter to glide over deep snow and decreasing your chances of dropping an armful of supplies over slippery ice.

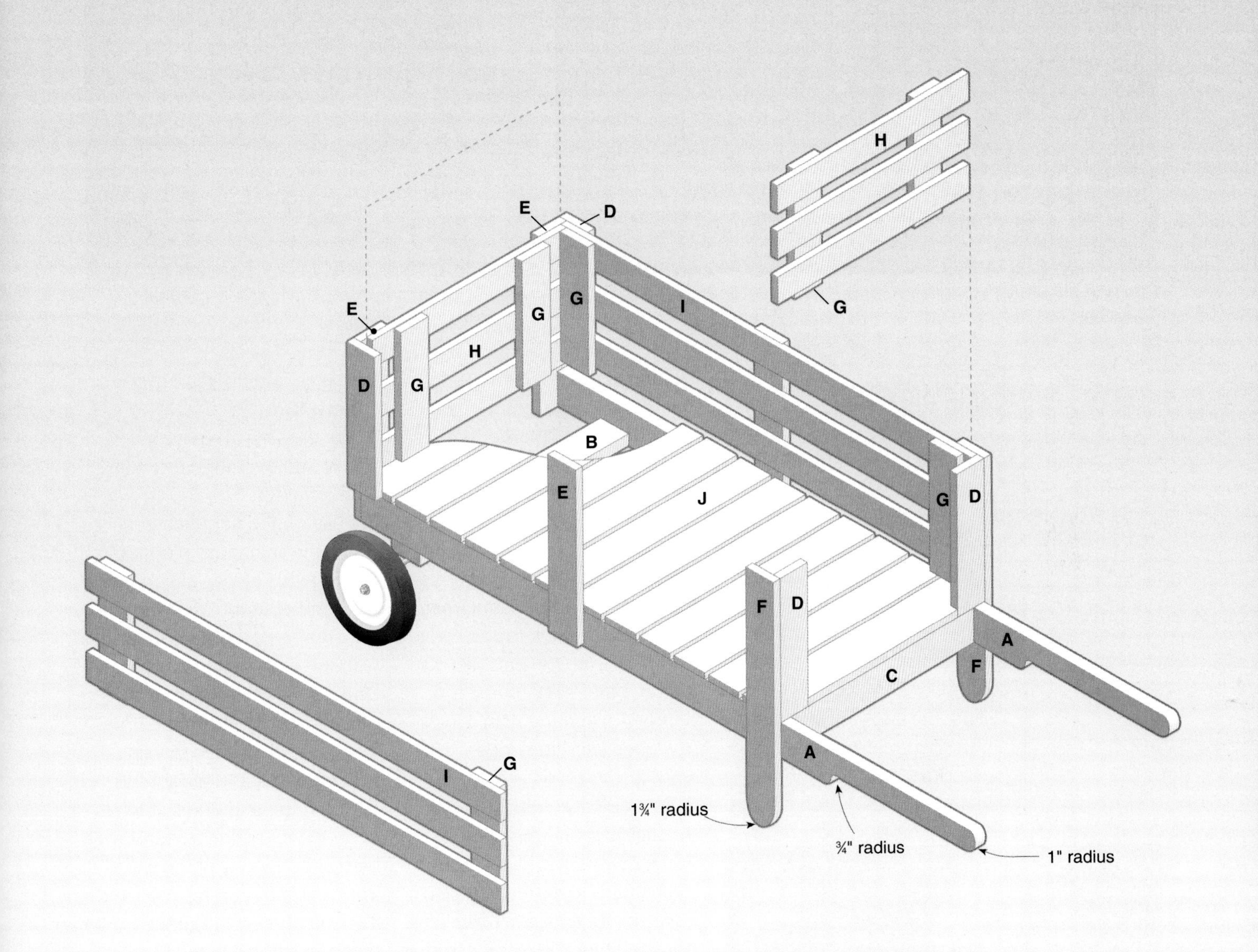

Cutting List

Key	Part	Dimension	Pcs.	Material
A	Handle	1½ × 3½ × 72⅞"	2	Cedar
B	Front stringer	1½ × 3½ × 24"	1	Cedar
C	Rear stringer	1½ × 3½ × 21"	1	Cedar
D	Short stile	⅞ × 3½ × 14⅜"	4	Cedar
E	Long stile	⅞ × 3½ × 17⅞"	4	Cedar

Cutting List

Key	Part	Dimension	Pcs.	Material
F	Rear stile	⅞ × 3½ × 24½"	2	Cedar
G	Gate stile	⅞ × 3½ × 13½"	8	Cedar
H	Gate rail	⅞ × 3½ × 22"	6	Cedar
I	Side rail	⅞ × 3½ × 46⅝"	6	Cedar
J	Slat	⅞ × 3½ × 24"	12	Cedar

Materials: 1½", 2", and 2½" deck screws, wood glue, 10"-dia. wheels (2), axle, ¾ × 4" metal straps (3), ¼× 1" lag screws, washers, crimp caps, finishing materials.
Note: Measurements reflect the actual size of dimension lumber.

Clamp the handles together and draw reference lines at the stringer locations.

When installing the stringers, make sure they are square with the handles.

Directions: Cabin Porter

ASSEMBLE THE HANDLES AND FRAMEWORK.

The framework for the cabin porter consists of handles connected by stringers at each end.

1. Cut the handles (A), front stringer (B) and rear stringer (C) to length. Sand the edges smooth.

2. Trim the back ends of the handles to create gripping surfaces. Draw a 16"-long cutting line on the face of each handle, starting at one end, 1½" up from the bottom edge. Set the point of a compass at the bottom edge, 14½" in from the end, and draw a 1½"-radius arc, creating a smooth curve leading up to the cut line. To round the ends of each handle, use a compass to draw a 1"-radius semicircle centered 1" below the top edge and 1" in from the end (see *Diagram,* page 203). Shape the handles by cutting with a jig saw, then sand the edges smooth.

3. Stringers and slats fit across the handles, creating the bottom frame of the porter. Clamp the handles together, edge to edge, so the ends are flush, and draw reference lines 25⅜" from the grip ends and 3½" from the square ends to locate the stringers **(photo A).** Place the front stringer flat across the bottom edges of the handles so the front edge of the stringer is flush with the 3½" reference lines. Attach it with glue and 2½" deck screws. Position the rear stringer between the handles so the back face of the stringer is flush with the 25⅜" reference lines. Attach it with glue and 2½" deck screws **(photo B).**

Apply glue and drive screws through the rails and into the corner pieces.

4. Cut the slats (J) to length, and round over their top edges with a sander.

5. Position the handle assembly so the shaped grip edges face down. Lay one slat over the handles at the front end so the corners of the slat are flush with the ends of the handles. Drill ⅛" pilot holes through the slat, and counterbore the holes to a ¼" depth. Fasten the slat with glue and 2" deck screws.

6. Notch the rear end slat to receive the rear short stiles. Draw lines at both ends of the slat ⅞" from a common long edge and 3½" from the ends. Cut the notches with a jig saw. Position the slat flush with the rear face of the rear stringer, and fasten it with glue and 2" screws.

7. Space the remaining slats evenly between the end slats with gaps of about ½". Fasten them with glue and 2" screws.

MAKE THE CORNERS.

Join the stiles to make the corners, which will support the side rails and end gates.

1. Cut the short stiles (D), long

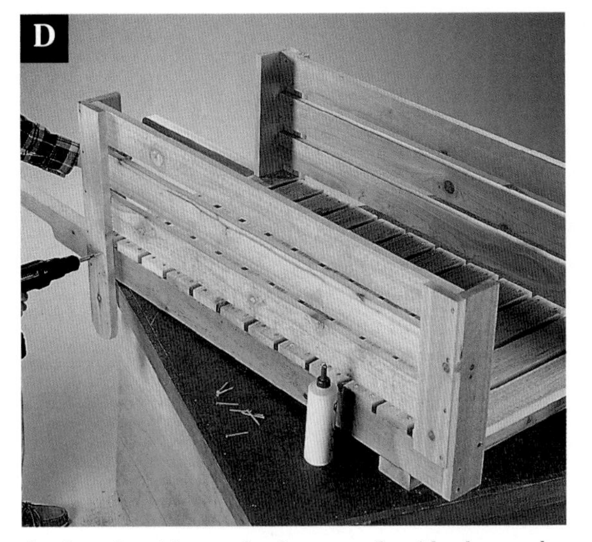

Anchor the sides to the framework with glue and screws driven through the stiles into the handles.

Attach the axles to the bottom of the front stringer with metal straps fastened with lag screws.

stiles (E) and rear stiles (F) to length. Use a compass to draw a 1¾"-radius semicircle at the bottom of each rear stile (see *Diagram*, page 203). Shape the ends with a jig saw, and sand the edges smooth.
2. Butt the edge of a short stile against the face of a rear stile so the pieces form a right angle. With the square ends flush, drill pilot holes every 2" through the rear stile and into the edge of the short stile. Counterbore the holes. Join the stiles with glue and 2" deck screws. Assemble the other rear corner.
3. Repeat this procedure to assemble the front corners, butting the edge of a long stile against the face of a short stile so the edges and tops are flush.

MAKE THE SIDES.
1. Cut the side rails (I) to length. Place three side rails tight between one front corner and one rear corner so the top of the upper rail is flush with the tops of the corners. Leave a 1" gap between rails. Fasten the rails to the corners with glue and 1½" deck screws **(photo C).**
2. Fasten the side assemblies to the handles with glue and 2" deck screws **(photo D).** Drive an additional screw through each stile and into the edge of an adjacent slat. Position the remaining stiles (E) on the outer sides of the rails, midway between the front and rear corners. Fasten the stiles to the rails with glue and 1½" deck screws.

MAKE THE GATES.
1. Cut the gate stiles (G) and gate rails (H) to length. Sand the short edges of the rails.
2. Lay the rails facedown together in groups of three with the ends flush. Draw reference lines across the rails 2" in from each end to locate the stiles.
3. Place two stiles on one rail with the tops flush and the outer stile edges on the reference lines. Fasten them with glue and 1½" deck screws driven through the stiles and into the rails.
4. Attach two more rails below the first one, leaving a 1" gap between the rails.
5. Follow the same procedure to make the other gate.
6. Set the gates in place between the porter sides to locate the four remaining gate stiles (G). These form the slots that keep the gates in place. Position the stiles flush with the tops of the top side rails and almost flush with the faces of the gate rails. Attach them with glue and 1½" deck screws driven through the stiles and into the rails. Slide the gates in and out of the slots to test for smooth operation.
7. Sand any rough areas, and apply the finish of your choice.

ATTACH THE WHEELS.
1. Cut the axle to length (24" plus the width of the two wheels plus 1½"). Attach the axle to the bottom of the front stringer with lag screws and metal straps bent in the center **(photo E).** Place one strap at each end of the stringer and one in the middle.
2. Slide three washers followed by a wheel over each end of the axle. Secure the wheels with crimp caps or by drilling a small hole in each end of the axle and installing an additional washer and a cotter pin.

PROJECT POWER TOOLS

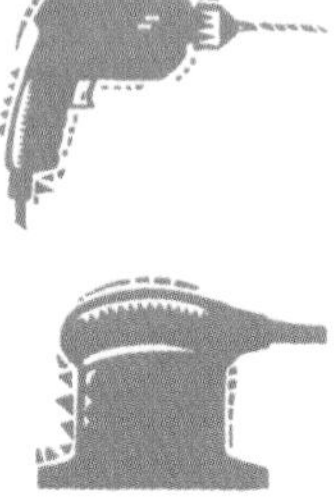

Compost Bin

Convert yard waste to garden fertilizer inside this simple and stylish cedar compost bin.

CONSTRUCTION MATERIALS

Quantity	Lumber
4	4 × 4" × 4' cedar posts
5	2 × 2" × 8' cedar
8	1 × 6" × 8' cedar fence boards

Composting yard debris is an increasingly popular practice that makes good environmental sense. Composting is the process of converting organic waste into rich fertilizer for the soil, usually in a compost bin. A well-designed compost bin has a few key features. It's big enough to contain the organic material as it decomposes. It allows cross-flow of air to speed the process. And the bin area is easy to reach whether you're adding waste, turning the compost or removing the composted material. This compost bin has all these features, plus one additional benefit not shared by most compost bins: it's very attractive.

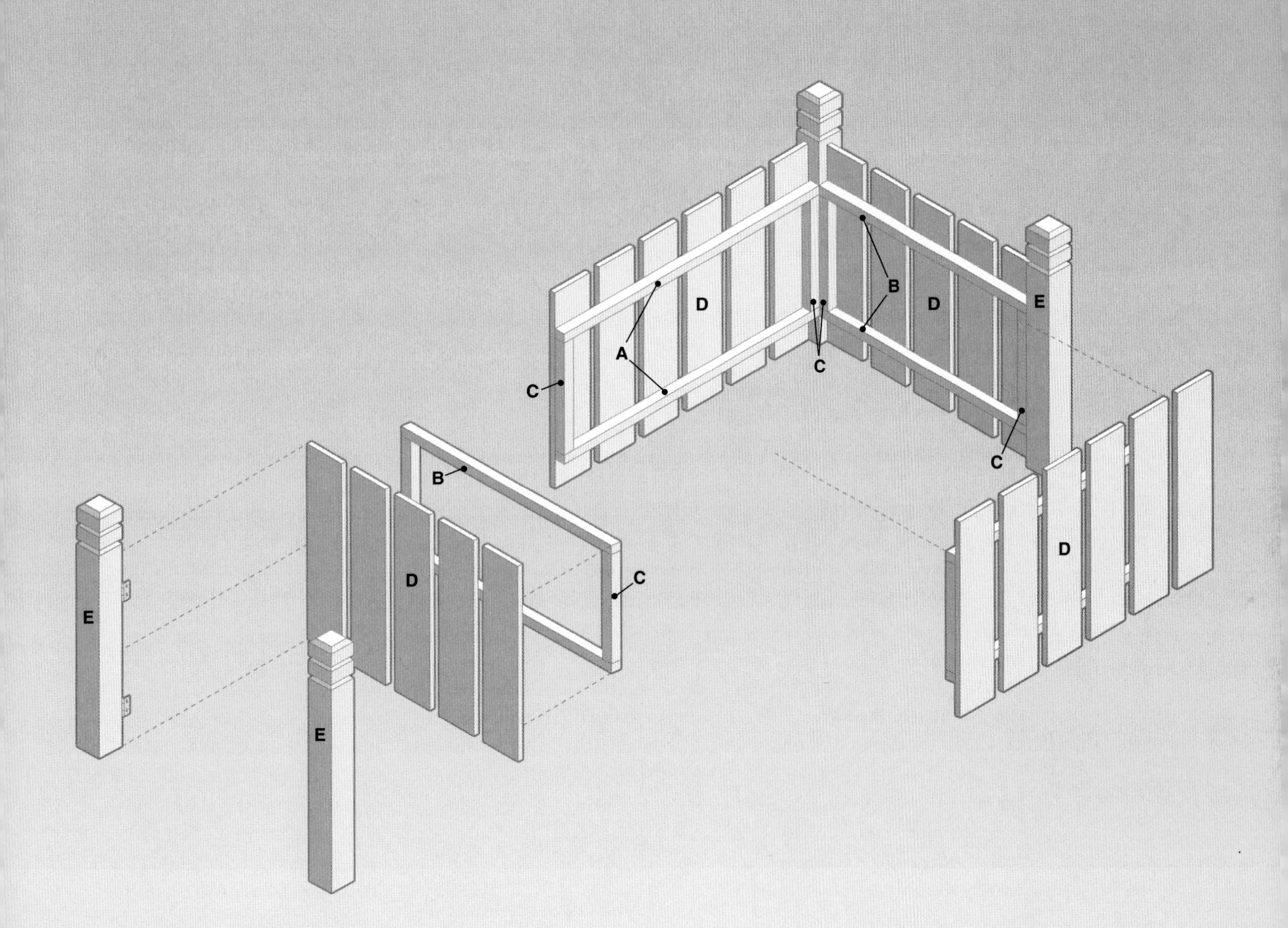

Cutting List

Key	Part	Dimension	Pcs.	Material
A	Side rail	1½ × 1½ × 40½"	4	Cedar
B	End rail	1½ × 1½ × 33½"	4	Cedar
C	Cleat	1½ × 1½ × 15"	8	Cedar
D	Slat	¾ × 5½ × 27"	22	Cedar
E	Post	3½ × 3½ × 30"	4	Cedar

Materials: 1½" and 3" galvanized deck screws, hook-and-eye latch mechanism, 3 × 3" brass butt hinges (one pair) and screws.

Note: Measurements reflect the actual size of dimension lumber.

Fasten the cleats between the rails to construct the panel frames.

Attach a slat at each end of the panel frame so the outer edges of the slats are flush with the outer edges of the frame.

Directions: Compost Bin

BUILD THE PANELS.

The four fence-type panels that make up the sides of this compost bin are cedar slats that attach to panel frames. The panel frames for the front and back of the bin are longer than the frames for the sides.

1. Cut the side rails (A), end rails (B) and cleats (C) to length. Group pairs of matching rails with a pair of cleats. Assemble each group into a frame—the cleats should be between the rails, flush with the ends. Drill ⅛" pilot holes into the rails. Counterbore the holes ¼" deep, using a counterbore bit. Fasten all four panel frames together by driving 3" deck screws through the rails and into each end of each cleat **(photo A).**

2. Cut all of the slats (D) to length. Lay the frames on a flat surface and place a slat at each end of each frame. Keep the edges of these outer slats flush with the outside edges of the frame and let the bottoms of the slats overhang the bottom frame rail by 4". Drill pilot holes in the slats. Counterbore the holes slightly. Fasten the outer slats to the frames with 1½" deck screws **(photo B).**

3. When you have fastened the outer slats to all of the frames, add slats between each pair of outer slats to fill out the panels. Insert a 1½" spacing block between the slats to set the correct gap. (This will allow air to flow into the bin.) Be sure to keep the ends of the slats aligned. Check with a tape measure to make sure the bottoms of all the slats are 4" below the bottom of the panel frame **(photo C).**

> **TIP**
>
> *Grass clippings, leaves, weeds and vegetable waste are some of the most commonly composted materials. Just about any formerly living organic material can be composted, but DO NOT add any of the following items to your compost bin:*
> - *animal material or waste*
> - *dairy products*
> - *papers with colored inks*
>
> *For more information on composting, contact your local library or agricultural extension office.*

ATTACH THE PANELS AND POSTS.

The four slatted panels are joined with corner posts to make the bin. Three of the panels are attached permanently to the posts, while one of the end panels is installed with hinges and a latch so it can swing open like a gate. You can use plain 4 × 4 cedar posts for the corner posts. For a more decorative look, you can buy prefabricated fence posts or deck rail posts with carving or contours at the top.

1. Cut the posts (E) to length. If you're using plain posts, you may want to do some decorative contouring at one end or attach post caps.

2. Stand a post upright on a flat work surface. Set one of the longer slatted panels next to the post, resting on the bottoms of the slats. Hold or clamp the panel to the post, with the back of the panel frame flush with

The inner slats should be 1½" apart, with the ends 4" below the bottom of the frame.

Stand the posts and panels upright, and fasten the panels to the posts by driving screws through the cleats.

one of the faces of the post. Fasten the panel to the post by driving 3" deck screws through the frame cleats and into the posts. Space screws at roughly 8" intervals.

3. Stand another post on end, and fasten the other end of the panel frame to it, making sure the posts are aligned.

4. Fasten one of the shorter panels to the adjoining face of one of the posts. The back faces of the frames should just meet in a properly formed corner **(photo D).** Fasten another post at the free end of the shorter panel.

5. Fasten the other longer panel to the posts so it is opposite the first longer panel, forming a U-shaped structure.

ATTACH THE GATE.

The unattached shorter panel is attached at the open end of the bin with hinges to create a swinging gate for loading and unloading material. Exterior wood stain will keep the cedar from turning gray. If you are planning to apply a finish, you'll find it easier to apply it before you hang the gate. Make sure all hardware is rated for exterior use.

1. Set the last panel between the posts at the open end of the bin. Move the sides of the bin slightly, if needed, so there is about ¼" of clearance between each end of the panel and the posts. Remove this panel gate and attach a pair of 3" butt hinges to a cleat, making sure the barrels of the hinges extend past the face of the outer slats.

2. Set the panel into the opening, and mark the location of the hinge plates onto the post. Open the hinge so it is flat, and attach it to the post **(photo E).**

3. Attach a hook-and-eye latch to the unhinged end of the panel to hold the gate closed.

Attach the hinges to the end panel frame, then fasten to the post.

Dock Box

This spacious dockside hold protects all your boating supplies, with room to spare.

PROJECT POWER TOOLS

Construction Materials

Quantity	Lumber
2	⅝" × 4 × 8' plywood siding
7	1 × 2" × 8' cedar
4	1 × 4" × 8' cedar
1	1 × 6" × 8' cedar
3	2 × 2" × 8' cedar

With its spacious storage compartment and appealing nautical design, this box is a perfect place for stowing water sports equipment. You won't have to haul gear inside anymore after offshore excursions. Life preservers, beach toys, ropes and even small coolers conveniently fit inside this attractive chest, which has ventilation holes to discourage mildew. Sturdy enough for seating, the large top can hold charts, fishing gear or a light snack while you await your next voyage. With a dock box to hold your gear, you can spend your energy carrying more important items—like the fresh catch of the day—up to your cabin.

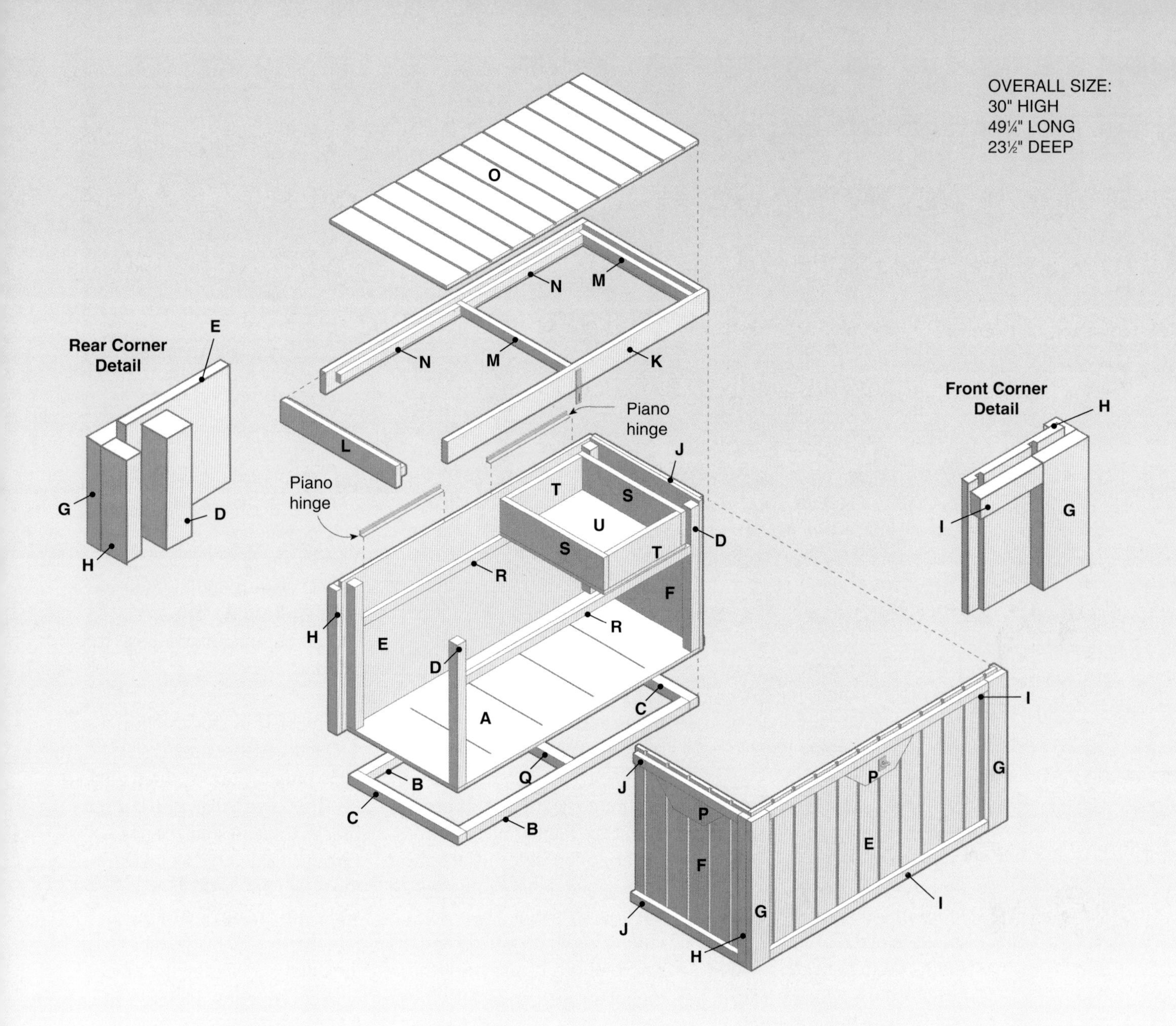

Cutting List

Key	Part	Dimension	Pcs.	Material
A	Bottom	⅝ × 46¼ × 20½"	1	Plywood siding
B	Bottom brace	1½ × 1½ × 43¼"	2	Cedar
C	End brace	1½ × 1½ × 20½"	2	Cedar
D	Corner brace	1½ × 1½ × 24⅜"	4	Cedar
E	Large panel	⅝ × 47½ × 27"	2	Plywood siding
F	Small panel	⅝ × 20½ × 27"	2	Plywood siding
G	Corner trim	⅞ × 3½ × 26½"	4	Cedar
H	Corner batten	⅞ × 1½ × 26½"	4	Cedar
I	Long trim	⅞ × 1½ × 42¼"	4	Cedar
J	End trim	⅞ × 1½ × 18¾"	4	Cedar
K	Lid side	⅞ × 3½ × 49¼"	2	Cedar

Cutting List

Key	Part	Dimension	Pcs.	Material
L	Lid end	⅞ × 3½ × 21¾"	2	Cedar
M	Top support	⅞ × 1½ × 21¾"	3	Cedar
N	Ledger	⅞ × 1½ × 22⅜"	4	Cedar
O	Top panel	⅝ × 47½ × 21¾"	1	Plywood siding
P	Handle	⅞ × 3½ × 13½"	4	Cedar
Q	Cross brace	1½ × 1½ × 17½	1	Cedar
R	Tray slide	⅞ × 1½ × 43¼"	2	Cedar
S	Tray side	⅞ × 5½ × 20¼"	2	Cedar
T	Tray end	⅞ × 5½ × 14"	2	Cedar
U	Tray bottom	⅝ × 15¾ × 20¼"	1	Plywood siding

Materials: 1¼" and 1⅝" deck screws, 6d finish nails, 1" wire brads, construction adhesive, 1½ × 30" or 36" piano hinge, hasp, lid support chains (2), finishing materials.

Note: Measurements reflect the actual size of dimension lumber.

For ventilation, cut slots into the bottom panel, using a straightedge as a stop block for the foot of your circular saw.

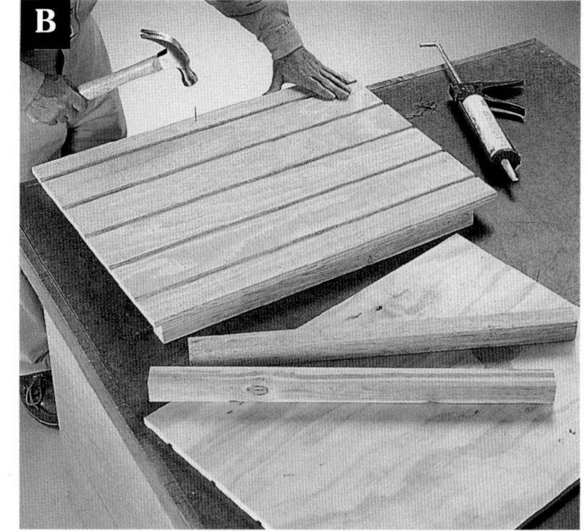

Position the corner braces beneath the small panels, and fasten them with adhesive and finish nails.

Directions: Dock Box

MAKE THE BOX BOTTOM. The box bottom is made of grooved plywood siding attached to a rectangular 2 × 2 box frame.

1. Cut the bottom (A), bottom braces (B), end braces (C) and cross brace (Q) to size. Apply construction adhesive or moisture-resistant wood glue to the ends of the bottom braces. Clamp them between the end braces so the edges are flush. Drill ⅛" pilot holes through each end brace into the bottom braces. Counterbore the holes ¼" deep, using a counterbore bit. Drive 1⅝" deck screws through the pilot holes to reinforce the joints.

2. Center the cross brace in the frame and attach it with adhesive and 1⅝" deck screws.

3. Attach the box bottom to the box frame with 1⅝" deck screws.

4. Cut six ventilation slots in the bottom panel. First, clamp a straightedge near one edge of the bottom panel. Then, set the cutting depth on your circular saw to about 1" and press the foot of the saw up against the straightedge. Turn on the saw, and press down with the blade in a rocking motion until you've cut through the bottom panel **(photo A).** The slots should be spaced evenly, 8" to 9" apart.

ATTACH THE BOX SIDES.

1. Cut the corner braces (D), large panels (E) and small panels (F) to size. Align two corner braces under a small panel (grooved side up). Make sure the edges are flush, with a ½"-wide gap at one end of the panel and a 2⅛"-wide gap at the other end. Fasten the braces with construction adhesive and 6d finish nails **(photo B).**

2. Repeat the procedure for the other small panel.

3. Attach the small panels, with the 2" space facing downward, to the end braces, using 6d nails and construction adhesive.

4. Place the large panels in position and drive nails through the panels into the bottom braces and corner braces.

MAKE THE TRIM PIECES.

1. Cut the corner trim (G) and corner battens (H) to length. Set the project on its side. Use construction adhesive and nails to attach the corner bat-

Attach the corner trim pieces flush with the edges of the corner battens to cover the plywood joints.

A handle block is attached to each face of the box, up against the bottom of the top trim piece.

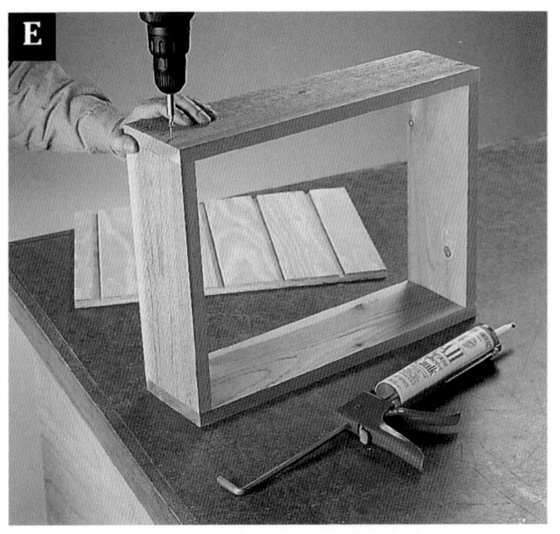

Counterbore the screw heads so they don't obstruct the movement of the tray on the tray slides.

tens flush with the bottom, covering the seam between panels. There should be a ½"-wide gap between the tops of the corner pieces and the top of the box. Then, attach the corner trim **(photo C).**

2. Cut the long trim (I) and the end trim (J) to length. Attach the lower trim flush with the bottom, using construction adhesive and finish nails.

3. Attach the upper trim pieces flush with the corner pieces, using adhesive. Drive 1¼" deck screws from inside the box into the trim pieces.

ATTACH THE HANDLES.

The handles (P) are trapezoid-shaped blocks cut from cedar.

1. Cut four handles to length. Mark each piece 3¾" in from each end along one long edge. Connect the marks diagonally to the adjacent corners to form cutting lines. Cut with a circular saw or a power miter box.

2. Center a handle against the bottom edge of the top trim piece on each face. Attach each handle with adhesive and 1¼" deck screws **(photo D).**

MAKE THE TRAY.

The tray rests inside the dock box on slides.

1. Cut the tray slides (R) to length. Mount the slides inside the box, 7" down from the top edge, using adhesive and 1¼" deck screws.

2. Cut the tray sides (S), tray ends (T) and tray bottom (U) to size. Drill pilot holes in the tray ends and counterbore the holes. Then, fasten the tray ends between the tray sides with adhesive and 1⅝" deck screws **(photo E).** Attach the tray bottom with adhesive and 1" wire brads.

MAKE THE LID.

1. Cut the lid sides (K) and lid ends (L) to length. Fasten them together with adhesive and drive 6d nails through the lid sides and into the ends.

2. Cut the top panel (O), top supports (M) and ledgers (N) to length. Attach two top supports to the inside edges of the frame, ⅝" down from the top edge, using adhesive and 1¼" screws **(photo F).** Attach the ledgers to the long sides of the lid—one at each corner—with adhesive and 1¼" deck screws. Place the remaining top support into the gap in the middle. Fasten it by driving 6d nails into the ends of the support.

3. Fit the top panel into the lid. Fasten with 6d nails and adhesive. Sand all exposed edges.

4. Attach the lid to the box with a piano hinge cut in two. Attach a pair of chains between the bottom of the lid and the front of the box to hold the lid upright when open. To lock the box, attach a hasp to the handle and lid at the front of the box.

5. Apply exterior stain or water sealer for protection. Caulk the gap around the top panel and lid frame with exterior caulk.

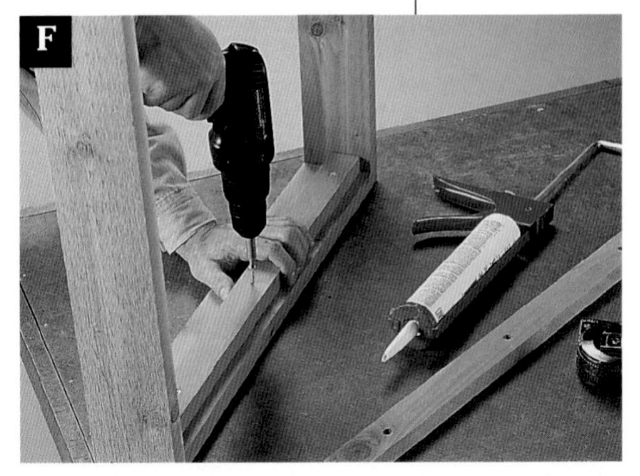

Top supports in the lid frame support the top panel.

Piranha
Carbide Teeth

Storage & Utility Projects

There just never seems to be enough space to store everything. Instead, we put up with messy woodpiles, scattered garbage cans, and jumbled boots in the back entryway. See how much easier everyday life can be by adding a few simple outdoor storage and utility solutions to your home. You'll also find stylish solutions to mailboxes and address markers, as well as grills and trash collection areas.

PROJECT POWER TOOLS

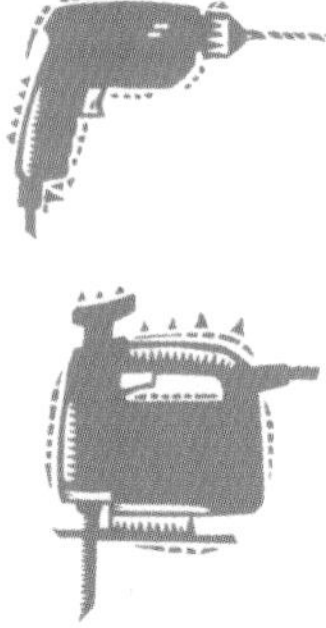

Outdoor Storage Center

Create additional storage space for backyard games and equipment with this efficient outdoor storage center.

Construction Materials

Quantity	Lumber
2	⅝" × 4 × 8' textured cedar plywood siding
2	¾" × 2 × 4' BC fir plywood handy panels
2	1 × 2" × 8' cedar
6	1 × 3" × 8' rough-sawn cedar
2	1 × 4" × 8' rough-sawn cedar
1	2 × 2" × 8' pine
1	1 × 2" × 8' pine

Sturdy cedar construction and a rustic appearance make this storage center an excellent addition to any backyard or outdoor setting. The top lid flips up for quick and easy access to the upper shelf storage area, while the bottom doors swing open for access to the lower storage compartments. The raised bottom shelf keeps all stored items up off the ground, where they stay safe and dry. Lawn chairs, yard games, grilling supplies, fishing and boating equipment, and much more can be kept out of sight and protected from the weather. If security is a concern, simply add a locking hasp and padlock to the top lid to keep your property safe and secure. If you have a lot of traffic in and out of the top compartment, add lid support hardware to prop the lid open.

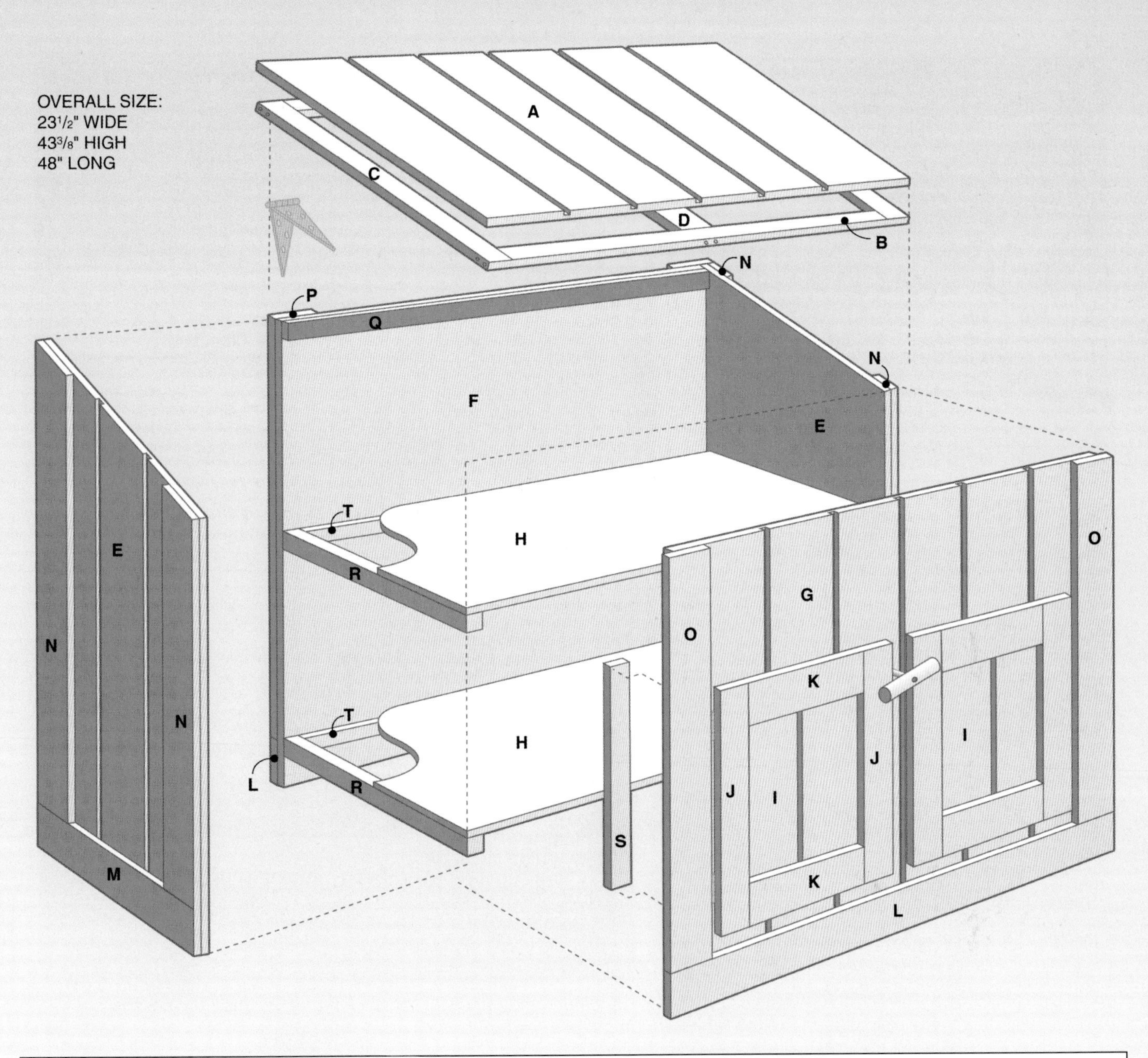

Key	Part	Dimension	Pcs.	Material
A	Lid	⅝ × 24 × 48"	1	Plywood siding
B	Lid edge	¾ × 1½ × 45"	2	Cedar
C	Lid end	¾ × 1½ × 24"	2	Cedar
D	Lid stringer	¾ × 2½ × 21"	1	Cedar
E	End panel	⅝ × 22 × 42"	2	Plywood siding
F	Back panel	⅝ × 44¾ × 42"	1	Plywood siding
G	Front panel	⅝ × 44¾ × 37½"	1	Plywood siding
H	Shelf	¾ × 20¾ × 44¾"	2	Fir plywood
I	Door panel	⅝ × 15¾ × 17¾"	2	Plywood siding
J	Door stile	¾ × 3½ × 21¼"	4	Cedar

Cutting List

Key	Part	Dimension	Pcs.	Material
K	Door rail	¾ × 3½ × 12¼"	4	Cedar
L	Kickboard	¾ × 2½ × 47½"	2	Cedar
M	End plate	¾ × 2½ × 22"	2	Cedar
N	End trim	¾ × 2½ × 39½"	4	Cedar
O	Front trim	¾ × 2½ × 35"	2	Cedar
P	Back trim	¾ × 2½ × 39½"	2	Cedar
Q	Hinge cleat	¾ × 1½ × 44¾"	1	Pine
R	Shelf cleat	1½ × 1½ × 20¾"	4	Pine
S	Back cleat	1½ × 1½ × 41¾"	2	Pine
T	Door cleat	¾ × 1½ × 18"	2	Pine

Materials: Moisture-resistant glue, butt hinges (4), 4" strap hinges (2), 1¼" and 2½" deck screws, door catches (2) or a 1"-dia. × 12" dowel and a ¼"-dia. × 4" carriage bolt, finishing materials.

Note: Measurements reflect actual size of dimension lumber.

Cut and fasten the lid to the lid framework with the grooves in the panel running back to front.

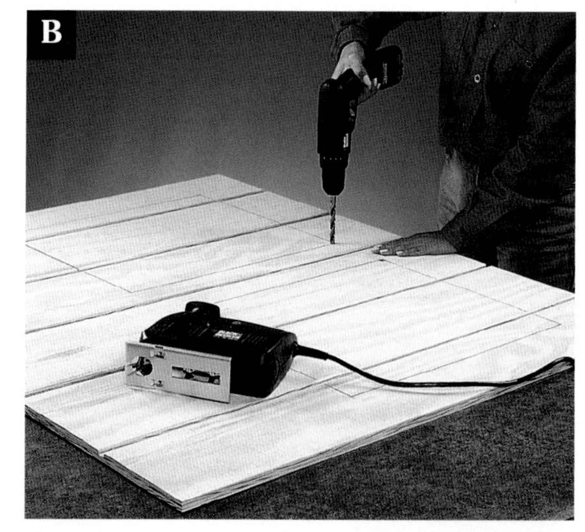

Drill a ⅜"-dia. starter hole at a corner of each door opening and cut out the openings with a jig saw.

Directions: Outdoor Storage Center

MAKE THE LID ASSEMBLY.
1. Use a circular saw and a straightedge to cut the lid (A).
2. Cut the lid edges (B), lid ends (C) and lid stringer (D).
3. Lay the lid ends and edges on their faces, smooth side up. Attach the lid ends flush with the outsides of the lid edges, using glue and 2½" deck screws. Attach the lid stringer midway between the lid ends in the same manner.
4. Apply glue to the top faces of the lid ends, stringer and lid edges. Set the lid on the frame assembly **(photo A)** and screw it in place with 1¼" deck screws.

MAKE THE PANELS.
1. Cut the back panel (F) and front panel (G) to size. On the inside face of the front panel, measure up from the bottom and draw straight lines at 5" and 23". Measure in 4" and 20" from each side and draw lines. These lines mark the cutout lines for the door openings.
2. Drill a ⅜"-dia. starter hole at one corner in each door opening **(photo B).** Cut out the door openings with a jig saw and sand the edges smooth.
3. Cut the end panels (E) to size. On the front edge of each panel, measure down 4½" and place a mark. Draw a line connecting each mark with the top corner on the back edge of the panel, creating cross-cutting lines for the back-to-front tapers. Cross-cut along the lines with a circular saw.

Attach the end panels to the back panel, keeping the back panel flush with the back edges of the end panels.

ASSEMBLE THE PANELS.
1. Stand the back panel on its bottom edge and butt it up between the end panels, flush with the back edges.
2. Fasten the back panel between the side panels with glue and 1¼" deck screws **(photo C).**

ATTACH THE SHELVES.
1. Cut the shelves (H) to size. Measure up 25" from the bottoms of the end panels and draw reference marks for positioning the top shelf. Cut the shelf cleats (R) and back cleats (S) to length. Attach the cleats just below the reference lines

Place the shelf on top of the cleats and fasten with glue and screws.

with glue. Drive 1¼" deck screws through the end panels and back panels and into the cleats.
2. Fasten the shelf to the cleats with 1¼" deck screws **(photo D).** Drive 1¼" deck screws through the back panel and into the shelf.
3. Mark reference lines for the bottom shelf, 4" from the bottoms of the side panels. Install the bottom shelves in the same manner as the top shelves.
4. Fasten the front panel (G) between the end panels with glue and 2½" deck screws.

CUT AND INSTALL TRIM.
1. Cut the kickboards (L), the end plates (M), the end trim (N), the front trim (O) and the back trim (P) to length. Sand the ends smooth. Attach the end plates at the bases of the side panels. Drill ⅛" pilot holes in the end plates. Counterbore the holes ¼" deep, using a counterbore bit. Drive 1¼" deck screws through the end plate and into the side panels.
2. Attach the front and back kickboards to the bases of the front and back panels.
3. Hold the end trim pieces against the side panels at both the front and back edges. Trace the profile of the tapered side panels onto the trim pieces to make cutting lines. The trim pieces at the fronts should be flush with the front panel. Cut at the lines with a circular saw.
4. Attach the end trim pieces to the side panels with 1¼" deck screws **(photo E).** Attach the front and back trim to the front and back panels, covering the edges of the end trim.

ATTACH THE DOORS AND LID.
1. Cut the door stiles (J) and door rails (K) to length. Attach them to the cutout door panels (I), forming a frame that extends 1¾" past the edges of the door panels on all sides.
2. Cut door cleats (T) to length. Screw them to the inside faces of the front panel directly behind the hinge locations at the outside edges of the openings. Mount two butt hinges on the outside edge of each door, using 1¼" deck screws.
3. Install a door catch for each door or use a 1" dowel bolted to the front panel as a turnbuckle.
4. Cut the hinge cleat (Q) to length and attach it to the inside face of the back panel, flush with the top edge.
5. Put the lid and strap hinges in place, with the upper hinge plates positioned between the back trim and lid ends. Drill pilot holes on the back trim for the lower hinge plate and mark the hinge pin location on the back edge of the lid end. Remove the lid and use the location marks to attach the upper hinge plate with 1¼" deck screws. Put the lid in place and attach the lower hinge plates in the same manner.

APPLY FINISHING TOUCHES.
Sand edges smooth. Apply a clear wood sealer or any other finish of your choice.

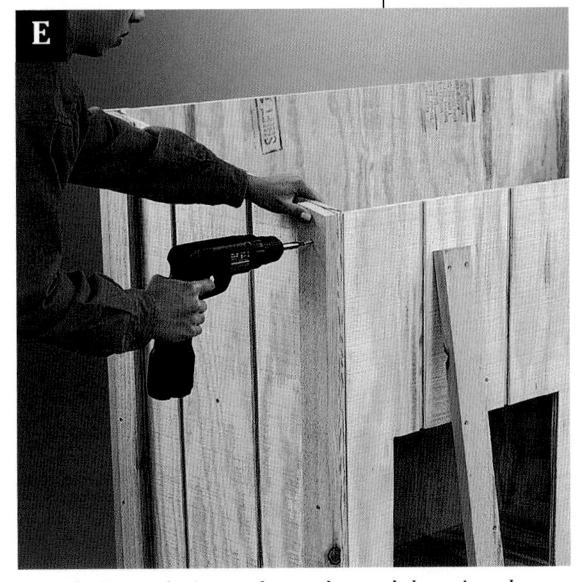

Attach the end trim to the end panel, keeping the front edge of the trim flush with the front edge of the front panel.

PROJECT
POWER TOOLS

Grill Garage

Eliminate mess and clutter, and shelter grilling appliances from the elements with this spacious grill garage.

CONSTRUCTION MATERIALS

Quantity	Lumber
2	½" × 4 × 8' textured cedar sheet siding
1	¾" × 2 × 2' plywood
10	1 × 2" × 8' cedar

Summer cookouts will be more enjoyable with this handy grill garage and storage unit. Unlike most prefabricated grill garages, this project is sized to store today's popular gas grills, as well as traditional charcoal grills. And while you're using your grill, the spacious top platforms of the grill garage can be used as convenient staging and serving areas. The walls of this grill garage are made from inexpensive, attractive rough cedar siding panels. Fitted with a cabinet-style door, the storage compartment can accommodate two large bags of charcoal, plus all your grilling accessories.

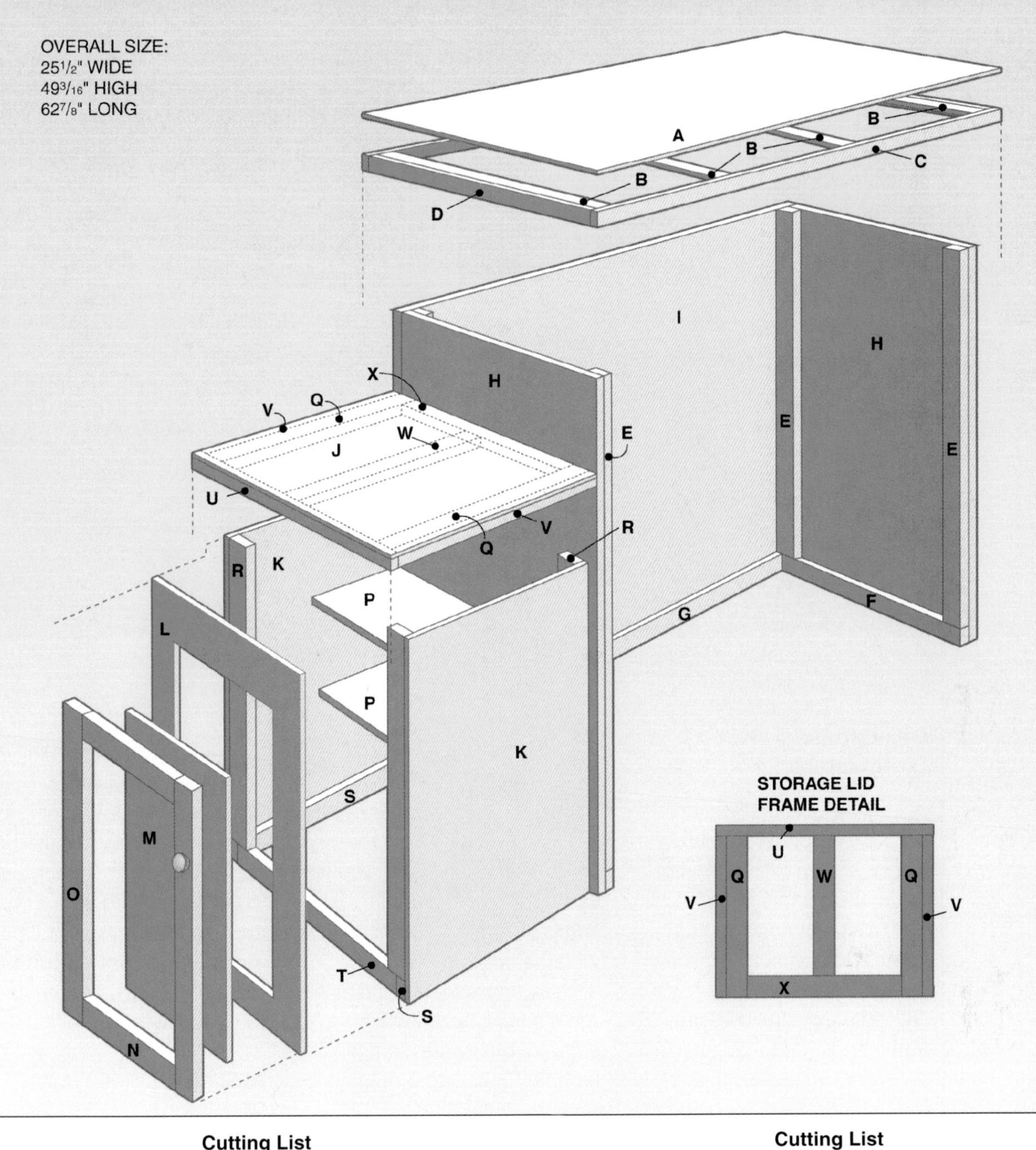

Cutting List

Key	Part	Dimension	Pcs.	Material
A	Garage lid	7/16 × 25 1/2 × 43 5/8"	1	Cedar siding
B	Lid stringer	3/4 × 1 1/2 × 24"	4	Cedar
C	Lid-frame side	3/4 × 1 1/2 × 43 5/8"	2	Cedar
D	Lid-frame end	3/4 × 1 1/2 × 24"	2	Cedar
E	Posts	3/4 × 1 1/2 × 46 1/2"	4	Cedar
F	End plate	3/4 × 1 1/2 × 22 13/16"	2	Cedar
G	Back plate	3/4 × 1 1/2 × 41 1/4"	1	Cedar
H	End panel	7/16 × 23 9/16 × 48"	2	Cedar siding
I	Back panel	7/16 × 42 1/8 × 48"	1	Cedar siding
J	Storage lid	7/16 × 20 × 24"	1	Cedar siding
K	Side panel	7/16 × 19 1/4 × 29 1/4"	2	Cedar siding
L	Face panel	7/16 × 22 1/2 × 29 1/4"	1	Cedar siding

Cutting List

Key	Part	Dimension	Pcs.	Material
M	Door panel	7/16 × 18 1/2 × 23 1/4"	1	Cedar siding
N	Door rail	3/4 × 1 1/2 × 17"	2	Cedar
O	Door stile	3/4 × 1 1/2 × 24 3/4"	2	Cedar
P	Shelf	3/4 × 10 × 21 3/8"	2	Plywood
Q	End stringer	3/4 × 1 1/2 × 19 1/4"	2	Cedar
R	Short post	3/4 × 1 1/2 × 27 3/4"	4	Cedar
S	Side plate	3/4 × 1 1/2 × 19 1/4"	2	Cedar
T	Front plate	3/4 × 1 1/2 × 20 1/8"	1	Cedar
U	Front lid edge	3/4 × 1 1/2 × 24"	1	Cedar
V	Storage lid end	3/4 × 1 1/2 × 19 1/4"	2	Cedar
W	Center stringer	3/4 × 1 1/2 × 17 3/4"	1	Cedar
X	Rear lid edge	3/4 × 1 1/2 × 19 1/2"	1	Cedar

Materials: Moisture-resistant glue, 1", 1 1/2", 2" and 3" deck screws, hinges, door pull, finishing materials.

Note: Measurements reflect actual size of dimension lumber.

Install stringers inside the garage-lid frame to strengthen the garage lid.

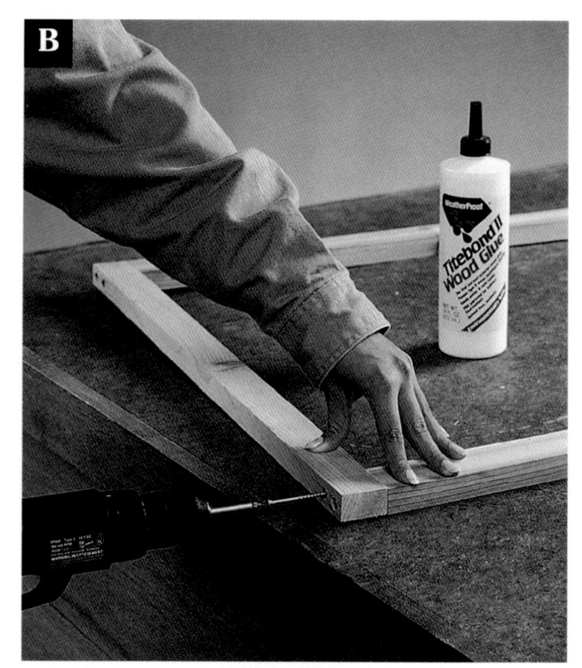

Use 1 × 2 posts to create the framework for the main garage compartment.

Directions: Grill Garage

MAKE THE GARAGE LID PANEL.

1. Cut the garage lid (A) to size. (Use a straightedge cutting guide whenever cutting sheet goods.) Cut the lid stringers (B), lid-frame sides (C) and lid-frame ends (D) to length. On a flat work surface, arrange the frame ends and sides on edge to form the lid frame. Fasten the lid sides and lid ends together with glue. Drive 1½" deck screws through the sides and into the ends of the lid-frame ends.

2. Position the lid stringers facedown in the frame, with one on each end and two spaced evenly in between. Attach the stringers and frame with glue and 1½" deck screws **(photo A).**

3. Turn the frame over so the side where the stringers are flush with the top edges of the frame is facing up. Lay the garage lid on top of the frame assembly and test the fit—the edges of the lid should be flush with the edges of the frame.

4. Remove the garage lid and run a bead of glue on the top edges of the frame. Drill ⅛" pilot holes in the lid. Counterbore the holes ¼" deep, using a counterbore bit. Reposition the lid on the frame assembly. Drive 1" deck screws through the lid and into the tops of the frame components.

BUILD THE GARAGE WALLS.

1. Cut the posts (E) and end plates (F) to length. Cut the end panels (H) to size. Assemble an end plate and two posts into an open-end frame on your work surface. Fasten the parts together with glue. Drive 1½" deck screws through the end plate and into the ends of the posts **(photo B).**

2. Test the fit. Drill pilot holes in the end panel. Counterbore the holes. Attach an end panel to the frame with glue. Drive 1" deck screws through the panel and into the frame **(photo C).**

3. Build the other end panel the same way.

ASSEMBLE THE GARAGE PANELS.

1. Cut the back plate (G) to length. Cut the back panel (I) to size.

2. Stand one end-panel assembly up so it rests on the plate. Place a bead of glue along the edge of the post that will join the back panel. Position one end of the back panel flush against the post, making sure the rough side of the cedar siding is facing out. Attach the back panel to the end-panel assembly with 1½" screws. Attach the other end-panel assembly to the other side of the back

TIP

Cedar siding panels, like most sheet goods, usually have a different actual thickness than the nominal thickness implies. For example, the project shown here uses ½" nominal siding, which has an actual thickness of 7⁄16".

Attach the end panel to the open-ended frame assembly, making sure that the rough side of the cedar siding is facing outward.

panel the same way **(photo D).**
3. Place a bead of glue along the outside face of the back plate. Position the plate at the bottom of the back panel, so the ends of the plate form butt joints with the end-panel assemblies. Secure by driving 1" deck screws through the back panel and into the back plate.
4. Fit the garage lid panel around the tops of the end and back panels, shifting the panels slightly to create a tight fit. Drill pilot holes in the lid frame. Counterbore the holes. Attach the lid panel with glue. Drive 2" deck screws through the lid frame and into the tops of the end and back panels and frame posts.

BUILD THE CABINET LID.
1. Cut the storage lid (J) to size. Cut the end stringers (Q), center stringer (W), front lid edge (U), rear lid edge (X) and storage lid ends (V) to length.
2. Lay the two storage lid ends and the front lid edge on edge on a flat surface. Position the storage lid ends so that they butt into the back face of the front lid edge. Fasten the ends and edge together with glue and 1½" deck screws.
3. Lay the rear lid edge on its face between the end stringers, which are facedown, flush with the ends of the stringers. Mounting the rear lid edge in this way provides a flush fit at the rear of the storage unit assembly while maintaining an overhang on the sides and front. Fasten the rear lid edge and end stringers together with glue and 3" deck screws.
4. Fasten the storage-lid end/edge assembly to the end stringer/rear-lid edge assembly with glue and 3" deck screws to form a frame. Position the center stringer midway between the end stringers and attach with glue and 3" deck screws.
5. Turn the storage lid frame over so the side with the stringers flush with the tops of the frame faces up. Lay the storage lid panel on top of the frame so the edges are flush. Drill pilot holes in the lid and counterbore the holes. Attach the lid panel with glue and drive 1" deck screws through the panel and into the frame.

BUILD THE CABINET WALLS.
1. Cut the short posts (R) and side plates (S) to length. Cut the side panels (K) to size.
2. Attach a side plate to the bottom, inside edge of a side panel, so the plate is flush with the front edge of the panel **(photo E).** Attach the short

Attach the back panel to the posts of the end panels to assemble the walls of the main grill garage compartment.

Attach the side plate, with the face against the panel, to the bottom edge of the side panel.

Drill a ⅜"-dia. hole on the inside of one of the corners of the door layout, then cut out the door opening with a jig saw.

posts upright, flush with the ends of the side plates and the the side panels, by driving 2½" deck screws through each plate and into the end of the corresponding post. Drive a 1" deck screw through each side panel and into the corresponding post. Build the second cabinet side panel the same way.

MAKE THE CABINET DOOR FACE FRAME.

1. Cut the face panel (L) to size. On the inside of the panel, mark a cutout for the cabinet door opening. First, measure down from the top 4", and draw a line across the panel. Then, measure in from both sides 2" and draw straight lines across the panel. Finally, draw a line 2" up from the bottom. The layout lines should form an 18½" × 23¼" rectangle.

2. Drill a ⅜"-dia. starter hole for a jig saw blade at one corner of the cutout area **(photo F).** Cut out the door opening with a jig saw. Sand the edges smooth. Save the cutout piece for use as the door panel (M).

ASSEMBLE THE CABINET.

1. Arrange the cabinet walls so they are 22½" apart. Attach the face frame to a short post on each wall, using glue and 1" deck screws. Make sure the face frame is flush with the outside faces of the cabinet walls, and that the wide "rail" of the face frame is at the top of the cabinet, where there are no plates **(photo G).**

2. Cut the front plate (T) and fasten it to the bottom, inside edge of the face frame, butted against the short posts.

3. Place the cabinet lid assembly onto the cabinet walls and face frame. Attach the cabinet lid with glue. Drive 1" deck screws through the insides of the cabinet walls and into the frame of the lid **(photo H).**

MAKE AND INSTALL THE SHELVES.

1. Cut the shelves (P) to size. Lay out ¾ × 1½" notches in the back corners of the shelves so they fit around the cabinet posts that attach the cabinet to the garage wall. Cut out the notches in the shelves, using a jig saw.

2. On the inside of each cabinet wall, draw lines 8" down from the top and 11" up from the bottom to mark shelf locations. Fit the shelf notches around the back posts, then attach the shelves by driving 1½" deck screws through the cabinet sides and into the edges of the shelves. Drive at least two screws into each shelf edge.

ATTACH THE CABINET TO THE GARAGE.

1. Push the cabinet flush against the left wall of the garage.

2. Fasten the cabinet to the garage by driving 3" deck screws through the garage

Fasten the cutout face frame to the cabinet sides.

Set the cabinet-lid assembly over the cabinet walls and face frame. Fasten them with glue and screws.

Fasten the door rails and door stiles to the door panel using glue and screws, leaving a ¾" overlap on all sides of the door panel.

posts and into the short posts of the cabinet. Three screws into each post will provide sufficient holding power.

BUILD AND ATTACH THE DOORS.

1. Cut the door rails (N) and stiles (O) to length. Using the cutout from the face frame panel for the door panel (M), fasten the rails and stiles to the door panel using glue and 1½" deck screws. Leave a ¾" overlap on all sides **(photo I).** Be sure to mount the rails between the stiles, but flush with the stile ends.

2. Attach door hinges 3" from the top and bottom of one door stile. Mount the door to the face frame. Install the door pull.

APPLY THE FINISHING TOUCHES.

Sand and smooth the edges of the grill garage and prepare it for the finish of your choice. Since it is constructed with cedar, you can chose a clear wood sealer that leaves the rich wood grain and color visible. If you prefer a painted finish, use a quality primer and durable exterior enamel paint.

TIP

The grill garage is designed as a handy storage center for your grill and such supplies as charcoal and cooking utensils. Do not store heavy items on top of the garage lid, and never light your grill while it is still in the grill garage. Do not store lighter fluid in the grill garage—always keep lighter fluid out of reach of children, in a cool, sheltered area, such as a basement.

Firewood Shelter

Those stacks of firewood won't be an eyesore anymore once you build this ranch-style firewood shelter for your yard.

CONSTRUCTION MATERIALS

Quantity	Lumber
10	2 × 4" × 8' cedar
5	2 × 6" × 8' cedar
10	⅝ × 8" × 8' cedar lap siding

This handsome firewood shelter combines rustic ranch styling with ample sheltered storage that keeps firewood off the ground and obscured from sight. Clad on the sides and roof with beveled cedar lap siding, the shelter has the look and feel of a permanent structure. But because it's freestanding, you can move it around as needed. It requires no time-consuming foundation work.

This firewood shelter is large enough to hold an entire face cord of firewood. And since the storage area is sheltered and raised to avoid ground contact and allow air flow, wood dries quickly and is ready to use when you need it.

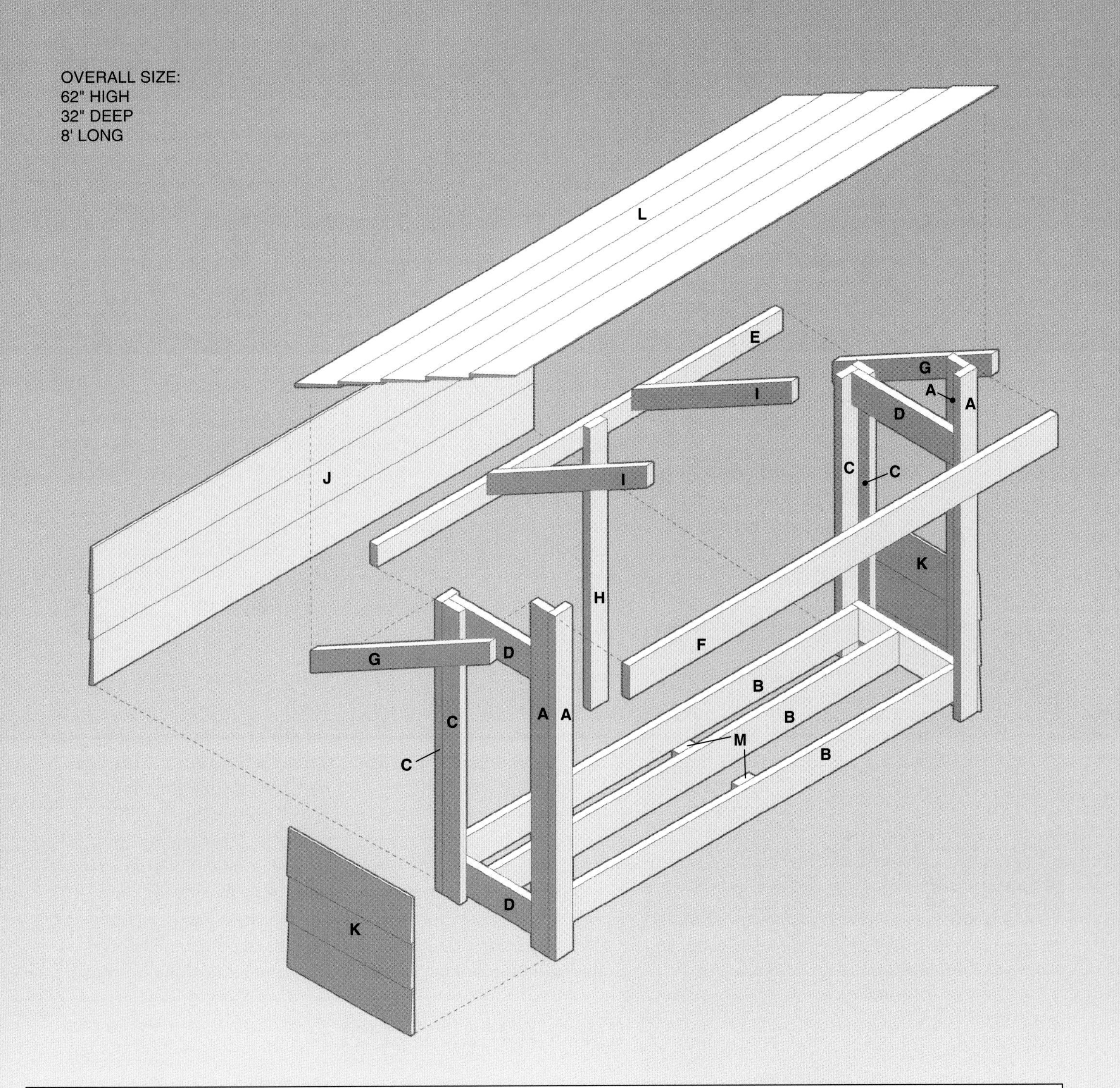

Cutting List

Key	Part	Dimension	Pcs.	Material
A	Front post	1½ × 3½ × 59"	4	Cedar
B	Bottom rail	1½ × 5½ × 82½"	3	Cedar
C	Rear post	1½ × 3½ × 50"	4	Cedar
D	End rail	1½ × 5½ × 21"	4	Cedar
E	Back rail	1½ × 3½ × 88½"	1	Cedar
F	Front rail	1½ × 5½ × 88½"	1	Cedar
G	Roof support	1½ × 3½ × 33¾"	2	Cedar

Cutting List

Key	Part	Dimension	Pcs.	Material
H	Middle post	1½ × 3½ × 50"	1	Cedar
I	Middle support	1½ × 3½ × 28"	2	Cedar
J	Back siding	⅝ × 8 × 88½"	3	Cedar siding
K	End siding	⅝ × 8 × 24"	6	Cedar siding
L	Roof strip	⅝ × 8 × 96"	5	Cedar siding
M	Prop	1½ × 3½ × 7½"	2	Cedar

Materials: ⅜ × 3½" lag screws (24), ⅜ × 4" lag screws (8), 1½" spiral siding nails, 2½" and 3" deck screws, finishing materials.

Note: Measurements reflect the actual size of dimension lumber.

Directions: Firewood Shelter

BUILD THE FRAME.

1. Cut the front posts (A) and rear posts (C) to length. Butt the edges of the front posts together in pairs to form the corner posts. Drill ⅛" pilot holes at 8" intervals. Counterbore the holes ¼" deep, using a counterbore bit. Join the post pairs with 2½" deck screws. Follow the same procedure to join the rear posts in pairs.

2. Cut the bottom rails (B) and end rails (D). Assemble two bottom rails and two end rails into a rectangular frame, with the end rails covering the ends of the bottom rails. Set the third bottom rail between the end rails, centered between the other bottom rails. Mark the ends of the bottom rails on the outside faces of the end rails. Drill two ⅜" pilot holes for lag screws through the end rails at each bottom rail position—do not drill into the bottom rails. Drill a ¾" counterbore for each pilot hole, deep enough to recess the screw heads. Drill a smaller, ¼" pilot hole through each pilot hole in the end rails, into the ends of the bottom rails **(photo A).** Drive a ⅜ × 3½" lag screw fitted with a washer at each pilot hole, using a socket wrench.

Use a smaller bit to extend the pilot holes for the lag screws into the ends of the bottom rails.

3. Draw reference lines across the inside faces of the corner posts, 2" up from the bottoms. With the corner posts upright and about 82" apart, set 2"-high spacers next to each corner post to support the frame. Position the bottom rail frame between the corner posts, and attach the frame to the corner posts by driving two 2½" deck screws through the corner posts and into the outer faces of the bottom rails. Drill pilot holes in the sides of the corner posts. Counterbore the holes. Drive a pair of ⅜ × 4" lag screws, fitted with washers, through the sides of the corner posts and into the bottom rails. The lag screws must go through the post and end rail, and into the end of the bottom rail. Avoid hitting the lag screws that have already been driven through the end rails.

4. Complete the frame by installing end rails at the tops of the corner posts. Drill pilot holes in the end rails. Counterbore the holes. Drive 2½" deck screws through the end rails and into the posts. Make sure the tops of the end rails are flush with the tops of the rear posts **(photo B).**

MAKE THE ROOF FRAME.

1. Cut the back rail (E), front rail (F), roof supports (G), middle post (H) and middle supports (I) to length. The roof supports and middle supports are mitered at the ends. To make the miter cutting lines, mark a point 1½" in from each end, along the edge of the board. Draw diagonal lines from each point to the opposing corner. Cut along the lines

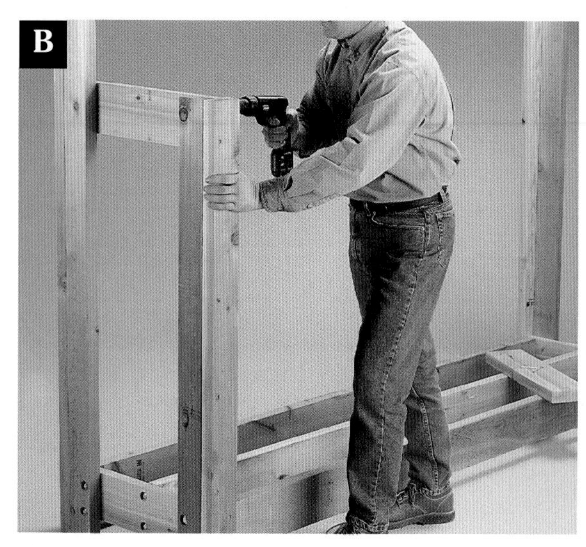

Attach end rails between front and rear corner posts.

Miter-cut the middle supports and roof supports with a circular saw.

Attach the front rail by driving screws through the outer roof supports, making sure the top of the rail is flush with the tops of the supports.

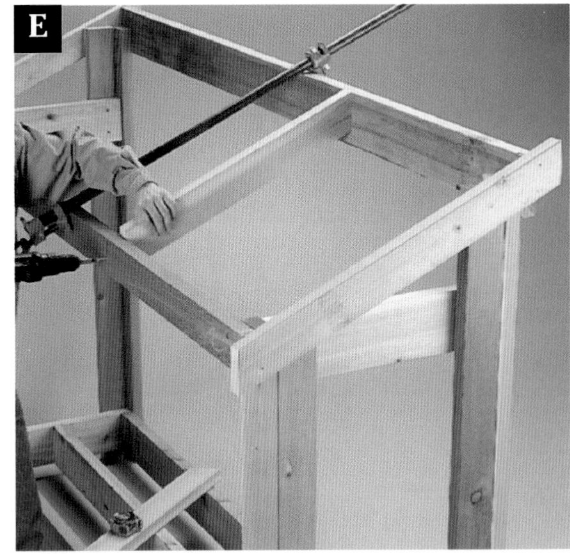
Attach the middle roof supports by driving screws through the front and back rails.

with a circular saw **(photo C).**
2. Drill pilot holes in the back rail. Counterbore the holes. Use 3" deck screws to fasten the back rail to the backs of the rear corner posts, flush with their tops and sides. Use the same procedure to fasten a roof support to the outsides of the corner posts. Make sure the top of each support is flush with the high point of each post end. The supports should overhang the posts equally in the front and rear.
3. Drill pilot holes in the roof supports. Counterbore the holes. Drive deck screws to attach the front rail between the roof supports **(photo D),** with the top flush with the tops of the roof supports. Attach the middle supports between the front rail and back rail, 30" in from each rail end. Drive 3" deck screws through the front and back rails into the ends of the middle supports **(photo E).** Use a pipe clamp to hold the supports in place as you attach them.
4. Drill pilot holes in the middle post (H). Counterbore the holes. Position the middle post so it fits against the outside of the rear bottom rail and the inside of the top back rail. Make sure the middle post is perpendicular and extends past the bottom rail by 2". Attach it with 2½" deck screws.
5. Cut a pair of props (M) to length. Attach them to the front two bottom rails, aligned with the middle post. Make sure the tops of the props are flush with the tops of the bottom rails.

ATTACH SIDING AND ROOF.
1. Cut pieces of 8"-wide beveled cedar lap siding to length to make the siding strips (J, K) and the roof strips (L). Starting 2" up from the bottoms of the rear posts, fasten the back siding strips (J) with two 1½" siding nails driven through each strip and into the posts, near the top and bottom edge of the strip. Work your way up, overlapping each piece of siding by ½", making sure the thicker edges of the siding face down. Attach the end siding (K) to the corner posts, with the seams aligned with the seams in the back siding.

Attach the roof strips with siding nails, starting at the back edge and working your way forward.

2. Attach the roof strips (L) to the roof supports, starting at the back edge. Drive two nails into each roof support. Make sure the wide edge of the siding faces down. Attach the rest of the roof strips, overlapping the strip below by about ½" **(photo F),** until you reach the front edges of the roof supports. You can leave the cedar wood untreated or apply an exterior wood stain to keep it from turning gray as it weathers.

Cabin Marker

Hidden driveways and remote roads won't escape first-time visitors if they are marked with a striking, personalized cabin marker.

PROJECT POWER TOOLS

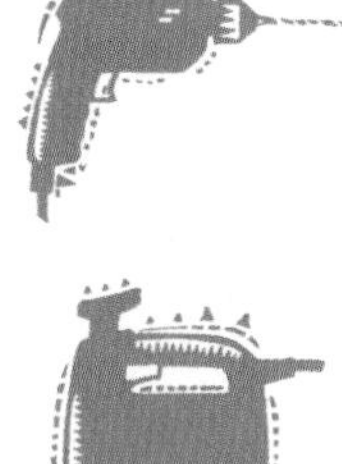

Trips to a friend's cabin or vacation home, though usually enjoyable, often start on a confusing note. "Do you have the address written down?" is a common refrain after the fourth left into a dead end in the woods. You can save your friends confusion and wasted time by displaying your name, address and mailbox at the head of your driveway. And on a safety note, emergency vehicles can spot your home more quickly with a well-marked name and address on it.

The simple design of this cabin marker and mailbox stand makes it suitable for almost any yard. Its height ensures a certain level of prominence, but the cedar material and basic construction allow it to fit right in with its natural surroundings.

One of the best features of the cabin marker may be the least noticed—the base section. The base is a multi-tiered pyramid of 4 × 4 cedar timbers. It provides ample weight and stability, so you won't need to go to the trouble of digging a hole or pouring concrete. Just position the marker wherever you want it, and stake it in place. Much more attractive than a simple mailbox stand, this project will provide just the touch of originality that your cabin or vacation home deserves.

Construction Materials

Quantity	Lumber
1	1 × 6" × 8' cedar
1	2 × 2" × 6' cedar
4	2 × 4" × 8' cedar
3	4 × 4" × 8' cedar

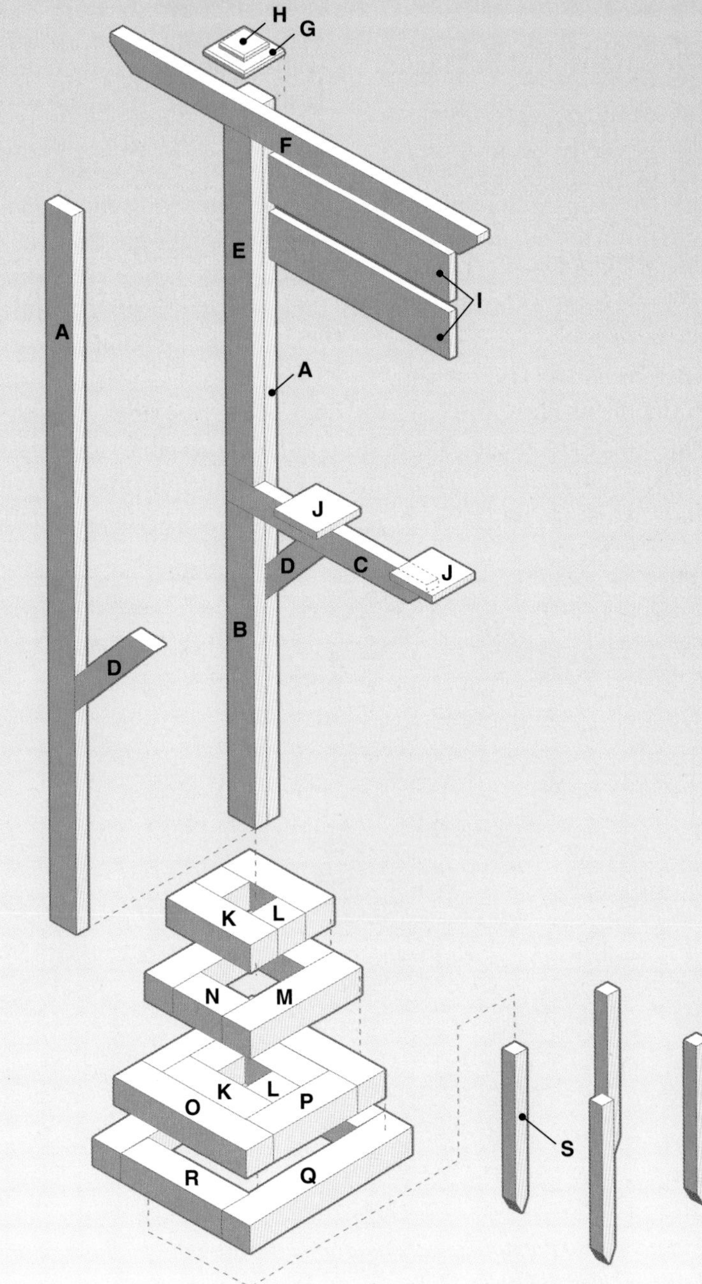

OVERALL SIZE:
85¾" HIGH
48½" WIDE
22" DEEP

Cutting List

Key	Part	Dimension	Pcs.	Material
A	Post side	1½ × 3½ × 84"	2	Cedar
B	Post section	1½ × 3½ × 36½"	1	Cedar
C	Mailbox arm	1½ × 3½ × 23½"	1	Cedar
D	Mailbox brace	1½ × 3½ × 17½"	2	Cedar
E	Post section	1½ × 3½ × 40½"	1	Cedar
F	Sign arm	1½ × 3½ × 48½"	1	Cedar
G	Top plate	⅞ × 5½ × 5½"	1	Cedar
H	Cap	⅞ × 3½ × 3½"	1	Cedar
I	Sign board	⅞ × 5½ × 24"	2	Cedar
J	Mailbox cleat	⅞ × 5½ × 5⅞"	2	Cedar

Cutting List

Key	Part	Dimension	Pcs.	Material
K	Base piece	3½ × 3½ × 10½"	4	Cedar
L	Base piece	3½ × 3½ × 4½"	4	Cedar
M	Base piece	3½ × 3½ × 15"	2	Cedar
N	Base piece	3½ × 3½ × 7"	2	Cedar
O	Base piece	3½ × 3½ × 17½"	2	Cedar
P	Base piece	3½ × 3½ × 11½"	2	Cedar
Q	Base piece	3½ × 3½ × 22"	2	Cedar
R	Base piece	3½ × 3½ × 14"	2	Cedar
S	Stake	1½ × 1½ × 18"	4	Cedar

Materials: Moisture-resistant glue, epoxy glue, 2", 2½" and 4" deck screws, #10 screw eyes (8), S-hooks (4), ⅜"-dia. × 5" galvanized lag screws with 1" washers (8), finishing materials.
Note: Measurements reflect the actual size of dimension lumber.

Directions: Cabin Marker

MAKE THE POST.

The post is made in three layers. Two post sections and two arms form the central layer, which is sandwiched between two post sides. The arms extend out from the post to support a mailbox and an address sign.

1. Cut the mailbox arm (C) and sign arm (F) to length. One end of the mailbox arm and both ends of the sign arm are cut with decorative slants on their bottom edges. To cut the ends of the arms to shape, mark a point on the three ends, 1" down from a long edge. On the opposite long edge, mark a point on the face 2½" in from the end. Draw a straight line connecting the points, and cut along it.

2. Cut the post sides (A) and post sections (B, E) to length. To assemble the post, you will sandwich the sections and the arms between the sides. Set one of the post sides on a flat work surface, and position the lower post section (B) on top of it, face to face, with the ends flush. Attach the lower post section to the side with wood glue and 2½" deck screws.

3. Position the mailbox arm on the side, making sure the square end is flush with the edge of the side. Use a square to make sure the mailbox arm is perpendicular to the side. Attach the mailbox arm, using glue and 2½" deck screws.

4. Butt the end of the upper post section (E) against the top edge of the mailbox arm, and attach it to the side in the same manner **(photo A).**

5. Position the sign arm at the top of the assembly so it extends 30" past the post on the side with the mailbox arm. Attach the sign arm to the post side with glue and deck screws.

6. Apply glue to the remaining side. Attach it to the post sections with glue and 4" deck screws, making sure all the ends are flush.

Butt an end of the upper section against the top edge of the mailbox arm, and fasten it to the side.

ATTACH THE MAILBOX CLEATS AND BRACES.

The cleats on the mailbox arm

Position a mailbox brace on each side of the mailbox arm, and fasten them to the post and arm.

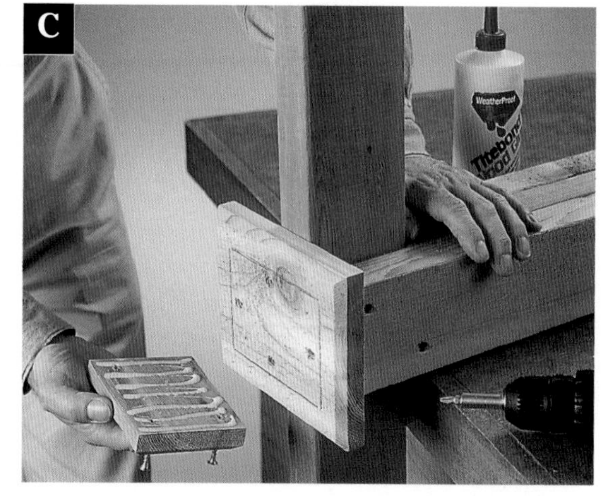

Apply glue to the bottom face of the cap, and center it on the top of the post.

provide a stable nailing surface for a "rural-style" mailbox. The mailbox braces fasten to the post and mailbox arm to provide support.

1. Cut the mailbox cleats (J) to length and sand smooth. Center the cleats on the top of the mailbox arm. The frontmost cleat should overhang the front of the mailbox arm by 1". Center the remaining cleat 12½" in from the front of the mailbox arm. Attach the cleats with glue and 2½" deck screws.

2. Cut the mailbox braces (D) to length. Their ends must be cut at an angle. Use a power miter box, or a backsaw and miter box, to miter-cut each end of each mailbox brace at a 45° angle. Make sure the cuts at either end slant toward each other (see *Diagram,* page 231).

3. Position a mailbox brace against the side of the mailbox arm so one end is flush with the top edge of the mailbox arm and the other rests squarely against the post. Drill ⅛" pilot holes. Counterbore the holes ¼" deep, using a counterbore bit. Attach the mailbox braces with glue and 2½" deck screws **(photo B).**

COMPLETE THE POST TOP.

The post assembly is capped with a post top and cap made of 1" dimension lumber.

1. Cut the top plate (G) and cap (H) to size. Using a power sander, make ¼"-wide × ¼"-deep bevels along the top edges of the top and cap.

2. Center the top on the post, and attach it with glue and 2" deck screws. Center the cap on the top and attach it **(photo C).**

MAKE THE BASE.

The base for the cabin marker is made from cedar frames that increase in size from top to bottom. The frames are stacked to create a four-level pyramid. A fifth frame fits inside one of the frames to make a stabilizer for the post. The bottom frame is fastened to stakes driven into the ground to provide a secure anchor that does not require digging holes and pouring concrete footings.

1. Cut the 4 × 4 base pieces (K, L, M, N, O, P, Q, R) to length for all five frames. Assemble them into five frames according to the *Diagram,* page 231. To join the pieces, use 4" deck screws driven into pilot holes that have been counterbored 1½" deep.

2. After all five frames are built, join one of the small frames and the two next-smallest frames together in a pyramid, using glue and 4" deck screws **(photo D).** Insert the other small frame into the opening in the third-smallest frame. Secure with deck screws.

3. Set the base assembly on top of the large frame; do not attach them. Insert the post into the opening, and secure it with lag screws, driven through the top frame and into the post. (NOTE: The bottom frame is anchored to the ground on site before being attached to the pyramid.)

MAKE THE SIGN BOARDS.

1. Cut the sign boards (I) to size. Sand them smooth.

2. Stencil your name and address onto the signs. Or, you can use adhesive letters, freehand painting, a router with a veining bit or a woodburner. Be sure to test the technique on a sanded scrap of cedar before working on the signs.

Attach the base tiers to each other, working from top to bottom.

APPLY FINISHING TOUCHES.

1. Join the two signs together with #10 screw eyes and S-hooks. Drill pilot holes for the screw eyes in the sign arm and signs. Apply epoxy glue to the threads of the screws before inserting them. Apply your finish of choice.

2. Position the bottom frame of the base in the desired location. The area should be flat and level so the post is plumb. Check the frame with a level. Add or remove dirt around the base to achieve a level base before installing.

3. Cut the stakes (S) to length, and sharpen one end of each stake. Set the stakes in the inside corners of the frame. Drive them into the ground until the tops are lower than the tops of the frame. Attach the stakes to the frames with 4" deck screws.

4. Center the cabin marker on the bottom frame. Complete the base by driving 5" lag screws through the tops of the base into the bottom frame.

Front-porch Mailbox

This cedar mailbox is a practical, good-looking project that is very easy to build. The simple design is created using basic joinery and mostly straight cuts.

PROJECT
POWER TOOLS

If you want to build a useful, long-lasting item in just a few hours, this mailbox is the project for you. Replace that impersonal metal mailbox you bought at the hardware store with a distinctive cedar mailbox that's a lot of fun to build. The lines and design are so simple on this project that it suits nearly any home entrance. The mailbox features a hinged lid and a convenient lower shelf that is sized to hold magazines and newspapers.

We used select cedar to build our mailbox, then applied a clear, protective finish. Plain brass house numbers dress up the flat surface of the lid, which also features a decorative scallop that doubles as a handgrip.

If you are ambitious and economy-minded, you can build this entire mailbox using just one 8'-long piece of 1 × 10 cedar. That means, however, that you'll have to do quite a bit of rip-cutting to make the parts. If you have a good straightedge and some patience, rip-cutting is not difficult. But you may prefer to simply purchase dimensional lumber that matches the widths of the pieces (see the *Construction Materials* list to the left).

If your house is sided with wood siding, you can hang the mailbox by screwing the back directly to the siding. If you have vinyl or metal siding, be sure that the screws make it all the way through the siding and into wood sheathing or wood wall studs. If you have masonry siding, like brick or stucco, use masonry anchors to hang the mailbox.

Construction Materials

Quantity	Lumber
1	1 × 10" × 4' cedar
1	1 × 8" × 3' cedar
1	1 × 4" × 3' cedar
1	1 × 3" × 3' cedar
1	1 × 2" × 3' cedar

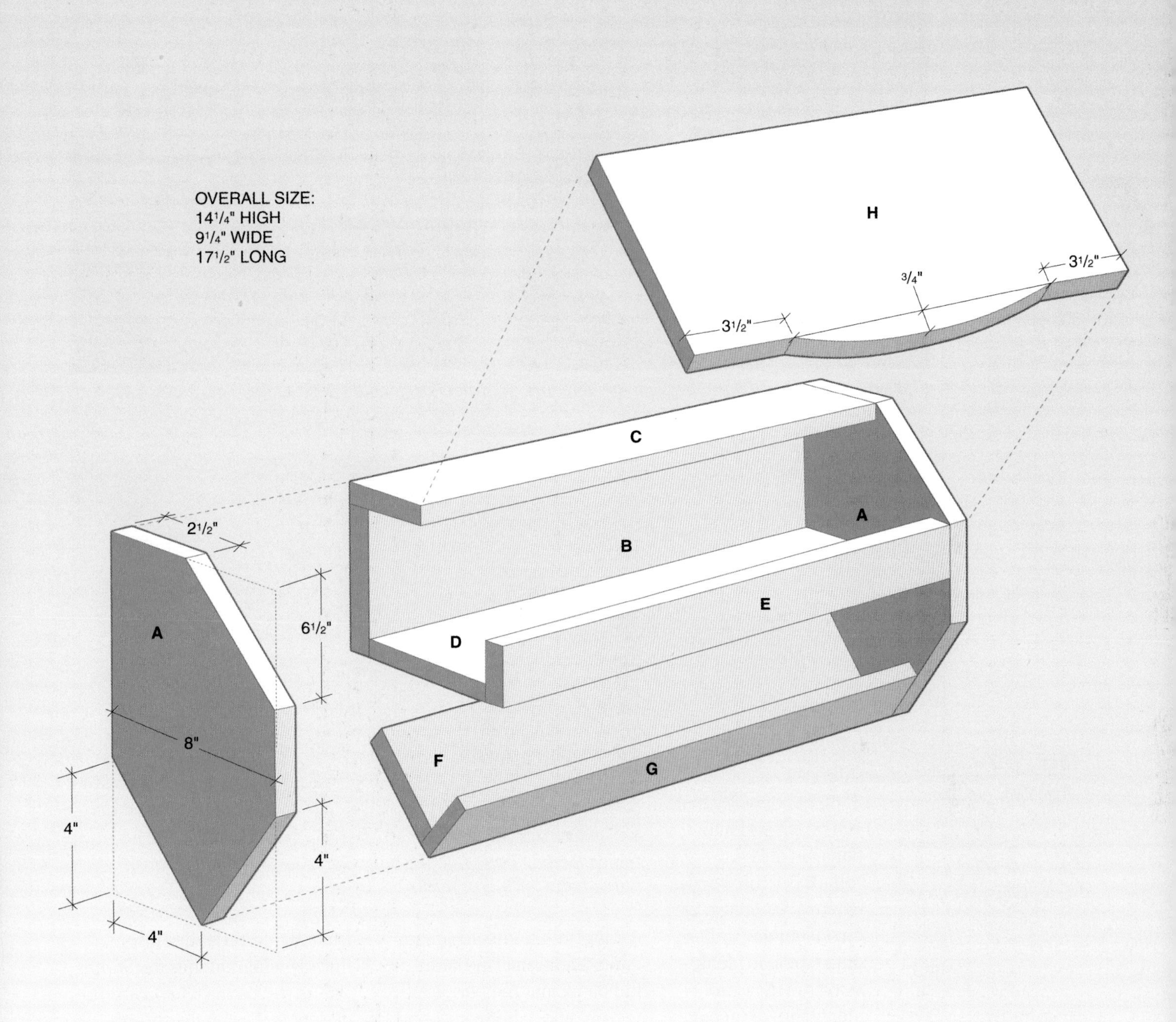

Key	Part	Dimension	Pcs.	Material
A	Side	¾ × 8 × 14¼"	2	Cedar
B	Back	¾ × 7¼ × 16"	1	Cedar
C	Top	¾ × 2½ × 16"	1	Cedar
D	Box bottom	¾ × 6½ × 16"	1	Cedar

Cutting List

Key	Part	Dimension	Pcs.	Material
E	Box front	¾ × 1½ × 16"	1	Cedar
F	Shelf bottom	¾ × 3½ × 16"	1	Cedar
G	Shelf lip	¾ × 2½ × 16"	1	Cedar
H	Lid	¾ × 9¼ × 17½"	1	Cedar

Materials: Moisture-resistant wood glue, 2" deck screws, masking tape, piano hinge with screws, finishing materials.

Note: Measurements reflect the actual size of dimension lumber.

Cutlines are drawn on the sides, and the parts are cut to shape with a jig saw.

After fastening the top between the sides, fasten the back with deck screws.

Directions: Front-porch Mailbox

BUILD THE SIDES.
The sides are the trickiest parts to build in this mailbox design. But if you can use a ruler and cut a straight line, you should have no problems.

1. Cut two 8 × 14¼" pieces to make the sides (A).

2. Pieces of wood that will be shaped into parts are called "blanks" in the woodworkers' language. Lay out the cutting pattern onto one side blank, using the measurements shown on page 235. Mark all of the cutting lines, then double-check the dimensions to make sure the piece will be the right size when it is cut to shape. Make the cuts in the blank, using a jig saw, to create one side. Sand edges smooth.

3. Use this side as a template to mark the second blank. Arrange the template so the grain direction is the same in the blank and the template. Cut out and sand the second side **(photo A).**

TIPS

If you are planning on stenciling your name or address on the front of the mailbox, remember these helpful hints:
- *Secure the stencil pattern to the surface with spray adhesive and drafting tape.*
- *Dip a stencil brush into the exterior latex paint, then wipe most of it off on a paper towel before painting.*
- *Let the paint dry thoroughly before removing the stencil.*

Attach the box bottom to the back with glue and screws driven through the back and sides.

ATTACH THE BACK AND TOP.
Fasten all the pieces on the mailbox with exterior wood glue and 2" deck screws. Although cedar is a fine outdoor wood, it can be quite brittle. To prevent splitting, drill ⅛" pilot holes and counterbore the holes ¼" deep, using a counterbore bit. Space the screws evenly when driving them.

1. Cut the back (B) and top (C) to length. Fasten the top between the 2½"-wide faces on the two sides with glue and 2" deck screws. Position the top so that the rear face is flush with the rear side edges, and the top face is flush with the top side edges.

2. Use glue and deck screws to fasten the back between the sides, flush with the 10¼"-long edges **(photo B),** and butted against the top.

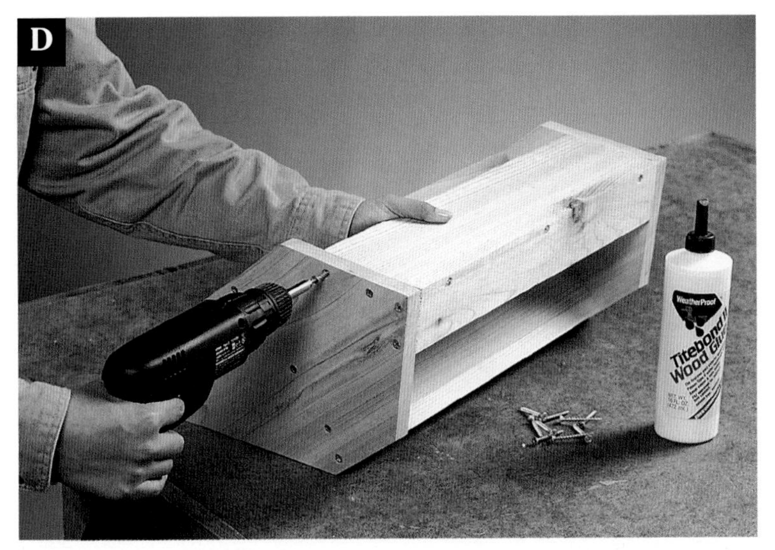

Keep the lip edges flush with the side edges to form the newspaper shelf.

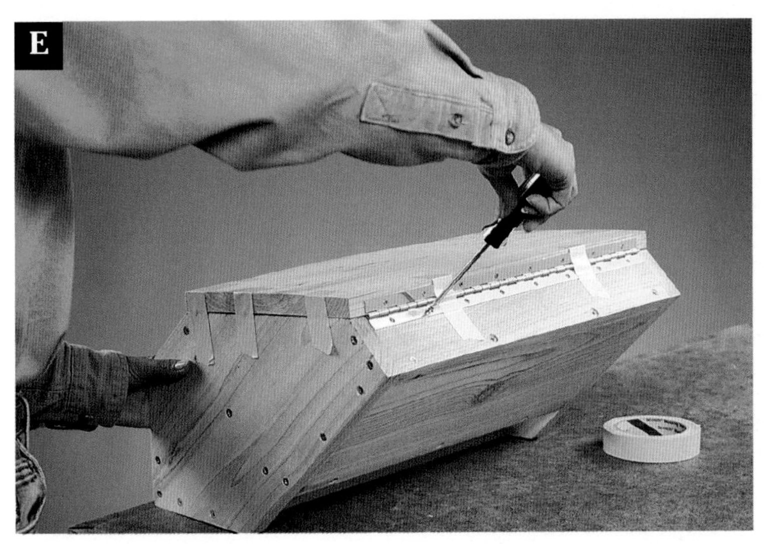

Once the pieces are taped in place, the continuous hinge is attached to join the lid to the top.

ATTACH THE BOX BOTTOM AND FRONT.

The bottom and front pieces form the letter compartment inside the mailbox.

1. Cut the bottom (D) and front (E) to length. Fasten the bottom to the back and sides, making sure the bottom edges are flush **(photo C).**

2. Once the bottom is attached, fasten the front to the sides and bottom, keeping the bottom edges flush.

ATTACH THE NEWSPAPER SHELF.

The lower shelf on the underside of the mailbox is designed for overflow mail.

1. To make the lower shelf, cut the shelf bottom (F) and shelf lip (G) to size. Fasten the shelf lip to the leg of the "V" formed by the sides that are closer to the front.

2. Fasten the shelf bottom to the sides along the back edges to complete the shelf assembly **(photo D).**

CUT AND ATTACH THE LID.

1. Cut the lid (H) to length (9¼" is the actual width of a 1 × 10). Draw a reference line parallel to and ¾" away from one of the long edges.

2. With a jig saw, make a 3½"-long cut at each end of the line. Mark the midpoint of the edge (8¾"), then cut a shallow scallop to connect the cuts with the midpoint. Smooth out the cut with a sander.

3. Attach a brass, 15"-long continuous hinge known as a piano hinge to the top edge of the lid. Then position the lid so the other wing of the hinge fits squarely onto the top of the mailbox. Secure the lid to the mailbox with masking tape. Attach the hinge to the mailbox **(photo E).**

APPLY FINISHING TOUCHES.

1. Sand all surfaces smooth with 150-grit sandpaper.

2. Finish the mailbox with a clear wood sealer or other finish of your choice. Add 3" brass house numbers on the lid. Or, stencil an address or name onto it (see *Tips*, page 236). Once the finish has dried, hang the mailbox on the wall by driving screws through the back.

TIP

Clear wood sealer can be refreshed if it starts to yellow or peel. Wash the wood with a strong detergent, then sand the surface lightly to remove flaking or peeling sealer. Wash the surface again, then simply brush a fresh coat of sealer onto the wood.

PROJECT
POWER TOOLS

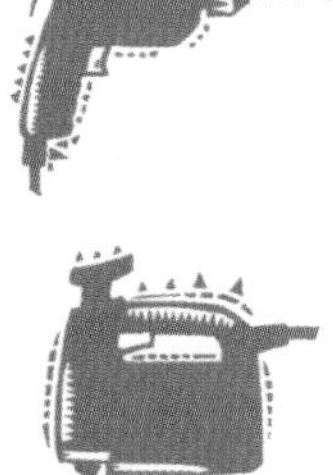

Trash Can Corral

This two-sided structure keeps trash cans out of sight but accessible from the curb or alley.

CONSTRUCTION MATERIALS

Quantity	Lumber
3	2 × 4" × 6' cedar
2	2 × 2" × 10' cedar
1	1 × 8" × 2' cedar
7	1 × 6" × 10' cedar
8	1 × 4" × 8' cedar

Nothing ruins a view from a favorite window like the sight of trash cans—especially as garbage-collection day draws near. With this trash can corral, you'll see an attractive, freestanding cedar fence instead of unsightly trash cans.

The two fence-style panels support one another, so you don't need to set fence posts in the ground or in concrete. And because the collars at the bases of the posts can be adjusted, you can position the can corral on uneven or slightly sloping ground. The staggered panel slats hide the cans completely, but still allow air to pass through to ensure adequate ventilation.

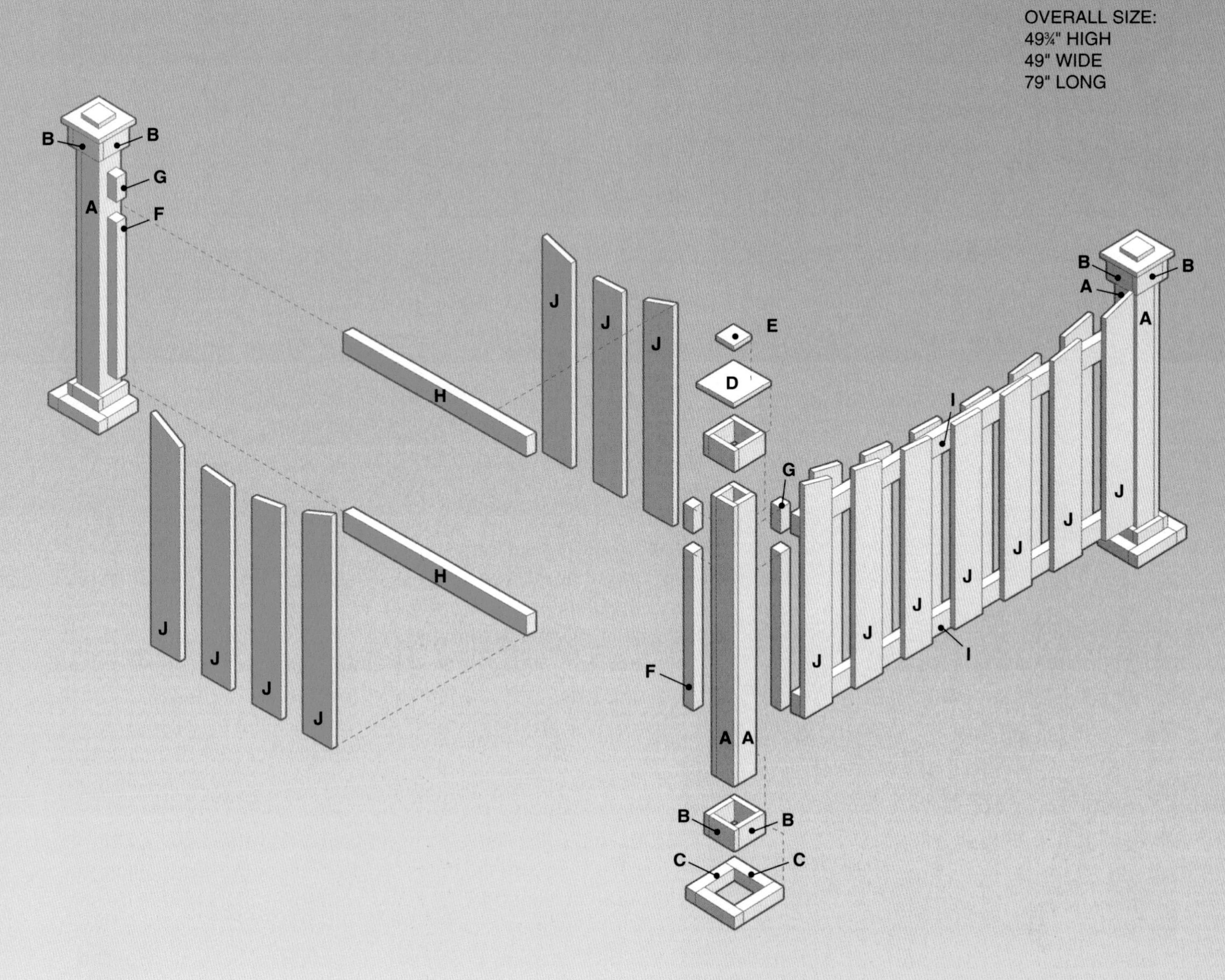

Key	Part	Dimension	Pcs.	Material
A	Post board	⅞ × 3½ × 48"	12	Cedar
B	Collar strip	⅞ × 3½ × 5¼"	24	Cedar
C	Foot strip	1½ × 1½ × 7⅝"	12	Cedar
D	Collar top	⅞ × 7¼ × 7¼"	3	Cedar
E	Collar cap	⅞ × 3½ × 3½"	3	Cedar

Cutting List

Key	Part	Dimension	Pcs.	Material
F	Long post cleat	1½ × 1½ × 26⅞"	4	Cedar
G	Short post cleat	1½ × 1½ × 4"	4	Cedar
H	Short stringer	1½ × 3½ × 35½"	2	Cedar
I	Long stringer	1½ × 3½ × 65½"	2	Cedar
J	Slat	⅞ × 5½ × 39½"	20	Cedar

Materials: 1½" and 2" deck screws, finishing materials.

Note: Measurements reflect the actual size of dimension lumber.

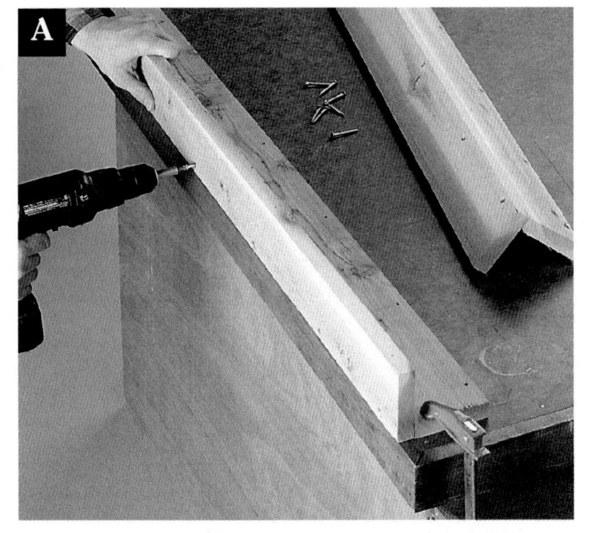

The post is made from four edge-joined boards.

Fit the top collar assemblies onto each post and attach them.

Attach the short post cleats 3⅝" up from the tops of the long post cleats. The top stringers on the panels fit between the cleats when installed.

Directions: Trash Can Corral

BUILD THE POSTS.
Each post is made of four boards butted together to form a square.
1. Cut the post boards (A) to length and sand them smooth.
2. Clamp one post board to your work surface. Then, butt another post board against it at a right angle. With the ends flush, drill ⅛" pilot holes at 8" intervals and counterbore the holes to a ¼" depth. Connect the boards by driving 2" deck screws through the pilot holes **(photo A).** Repeat the step until the post boards are fastened together in pairs. Then, fasten the pairs together to form the three posts.

MAKE AND ATTACH THE COLLARS.
Each post is wrapped at the top and bottom by a four-piece collar. The top collars have two-piece flat caps, and the base collars are wrapped with 2 × 2 strips for stability.
1. Cut the collar strips (B), collar tops (D) and collar caps (E) to size.
2. Join the collar strips together to form square frames, using 1½" deck screws.
3. Center each collar cap on top of a collar top, and attach the caps to the tops with 1½" deck screws.
4. To center the tops on the frames, mark lines on the bottoms of the top pieces, 1⅛" in from the edges. Then, drill pilot holes ½" in from the lines, and counterbore the holes.
5. Use the lines to center a frame under each top. Then, drive 1½" deck screws through the holes and into the frames.
6. Slip a top collar assembly over one end of each post **(photo B),** and drill centered pilot holes on each side of the collar. Counterbore the holes and drive 1½" deck screws into the posts.
7. Attach the remaining frames to the other ends of the posts, with the bottom edges flush.
8. Cut the foot strips (C) to length. Lay them around the bottoms of the base collars, and screw them together with 2" deck screws. Make sure the bottoms of the frames are flush with the bottoms of the collars. Then, attach the frames to the collars with 2" deck screws.

ATTACH THE CLEATS AND SUPPORTS.
The long and short post cleats (F, G) attach to the posts between and above the stringers.
1. Cut the post cleats to length. Center a long post cleat on one face of each post and attach it with 2" deck screws so the bottom is 4" above the top of the base collar on each post.
2. For the corner post, fasten a second long post cleat on an adjacent post face, 4" up from the bottom collar.

Use 4½"-wide spacers to set the gaps between panel slats.

Use a flexible guide to mark the top contours.

3. Center the short post cleats on the same post faces, 3⅝" up from the tops of the long post cleats. Attach the short post cleats to the posts, with 2" deck screws, making sure the short cleats are aligned with the long cleats **(photo C).**

BUILD THE FENCE PANELS.

1. Cut the short stringers (H), long stringers (I) and slats (J).

2. Position the short stringers on your work surface so they are parallel and separated by a 26⅞" gap. Attach a slat at each end of the stringers, so the ends of the stringers are flush with the outside edges of the slats. Drive a 1½" deck screw through each slat and into the face of each stringer.

3. Measure diagonally from corner to corner to make sure the fence panel is square. If the measurements are equal, the fence is square. If not, apply pressure to one side of the assembly until it is square. Drive another screw through each slat and into each stringer.

4. Cut 4½" spacers to set the gaps between panel slats, and attach the remaining slats on the same side of the stringers by driving two 1½" deck screws at each end. Check that the bottoms of the slats are flush with the stringer **(photo D).**

5. Turn the panel over and attach slats to the other side, starting 4½" from the ends so slats on opposite sides are staggered—there will only be three slats on this side. Build the long panel the same way.

CONTOUR THE PANEL TOPS.

To lay out the curve at the top of each fence panel, you will need to make a marking guide.

1. Cut a thin, flexible strip of wood at least 6" longer than the long fence panel. On each panel, tack nails at the top outside corner of each end slat, and another nail midway across each panel, ½" above the top stringer.

2. Form a smooth curve by positioning the guide with the ends above the outside nails, and the midpoint below the nail in the center. Trace the contour onto the slats **(photo E),** and cut along the line with a jig saw. Use a short blade to avoid striking the slats on the other side. Use the same procedure on the other side of each panel. Sand the cuts smooth.

Set the completed fence panels between the cleats on the faces of the posts.

3. Position the fence panels between the posts so the top stringer in each panel fits in the gap between the long and short post cleats **(photo F).** Drive 2" deck screws through the slats and into the cleats.

APPLY FINISHING TOUCHES.

Apply exterior wood stain to protect the cedar. You can increase the height of any of the posts slightly by detaching the base collar, lifting the post and reattaching the collar.

PROJECT
POWER TOOLS

Trash Can Protector

Store that unsightly garbage can in this little structure to keep it hidden and safe from pests.

Construction Materials

Quantity	Lumber
4	2 × 4" × 8' pine
1	¾" × 2 × 4' plywood
3	⅜" × 4 × 8' grooved siding
11	1 × 2" × 8' cedar

Anyone who has spent time at a cabin or a vacation home knows that garbage can storage can be a problem. Trash cans are not only ugly, they attract raccoons and other troublesome pests. Keep the trash out of sight and away from nighttime visitors with this simple trash can protector. It accommodates a 44-gallon can and features a front and top that swing open for easy access. The 2 × 4 frame is paneled with grooved plywood siding. A cedar frame and cross rail stiffens each side panel and adds a decorative touch to the project. We used construction adhesive on this project because it adapts well to varying rates of expansion.

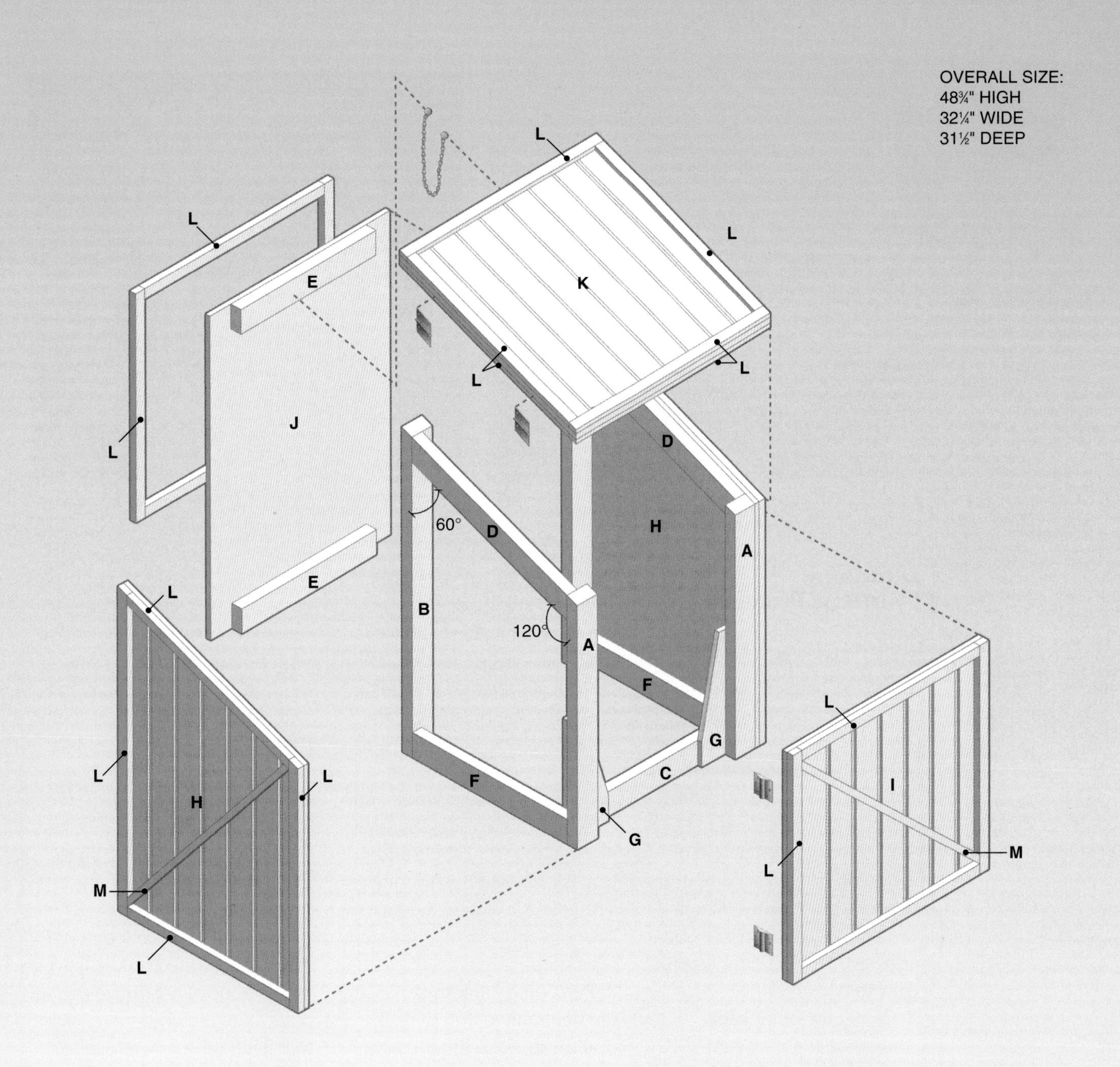

Key	Part	Dimension	Pcs.	Material
A	Front strut	1½ × 3½ × 36"	2	Pine
B	Back strut	1½ × 3½ × 46"	2	Pine
C	Front stringer	1½ × 3½ × 26¼"	1	Pine
D	Top rail	1½ × 3½ × 29¼"	2	Pine
E	Back stringer	1½ × 3½ × 22¼"	2	Pine
F	Bottom stringer	1½ × 3½ × 25½"	2	Pine
G	Brace	¾ × 6 × 16"	2	Plywood

Cutting List

Key	Part	Dimension	Pcs.	Material
H	Side panel	⅝ × 29⅛ × 45¾"	2	Plywood siding
I	Front panel	⅝ × 32¼ × 34"	1	Plywood siding
J	Back panel	⅝ × 29¼ × 45¾"	1	Plywood siding
K	Top panel	⅝ × 33¾ × 32¼"	1	Plywood siding
L	Side batten	⅞ × 1½ × *	20	Cedar
M	Cross rail	⅞ × 1½ × *	3	Cedar

Materials: Construction adhesive, 1¼", 2" and 3" deck screws, 1" panhead screws (2), 6d galvanized finish nails, 3 × 3" utility hinges (2), 3 × 3" spring-loaded hinges (2), galvanized steel or plastic door pulls (2), steel chain (18"), finishing materials.

Note: Measurements reflect the actual size of dimension lumber.
*Cut to fit.

Center the back stringer flush with the bottom edge of the back before attaching it with screws and construction adhesive.

Attach side panels to the frame, making sure the front and top edges are flush with the frame.

Directions: Trash Can Protector

MAKE THE BACK PANEL.

1. Cut the back panel (J) to size. Cut the back stringers (E) to length. Position one back stringer on the ungrooved face of the back panel, flush with the bottom edge. Center this back stringer so each end of the back stringer is 3½" in from the back panel sides. Attach the stringer with construction adhesive and 1¼" deck screws **(photo A).**

2. Center the other back stringer on the back panel, 1" down from the top edge. Attach it to the panel.

MAKE THE SIDE FRAMES.

1. Set a circular saw or a power miter box to cut at a 30° angle. Cut the front struts (A) and back struts (B) to length, making sure one end of each front and back strut is cut at a 30° angle. This slanted end cut should not affect the overall length of the struts (see *Diagram,* page 243).

2. Cut the bottom stringers (F) to length. Position the front struts on edge on your work surface. Butt the end of a bottom stringer against each front strut. The outside faces of the bottom stringers should be flush with the outside edges of the front struts, and flush with the square ends of the front struts. Apply construction adhesive, and drive 3" deck screws through the front struts and into the ends of the bottom stringers. Make sure the tops of the front struts slant in the same direction.

3. Position a back strut against the unattached end of each bottom stringer. Attach them, making sure the slanted top ends of the back struts are facing the front struts.

4. Cut the top rails (D) to length. For the top rails to fit between the front and back struts, their ends must be cut at a 30° angle. When you cut the ends, make sure they are slanted in the correct directions. (First, check the struts and bottom stringer for square. Then, hold the top rails in place against the front and back struts, and trace the angle onto the top rails.)

5. Position the top rails between the front struts and back struts, making sure the outside faces are flush with the outside edges of the front struts and back struts. Attach the top rails with construction adhesive and 3" deck screws.

MAKE AND ATTACH THE SIDE BRACES.

1. Cut the braces (G) to size. Cut a slanted profile on the braces, so the top edge is 3" long and the bottom edge is 6" long. This slanted profile leaves room to move a garbage can in and out of the unit.

2. Position the braces against the inside faces of the front struts, making sure the braces butt against the bottom stringers. Fasten the braces with construction adhesive and 2" deck screws.

JOIN THE SIDES.

1. Cut the front stringer (C) to length. Stand the side frames up on your work surface so the front struts face the same direction. Position the front stringer between the side frames with one face abutting each brace.

Mark the ends of the cross rails before cutting them to fit between the frame corners.

2. Make sure the ends of the front stringer are in contact with the bottom stringers. Attach the front stringer, using construction adhesive and 3" deck screws.

COMPLETE THE FRAME.

1. Position the back panel against the back of the frame sides. Make sure the back stringers are flush between the back struts, and attach the back panel to the frame with construction adhesive and 1¼" deck screws. The side edges of the back should be flush with the frame sides.

2. Cut the side panels (H) to size. Set them against the frame sides, making sure the front and back edges align and the panel rests squarely on the ground.

3. Trace the top edge of the frame onto the back (ungrooved) face of the side panels. Cut along the line with a circular saw. Attach the side panels with construction adhesive and 1¼" deck screws **(photo B).**

ATTACH BATTENS AND CROSS RAILS.

The battens are pieces of trim that frame the panels. The cross rails fit diagonally from corner to corner on the side panels.

1. Measure the sides carefully before cutting the battens (L) to fit along the side and back panel edges. Be sure to cut side battens with the appropriate angle cuts. Fasten the battens to the side panels with construction adhesive and 6d finish nails.

2. Position the diagonal cross rails on the sides as shown so they span from corner to corner. Mark the angles required for the ends to fit snugly into the corners **(photo C).** Cut the cross rails to size and attach them.

3. Attach the battens to the back panel, making sure their outside and top edges are flush with the side faces and top edges of the rear side battens.

MAKE THE TOP AND FRONT PANELS.

1. Cut the top panel (K) and front panel (I) to size. Cut these parts from the same sheet of siding to make sure the grooves align on the finished project.

2. Cut and attach battens to frame the edges of the top panel on both faces.

3. Cut and attach battens and a cross rail on the front panel. Like the side panels, the front panel is framed on one face only and has a cross rail stretching from corner to corner.

4. Attach two spring-loaded, self-closing hinges to the front panel, and mount it on the front of the project. The bottom edge of the front panel should be 1" up from the bottoms of the stringers. Mount the top panel with two 3 × 3" utility hinges **(photo D).**

5. Attach handles or pulls on the front and top of the garbage can holder. Use two panhead screws to attach an 18"-long safety chain to the front face of the top back stringer and top panel to keep the top panel from swinging open too far. Fasten the ends of the chain 10" in from the hinged edge of the top panel.

6. Finish the trash can protector with an exterior-rated stain. If pests are a problem, attach latches to the front and top to hold them in place.

Mount the top to the top of a side panel with 3 × 3" utility hinges.

PROJECT
POWER TOOLS

Boot Butler

A traditional piece of home furniture, the boot butler combines shoe storage and seating in one dependable unit.

CONSTRUCTION MATERIALS

Quantity	Lumber
1	¾" × 4 × 8' plywood
2	4" × 4' pine ranch molding

This boot butler was designed for an enclosed front porch, but it can be used near the entrance to your home or cabin. It provides plenty of storage space and gives you a place to sit when you're putting on or removing shoes and boots. It's a classic piece of household furniture modernized and simplified for your home.

To help keep things tidy and clean, just fit some plastic boot trays neatly onto the bottom shelf. When the trays get dirty, simply take them out and clean them. The boot butler can store footwear of the entire family. It fits conveniently against the wall to save space and keep unsightly boots and shoes out of busy traffic lanes.

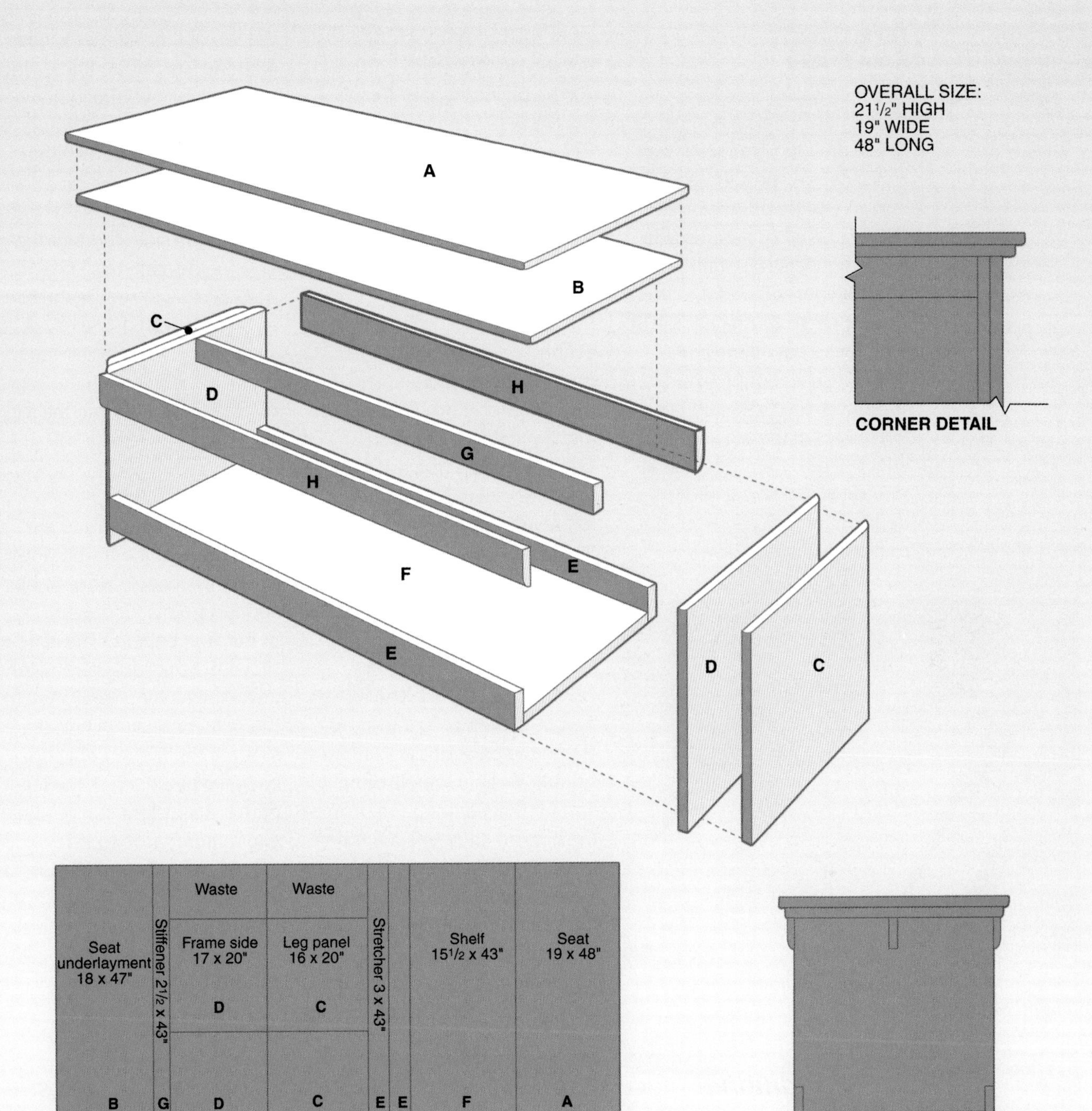

Cutting List

Key	Part	Dimension	Pcs.	Material
A	Seat	¾ × 19 × 48"	1	Plywood
B	Underlayment	¾ × 18 × 47"	1	Plywood
C	Leg panel	¾ × 16 × 20"	2	Plywood
D	Frame side	¾ × 17 × 20"	2	Plywood

Cutting List

Key	Part	Dimension	Pcs.	Material
E	Stretcher	¾ × 3 × 43"	2	Plywood
F	Shelf	¾ × 15½ × 43"	1	Plywood
G	Stiffener	¾ × 2½ × 43"	1	Plywood
H	Apron	½ × 3½ × 43¾"	2	Pine

Materials: Moisture-resistant glue, 1¼" and 2" deck screws, 8d finish nails, 15 × 21" plastic boot trays, finishing materials.

Note: Measurements reflect the actual size of dimension lumber.

Directions: Boot Butler

CUT THE PLYWOOD PARTS.
1. Cut the following parts with a circular saw and straightedge: seat (A), seat underlayment (B), leg panel (C), frame side (D), stretchers (E), shelf (F) and stiffener (G). Refer to the *Cutting Diagram* on page 247 to see how to cut all the parts from one sheet of plywood.
2. Smooth the sides and bottom edges of the legs, the top edges of the stretchers, and all the edges of the seat and underlayment with a sander or a router and ¼" roundover bit.

Use glue to reinforce the joints between the stretchers and the shelf.

ASSEMBLE THE SHELF AND STRETCHERS.
All the plywood parts are connected with screws and glue. Before you drive any screws, drill counterbores for the screw heads that are just deep enough to be filled with wood filler or putty.
1. Attach the stretchers (E) to the shelf (F) by drilling four evenly spaced ⅛" pilot holes through the outside edges of the stretchers and into the front and back edges of the shelf. Keep the screw holes at least 2" from the ends of the stretchers to prevent splitting.
2. Glue the joints **(photo A),** and drive 2" deck screws through the pilot holes and into the shelf.

BUILD THE BOX FRAME.
1. Attach the two frame sides (D) to the ends of the shelf assembly. Begin by measuring and marking a line 2" up from the bottom edge of each frame side. The stretcher's lower edges will fit here.
2. Position the stiffener (G) between the frame sides at the top centerpoints. Mark the stiffener's position, making sure the top of the stiffener is flush with the tops of the frame sides. Apply glue to all the joints.
3. Clamp the stretchers and stiffener in position with bar clamps. Drill two evenly spaced ⅛" pilot holes through each frame side and into the ends of the stiffener. Drive in 2" deck screws to secure the stiffener **(photo B).** For extra shelf support, drill pilot holes and drive a screw through the center of each frame side into the shelf.

COMPLETE THE LEG ASSEMBLY.
Attach leg panels (C) to the outer faces of the frame sides to provide wider, more stable support points for the seat.
1. Put wax paper or newspaper on your work surface to catch any excess glue. Apply glue to the outer face of each frame side and to the inner face of each leg panel. Press the leg panels against the frame panels, centered side to side to create a ½" reveal on each side of each frame panel. All top and bottom edges should be flush. Clamp each panel pair together **(photo C).**

The stiffener is screwed in place between the sides to keep the boot butler square.

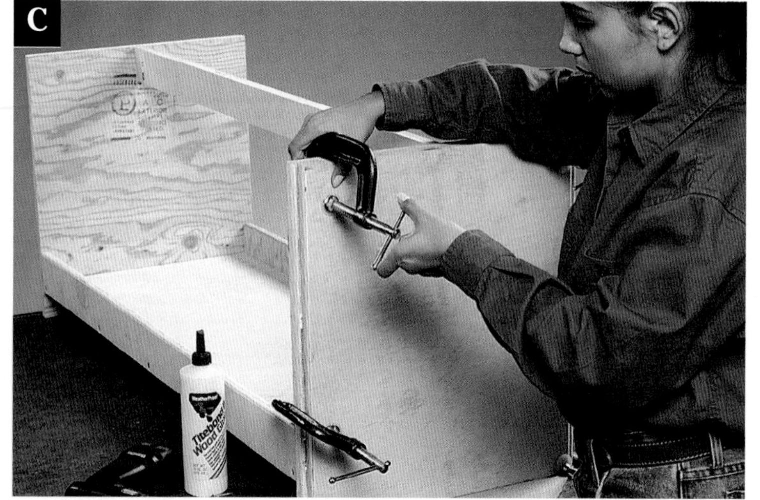

A pair of plywood panels are fastened together to create each leg assembly.

If the frame isn't square, fasten a pipe clamp diagonally across it and tighten.

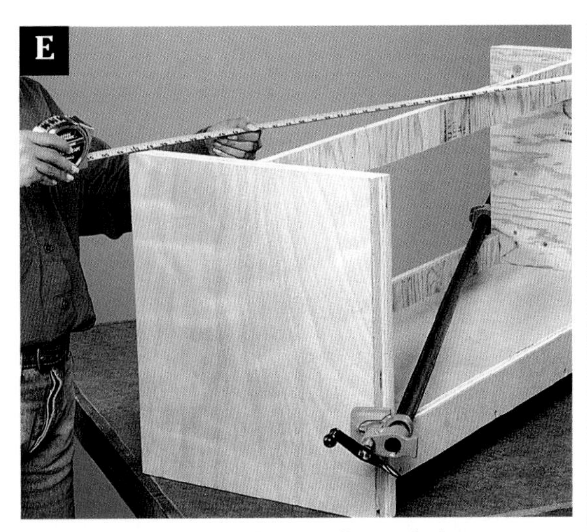

Set the frame assembly onto the underlayment. Center and trace the outline on the underlayment.

2. Drill pilot holes in the frame sides and counterbore the holes. Drive 1¼" deck screws through the frame sides and into the leg panels.
3. Measure the box frame diagonally from corner to corner, across the tops of the leg assembly to make sure it is square. Use a pipe clamp to draw the frame together until it is square and the diagonal measurements are equal **(photo D).**

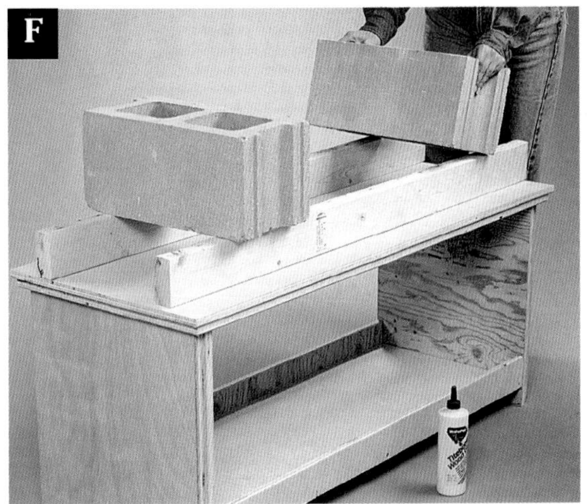

Apply weight on the seat to ensure a solid bond with the underlayment.

ATTACH THE SEAT.

The seat for the boot butler is made of an underlayment (B) layered with a plywood seat (A).
1. Lay the underlayment on a flat surface. Turn the leg and shelf assembly upside down and center it on the bottom face of the underlayment. Outline edges of the frame onto the underlayment **(photo E).**
2. Flip the leg and shelf assembly upright. Apply glue to the tops of the legs and stiffener. Position the underlayment on the assembly according to the alignment marks.
3. Drill ⅛" pilot holes through the underlayment into the legs and stiffeners. Counterbore the holes. Drive in 2" deck screws.
4. Apply a thin layer of glue to the top of the underlayment and the underside of the seat. Position the seat on the underlayment so the overhang is equal on all sides. Set heavy weights on top of the seat to create a solid glue bond **(photo F).**
5. Drill evenly spaced pilot holes in the underlayment and counterbore the holes. Drive 1¼" deck screws through the underlayment into the seat.
6. Cut the aprons (H) from 4"-wide pine ranch molding. Position them so the tops of the aprons are flush against the bottom edges of the underlayment, overlapping the edges of the frame panels slightly. Attach them with 8d finish nails.

APPLY FINISHING TOUCHES.

1. Fill all of the counterbored screw holes and plywood edges with wood putty and sand smooth.
2. Apply primer and paint. For a decorative touch, stencil or sponge-paint the surfaces.

TIP

Rigid, clear plastic boot trays are sold at most discount stores or building centers. The Boot Butler project shown here is designed to hold 15 × 21" plastic boot trays.

Conversion Charts

Drill Bit Guide

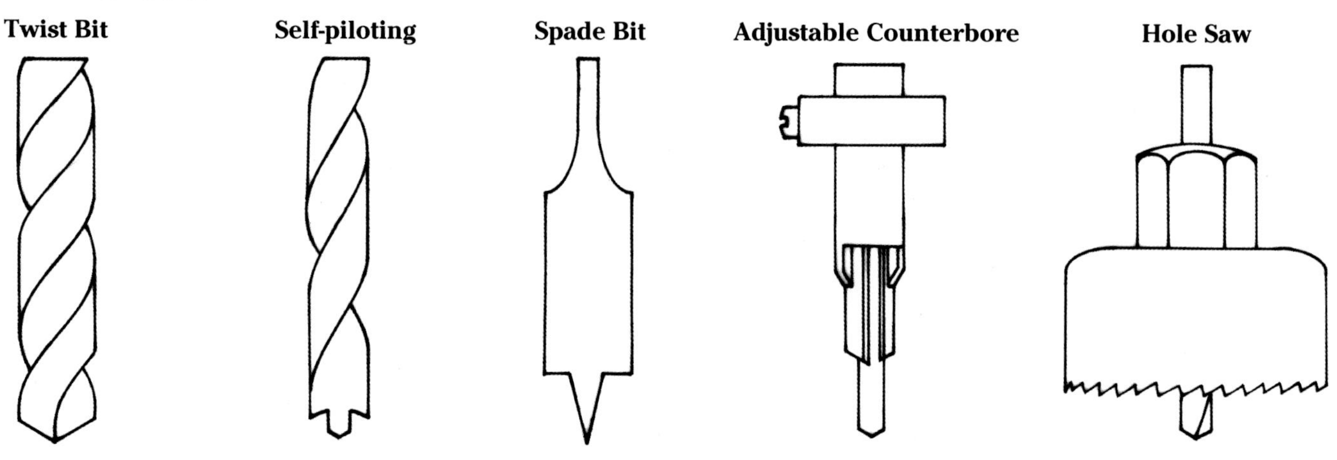

Counterbore, Shank & Pilot Hole Diameters

Screw Size	Counterbore Diameter for Screw Head	Clearance Hole for Screw Shank	Pilot Hole Diameter	
			Hard Wood	Soft Wood
#1	.146 (9/64)	5/64	3/64	1/32
#2	1/4	3/32	3/64	1/32
#3	1/4	7/64	1/16	3/64
#4	1/4	1/8	1/16	3/64
#5	1/4	1/8	5/64	1/16
#6	5/16	9/64	3/32	5/64
#7	5/16	5/32	3/32	5/64
#8	3/8	11/64	1/8	3/32
#9	3/8	11/64	1/8	3/32
#10	3/8	3/16	1/8	7/64
#11	1/2	3/16	5/32	9/64
#12	1/2	7/32	9/64	1/8

Abrasive Paper Grits - (Aluminum Oxide)

Very Coarse	Coarse	Medium	Fine	Very Fine
12 - 36	40 - 60	80 - 120	150 - 180	220 - 600

Saw Blades

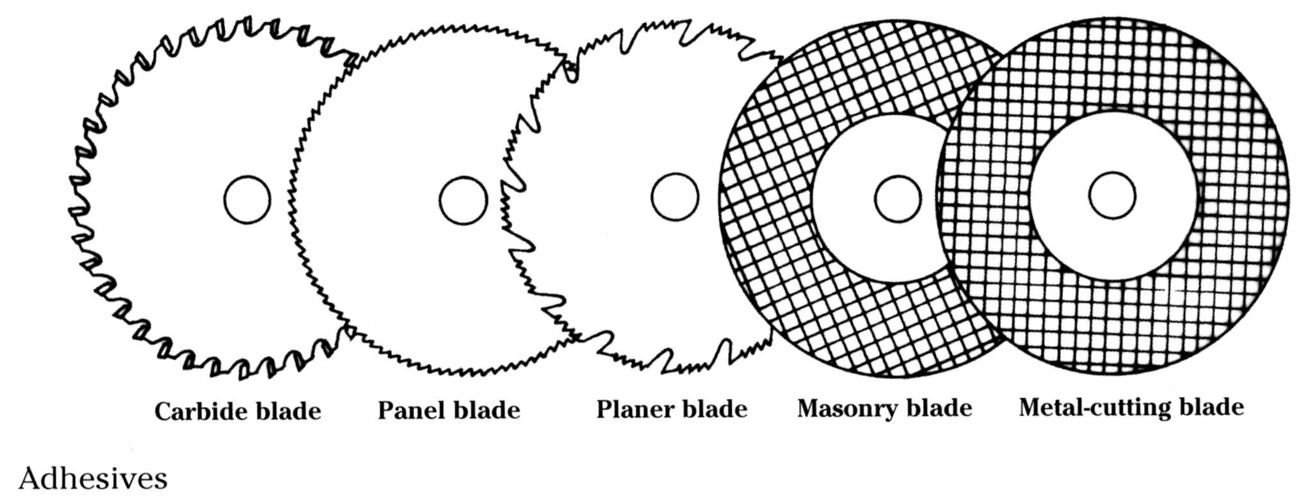

Adhesives

Type	Characteristics	Uses
White glue	**Strength:** moderate; rigid bond **Drying time:** several hours **Resistance to heat:** poor **Resistance to moisture:** poor **Hazards:** none **Cleanup/solvent:** soap and water	**Porous surfaces:** Wood (indoors) Paper Cloth
Yellow glue (carpenter's glue)	**Strength:** moderate to good; rigid bond **Drying time:** several hours; faster than white glue **Resistance to heat:** moderate **Resistance to moisture:** moderate **Hazards:** none **Cleanup/solvent:** soap and water	**Porous surfaces:** Wood (indoors) Paper Cloth
Two-part epoxy	**Strength:** excellent; strongest of all adhesives **Drying time:** varies, depending on manufacturer **Resistance to heat:** excellent **Resistance to moisture:** excellent **Hazards:** fumes are toxic and flammable **Cleanup/solvent:** acetone will dissolve some types	**Smooth & porous surfaces:** Wood (indoors & outdoors) Metal Masonry Glass Fiberglass
Hot glue	**Strength:** depends on type **Drying time:** less than 60 seconds **Resistance to heat:** fair **Resistance to moisture:** good **Hazards:** hot glue can cause burns **Cleanup/solvent:** heat will loosen bond	**Smooth & porous surfaces:** Glass Plastics Wood
Cyanoacrylate (instant glue)	**Strength:** excellent, but with little flexibility **Drying time:** a few seconds **Resistance to heat:** excellent **Resistance to moisture:** excellent **Hazards:** can bond skin instantly; toxic, flammable **Cleanup/solvent:** acetone	**Smooth surfaces:** Glass Ceramics Plastics Metal
Construction adhesive	**Strength:** good to excellent; very durable **Drying time:** 24 hours **Resistance to heat:** good **Resistance to moisture:** excellent **Hazards:** may irritate skin and eyes **Cleanup/solvent:** soap and water (while still wet)	**Porous surfaces:** Framing lumber Plywood and paneling Wallboard Foam panels Masonry
Water-base contact cement	**Strength:** good **Drying time:** bonds instantly; dries fully in 30 minutes **Resistance to heat:** excellent **Resistance to moisture:** good **Hazards:** may irritate skin and eyes **Cleanup/solvent:** soap and water (while still wet)	**Porous surfaces:** Plastic laminates Plywood Flooring Cloth
Silicone sealant (caulk)	**Strength:** fair to good; very flexible bond **Drying time:** 24 hours **Resistance to heat:** good **Resistance to moisture:** excellent **Hazards:** may irritate skin and eyes **Cleanup/solvent:** acetone	**Smooth & porous surfaces:** Wood Ceramics Fiberglass Plastics Glass

Conversion Charts (continued)

Lumber Dimensions

Nominal - U.S.	Actual - U.S.	Metric
1 × 2	¾" × 1½"	19 × 38 mm
1 × 3	¾" × 2½"	19 × 64 mm
1 × 4	¾" × 3½"	19 × 89 mm
1 × 5	¾" × 4½"	19 × 114 mm
1 × 6	¾" × 5½"	19 × 140 mm
1 × 7	¾" × 6¼"	19 × 159 mm
1 × 8	¾" × 7¼"	19 × 184 mm
1 × 10	¾" × 9¼"	19 × 235 mm
1 × 12	¾" × 11¼"	19 × 286 mm
1¼ × 4	1" × 3½"	25 × 89 mm
1¼ × 6	1" × 5½"	25 × 140 mm
1¼ × 8	1" × 7¼"	25 × 184 mm
1¼ × 10	1" × 9¼"	25 × 235 mm
1¼ × 12	1" × 11¼"	25 × 286 mm

Nominal - U.S.	Actual - U.S.	Metric
1½ × 4	1¼" × 3½"	32 × 89 mm
1½ × 6	1¼" × 5½"	32 × 140 mm
1½ × 8	1¼" × 7¼"	32 × 184 mm
1½ × 10	1¼" × 9¼"	32 × 235 mm
1½ × 12	1¼" × 11¼"	32 × 286 mm
2 × 4	1½" × 3½"	38 × 89 mm
2 × 6	1½" × 5½"	38 × 140 mm
2 × 8	1½" × 7¼"	38 × 184 mm
2 × 10	1½" × 9¼"	38 × 235 mm
2 × 12	1½" × 11¼"	38 × 286 mm
3 × 6	2½" × 5½"	64 × 140 mm
4 × 4	3½" × 3½"	89 × 89 mm
4 × 6	3½" × 5½"	89 × 140 mm

Index

Glossary

Belt sander — a tool used to resurface rough wood.

Breezeway — an open passageway, featuring a roof, that connects two structures.

Carpenter's level — a tool that features a bubble within a liquid that centers to indicate when a surface is properly aligned.

Carriage bolts — a bolt featuring a domed head and a square neck to securely fasten wood.

Casing nail — similar to a finishing nail, but with a slightly larger dimpled head for better holding power.

Caster wheels — wheels that swivel for easy positioning.

Caulk — a mastic substance, usually containing silicone, used to seal joints. Caulk is waterproof and flexible when dry and adheres to most dry surfaces.

Cedar — a lightweight, aromatic softwood with a natural resistance to moisture and insects.

Cedar lap siding — cedar boards cut with a bevel.

Cementboard — a rigid material with a fiberglass facing and a cement core that is undamaged by water.

Ceramic floor tile — sturdy tile suitable where durability is required.

Ceramic tile adhesive — multipurpose thin-set mortar applied with a V-notch trowel.

Circular saw — a type of saw used to make straight cuts.

Combination square — a measuring tool that combines a straight edge with a sliding head to measure both 45 ° and 90° angles.

Common nails — nails that have wide, flat heads.

Concrete — a mixture of Portland cement, sand, coarse gravel, and water.

Core box bit — a straight bit with a rounded bottom.

Cotter pin — a split pin whose shaft is inserted through a hole, then bent to fasten separate pieces together.

Counterboring — a technique for drilling holes that allows nail heads to be set below the surface of the wood when finished.

CPVC — chlorinated polyvinyl chloride; rigid plastic material used for high-quality water-supply pipe products.

Deck screws — screws with a light shank and coarse threads that are ideal for fastening soft woods.

Dimension lumber — the nominal size by which lumber is sold—usually larger than the actual size.

Dowels — round wooden rods.

Drill — a tool used to drill holes and drive screws.

Epoxy — a two-part glue that bonds powerfully and quickly.

Exterior plywood — plywood made with 100% waterproof glue.

Fencing staples — barbed staples that grab tightly onto wood.

Finish nails — nails with a thin shank and a cup-shaped head that are driven below the surface with a nail set.

Flagstone — large slabs of quarried stone cut into pieces up to 3" thick.

Flashing — aluminum or galvanized steel sheeting cut and bent into various sizes and shapes; used to keep water from entering joints between roof elements and to direct water away from structural elements.

Gang-sanding — a technique whereby matching pieces are clamped together before sanding to ensure identical shaping.

GFCI receptacle — a receptacle outfitted with a ground-fault circuit-interrupter. Also used on some extension cords to reduce the possibility of electric shock when operating an appliance or power tool.

Gnomon — the triangular part of a sundial face that casts a shadow to tell time.

Grommet — a metal ring used to line a hole cut into wood.

Grout — a fluid cement product used to fill spaces between ceramic tiles or other crevices.

Hardware cloth — galvanized, welded, and woven screening in a variety of mesh sizes; also called welded wire fabric.

Hasp — a two-part, hinged metal fastener featuring a loop, through which a lock is placed for security.

Heartwood — wood cut from the center of the tree; valued for its density, straightness, and resistance to decay.

Hot glue — a glue that is melted for application, then hardens as it cools to create a strong, durable bond; typically used for crafts, but very effective for lightweight exterior projects and indoor patching jobs.

Jig saw — a type of saw used to make contours, internal cuts, and short, straight cuts.

Lag screw — a heavy screw with a square or hexagonal head.

Lattice — panels woven from ¼" or ⅜" strips of cedar or treated lumber.

Linseed oil — a nontoxic drying oil obtained from the flax plant that is used to obtain a natural, protective finish on wood projects.

Locking nut — a nut that holds two pieces together securely, while still allowing the joint to pivot.

Miter — a 45° angled end-cut for fitting two pieces together at a corner.

Moisture-resistant glue — any exterior wood glue, such as plastic resin glue.

Mortar — a mixture of Portland cement, lime, and sand used to bond the bricks or blocks of a masonry wall.

Pine — a basic softwood used for interior projects.

Pipe clamp — a clamp that slides on a pipe to adjust to various sizes.

Plunge router — a router with a bit chuck that can be raised or lowered to start internal cuts.

Plywood — a wood product made of layers of thin veneer glued together.

Plywood siding — decorative exterior plywood, faced with either Douglas fir, cedar, or redwood, in a variety of textures, such as smooth, rough sawn, and grooved patterns.

Power miter saw — a type of saw used to make angled cuts in narrow stock.

Power sander — a tool used to prepare wood for a finish and to smooth sharp edges.

Prefabricated concrete pavers — decorative concrete blocks, available in a variety of shapes and sizes, that can be used to make walkways, patios, and walls.

Primer — a sealer that keeps wood resins from bleeding through the paint layer.

Redwood — a lightweight, moisture-resistant, rot-resistant, and insect-resistant wood.

Rip-cut — to cut a piece of wood parallel to the grain.

Router — a tool used to cut structural grooves, decorative edges, and roundover cuts in wood.

Sheet acrylic — clear plastic product available in thicknesses from 1/16" to 1".

Sheet goods — manufactured products generally sold in 4 ft. x 8 ft. sheets of various thicknesses.

Shim — a thin wedge of wood used to make a slight adjustment to achieve square alignment.

Squaring — a technique that ensures straight assembly by measuring diagonally from corner to corner in a frame.

Stain — a product that seals and adds color to wood.

Treated lumber — construction-grade pine that has been treated with chemical preservatives and insecticides.

Wood moldings — decorative trims for finishing projects.

Wood plugs — ⅜"-dia. x ¼"-thick disks with a slightly conical shape used to plug screw holes.

Wood sealer — a product that partially or totally blocks the wood's pores, preventing water penetration.

CREATIVE PUBLISHING INTERNATIONAL

- Complete Guide to Home Wiring
- Complete Guide to Home Plumbing
- Complete Guide to Easy Woodworking Projects
- Complete Guide to Building Decks
- Complete Guide to Home Carpentry
- Complete Guide to Painting & Decorating
- Complete Guide to Windows & Doors
- Complete Guide to Creative Landscapes
- Complete Guide to Home Storage
- Complete Guide to Bathrooms
- Complete Guide to Ceramic & Stone Tile
- Complete Guide to Flooring
- Complete Guide to Home Masonry
- Complete Guide to Roofing & Siding
- Complete Guide to Kitchens

CREATIVE PUBLISHING INTERNATIONAL
18705 LAKE DRIVE EAST
CHANHASSEN, MN 55317